# PV System Design and Performance

# PV System Design and Performance

Special Issue Editor
**Wilfried van Sark**

MDPI • Basel • Beijing • Wuhan • Barcelona • Belgrade

*Special Issue Editor*
Wilfried van Sark
Utrecht University
The Netherlands


*Editorial Office*
MDPI
St. Alban-Anlage 66
4052 Basel, Switzerland


This is a reprint of articles from the Special Issue published online in the open access journal *Energies* (ISSN 1996-1073) from 2017 to 2018 (available at: https://www.mdpi.com/journal/energies/special_issues/PV_system)

For citation purposes, cite each article independently as indicated on the article page online and as indicated below:

LastName, A.A.; LastName, B.B.; LastName, C.C. Article Title. *Journal Name* **Year**, *Article Number*, Page Range.

**ISBN 978-3-03921-622-2 (Pbk)**
**ISBN 978-3-03921-623-9 (PDF)**

Cover image courtesy of Wilfried van Sark.

# Contents

# About the Special Issue Editor

**Wilfried van Sark** has over 35 years of experience in the field of photovoltaic solar energy research and development. Starting in the area of solar cell characterization at AMOLF Amsterdam, his work entails III-V solar cell and processing development, thin film silicon cell and processing development, and, more recently, basic spectral shifting processes for next generation photovoltaics (up and down conversion) applied to luminescent solar concentrators. He is also a recognized expert in performance analysis of photovoltaic modules and systems as well as in life cycle and market analysis. In addition, his work in the deployment of photovoltaics in residential areas has led to research in building integrated photovoltaics, smart grids with electric mobility and vehicle-to-grid technology, as well as solar forecasting.

# Preface to "PV System Design and Performance"

This Special Issue, titled "Photovoltaic System Design and Performance," was originally published in MDPI's Energies journal. After a call for papers in 2017, the collection received a total of 21 papers on the topic. Main subtopics include the following: data analysis for optimal performance and fault analysis of photovoltaic systems, causes for energy loss, and design and integration issues. The papers in this book demonstrate the importance of designing and properly monitoring photovoltaic systems in the field in order to ensure continued good performance.

**Wilfried van Sark**
*Special Issue Editor*

*Editorial*

# Photovoltaic System Design and Performance

**Wilfried van Sark**

Utrecht University, Copernicus Institute, Princetonlaan 8a, 3584 CB Utrecht, The Netherlands;
w.g.j.h.m.vansark@uu.nl

Received: 7 February 2019; Accepted: 8 May 2019; Published: 14 May 2019

**Abstract:** This editorial summarizes the collection of papers in the Special Issue entitled Photovoltaic System Design and Performance, which was published in MDPI's Energies journal. Papers on this topic were submitted in 2017 and 2018, and a total of 21 papers were published. Main topics included data analysis for optimal performance and fault analysis, causes for energy loss, and design and integration issues. The papers in this Special Issue demonstrate the importance of designing and properly monitoring photovoltaic systems in the field in order to ensure maintaining good performance.

**Keywords:** photovoltaic (PV) system design; monitoring of PV systems; performance analysis; mapping performance; performance variability; system malfunction detection

---

## 1. Introduction

Photovoltaic (PV) solar technology grew rapidly and continuously in the past decades, leading to ~400 GWp installed capacity globally [1], and this led to enormous price reductions. The strength of the technology is its modular design, and PV power plants range from a few PV modules (~1 kWp) to millions (~250 MWp). Design of such systems depends on the scale level; residential systems are typically roof-based, either flat or tilted, while large systems allow to design for maximum annual yield but also require intricate electrical layouts with multiple inverters and connections to medium-voltage transmission networks. Additionally, operation is scale-dependent. In residential areas, non-ideal tilts and potential shading will lower annual yields, and designs to minimize these losses were developed, using, e.g., power optimization per module. Monitoring of such small systems is rare, as economic loss due to malfunction is relatively low. Large-scale systems have proper supervisory control and data acquisition (SCADA) systems to ensure maximum economic benefit, which is of importance to investors.

This Special Issue solicited papers with original research and studies related to the abovementioned topics, including, but not limited to, PV system design on residential and larger scales, methods for operational control and analysis, failure detection, performance analysis of systems, mapping performance differences, performance variability, and degradation of systems and modules. Papers selected for this Special Issue were subject to a rigorous peer review procedure with the aim of rapid and wide dissemination of research results, developments, and applications.

The response to the call for papers led to 28 submitted papers, of which 21 (75%) were accepted and seven (25%) rejected. The geographical distribution of the (first) author covers all continents, and is built as follows: Austria (one paper), Brazil (one), China (three), Colombia (two), Germany (one), Italy (one), Korea (two), Romania (one), Portugal (one), Spain (one), Slovenia (one), the Netherlands (four), United Kingdom (UK; one), and United States of America (USA; one).

## 2. Content of the Special Issue

The 21 papers collected in this Special Issue can broadly be divided into the following fields: (a) data analysis for optimal performance and fault analysis, (b) causes for energy loss, and (c) design and integration issues. They will be described in the following subsections.

### 2.1. Data Analysis for Optimal Performance

In the design phase of PV systems, yield predictions are made based on the typical weather data available or derived for the specific installation site. During operation, monitoring of yield is a necessity to demonstrate that the yield predictions were correct, which will satisfy investors. In addition to proper design and quality-controlled installation of the PV system, operation and maintenance (O&M) procedures should be in place that include adequate intervention after a yield loss is detected. Automatic detection of the cause of the yield loss will decrease downtime of a system, thus lowering potential economic loss. Several papers in this Special Issue addressed various aspects of O&M procedures. Santiago et al. [2] proposed a graphical procedure that supports performance analysis of PV plant operations, without additional costs. It is based on inverter data, which are automatically processed. Visualization using color maps or heat maps of direct current (DC) energy generated in the PV plant are shown as a function of day in the year and time of day, which is extremely insightful, as anomalies can be detected clearly. The usefulness of the method is demonstrated in a small PV plant of 17.8 kWp in size located in Córdoba, Spain, for a period of three years. The average daily performance ratio varied between 0.6 and 0.9, with highest values in winter. Detailed analysis of inverter data further showed that the presence of shadows could be easily detected, in particular the hours and days for which the system was mainly affected.

On a larger scale, Kausika et al. [3] and Moraitis et al. [4] demonstrated the usefulness of applying geographical information systems (GIS) in PV performance analysis on a multi-country and regional level. Maps were constructed for several European countries based on annual yield data from more than 30,000 PV systems. Color mapping and mapping differences between yields in different years assisted in identifying potential problems in PV systems across countries or regions. Moraitis et al. [4] provided an even more detailed analysis, showing the performance differences between PV system installed in rural and urban areas, and related these differences to the urban compactness causing more shading on systems than in rural areas. A seasonal dependence further corroborated the finding that shading is the cause for lower performance in urban areas.

Zhao et al. [5] proposed a novel fault diagnosis method for PV systems based on several fuzzy algorithms. The method was shown by simulation analyses to effectively detect short-circuit, open-circuit, partial occlusion, and other faults. Actual experiments with a 9.6-kWp PV system in the field further demonstrated that the method yields correct and effective fault detection, confirming the correctness and effectiveness of the proposed method, and that the method can complete the PV array diagnosis. The method was inspired by the notion that the unpredictability of faults can best be addressed using fuzzy theory.

At normal operation, the energy yield of PV systems is (nearly) linearly related to reference data such as irradiance or yield of neighboring PV systems. Tsafarakis et al. [6] described a new data analysis method that automatically distinguishes measurements that fit to a near-linear relationship (defined as inliers) from those which do not (outliers). The method, thus, can be used to detect and exclude any data input anomalies, and to detect and separate measurements where the PV system is functioning properly from the measurements characteristic for malfunctioning. The method was demonstrated in actual experimental data, showing that peer-to-peer comparison of yield is better than using irradiance data, whether from a pyranometer or derived from satellite images.

Although PV energy yield depends directly on solar irradiance and module temperature, Vergura [7] ignored these parameters and compared the energy performance of PV systems, or rather arrays of panels in the same 49-kWp plant in southern Italy, using several statistical methods including analysis of variance (ANOVA) in combination with various tests. This actually is a form of peer-to-peer comparison, and it was designed since the cost of advanced monitoring equipment is considered too high for the size of the PV plant. It was shown that these methods are effective in finding abnormal operating conditions.

Baschel et al. [8] reported on component reliability in large-scale PV systems and the effect on performance, based on a large operational dataset of failure rates for periods of 3–5 years. Impact assessment was performed using a fault tree analysis, as well as a failure mode and effects analysis, which was used to rank failures in term of occurrence and severity. Reduced energy yields were estimated based on actual failure probabilities. For example, transformer failures, while rare, lead to energy loss due to the long repair periods. Thus, transformer and inverter problems are responsible for two-thirds of the total energy loss.

Detailed analysis of performance of PV systems may require knowledge of detailed PV module parameters, such as diode parameters. As the current–voltage (I–V) characteristics involve an implicit function, fitting is needed to find parameters. For many decades, methods to solve this ere proposed. Kang et al. [9] newly proposed methods based on cuckoo search algorithms. An improved version for both single- and double-diode models was developed and tested using experimental I–V data. It was shown that the new method outperforms state-of-the-art algorithms, and finds parameters effectively.

*2.2. Causes for Energy Loss*

Several causes for energy loss can be identified. This subsection shows the effects of shadows and a solution to mitigate them, as well as the effects of soiling and degradation. Identification of the latter is possible using various imaging techniques.

Gutiérrez Galeano et al. [10] presented a simplified approach using a shading ratio to model and analyze the performance of partially shaded PV modules. A model was derived using well-known current–voltage characteristics, in which the shape and opacity of a shaded area are integrated. The method was developed to improve the detection of shaded PV systems, and it was experimentally validated. In order to mitigate shadow-induced losses in PV modules, Mirbagheri Golroodbari et al. [11] developed a smart PV module architecture consisting of 60 silicon solar cells. This module optimally consists of ten groups of six series-connected solar cells in parallel to a DC–DC buck converter. A model was developed that allows time-dependent simulations of moving shadows of poles and random shading patterns over the module. Compared to an ideal module, as well as standard series-connected or parallel-connected modules, the smart architecture outperforms the series-connected one by nearly 50%.

Another problem in PV system performance is soiling on the modules. Conceição et al. [12] presented the effect of soiling in Alentejo and Évora in Portugal on PV performance, and discussed the seasonal variation and the type of soiling. Sand from the Sahara was identified as soiling, in addition to pollen, especially in spring. A so-called soiling ratio index was defined using maximum power and short-circuit current of PV panels, and a detailed materials analysis was performed using scanning electron microscopy and energy-dispersive x-ray spectroscopy. The highest soiling rate of 4.1%/month was found in spring, while the summer and fall months showed 1.9%/month and 1.6/% month, respectively. This spring increase was due to the presence of pollen, in addition to inorganic materials. The effect of pollen was larger than that of organic materials, due to their larger particle size.

Degradation of solar cells as a result of various causes is known to affect the capacitance of solar cells. Cotfas et al. [13] demonstrated a simple technique to measure solar cell capacitance using an inductor, thus forming a resistor/inductor/capacitor (RLC) circuit. When connecting the inductor to the cell, oscillations in voltage and current appear, which are damped after some time. The frequency of these oscillations can be found using RLC circuit behavior, from which the capacitance of the cell is derived. Additional experiments showed that the capacitance increases with irradiance, as well as with temperature.

Eder et al. [14] studied the aging of PV test modules by comparing intact modules with modules with deliberately generated failures, i.e., micro-cracks, cell cracks, glass breakage, and connection defects. These modules were tested under different conditions in climate chambers and outdoors to be able to follow the propagation of stress-induced defects and their effect on performance. Module characterization was performed using electrical measurements and electroluminescence (EL), as well as using fluorescence imaging for the detection of aging effects in encapsulation materials of the modules. It was found that modules with mechanical failures were unaffected, while the pre-test presence of micro-cracks led to a higher rate of degradation after testing. Elevated temperature and irradiation were found to induce fluorescence effects in polymeric encapsulants after about one year of outdoor testing. Fluorescence imaging is a useful tool in detecting permeation of oxygen through the back sheet, which can be identified as bleaching of the fluorescence of the encapsulant top layer between the cells, above cell cracks.

While electroluminescence (EL) is widely used to qualitatively map electronic properties of PV modules, Kropp et al. [15] presented a new approach which allows for a quantitative assessment. The so-called "EL power prediction of modules" (ELMO) method is based on using two electroluminescence images to determine the electrical loss of damaged PV modules. The EL images are converted into spatially resolved series resistance images, and color-coded maps reveal the location of damages in the module. The method is also applicable for parallel resistance mapping, which was demonstrated in the analysis of potential-induced degradation.

*2.3. Design and Integration Issues*

Maximum power point tracking (MPPT) devices and algorithms are essential in maximizing power output of PV modules and systems. Robles Algarín et al. [16] discussed the common approach based on the so-called perturb-and-observer (P&O) algorithm, and showed that this may lead to oscillation issues around the operating point. They suggested an alternative way, i.e., to use a fuzzy controller for finding the maximum power point. A simulation model was designed for a PV panel, a buck converter, and a fuzzy controller, and the authors showed that their method with the fuzzy controller outperforms the common P&O approach, with faster settling time, lower power loss, and oscillations around the operating point.

Forecasting of energy generation of PV systems is important for integration of PV in the electricity grid, as it facilitates power management in the local grid. Accurate forecasting requires information about expected solar irradiance, which is not readily available. Therefore, Brecl and Topič [17] developed a method based on only weather forecast data, and solar irradiance was simulated based on discrete weather class values. The simple approach led to a root-mean-square error of measured and forecasted power data of 65%, while the correlation in terms of $R^2$ was high at 0.85.

Grid connection of a PV system may suffer from stability issues, as Huang et al. [18] argued. They provided a stability analysis based on a single-phase two-stage (modules, inverter) PV system connected to the grid. The nonlinearity of PV system response to irradiation poses a problem in linear system theory, which is usually used in grid stability analyses. Therefore, Huang et al. [18] developed a mathematical model addressing this issue. As the time dependence of current and voltage is the main problem in the stability analysis, a transformation was used to make the problem time-invariant. Results revealed that the mathematical model performs well compared to standard power simulation models (PSIM [19]).

Large PV power plants will not always operate at maximum power, and this allows inverters to provide reactive power support to the grid. Lourenço et al. [20] presented an evaluation method for reactive power support and associated cost. The method helps PV power plant owners optimize their bidding strategies in the reactive power market. At low irradiance, reactive power support can be large, while, at high irradiance, this is limited. Costs for providing reactive power were analyzed, as well as potential revenues in the Brazilian market, which are about 25% more than costs.

Kim et al. [21] described the design and construction of a floating PV system using fiber-reinforced polymer parts. The system consisted of modular structures connected by means of hinges, and anchored to shore. The 1-MWp plant is located in Dangjin City, Korea, at the waterway of the cooling water intake channel of a thermoelectric power plant, and it covers an area of about 21,000 m$^2$. Safety aspects were designed such that they comply with relevant design codes. The design also contains parts for electrical equipment (inverters, etc.) and pontoons that allow maintenance personnel to access the various areas of the system. Cost are expected to be lower compared to steel- or aluminum-based constructions.

Another contribution to floating PV was provided by Charles Lawrence Kamuyu et al. [22], who presented a multiple linear regression method using independent parameters of irradiance, ambient temperature, and wind speed to explain observed module temperature in a floating PV system. Interestingly, adding the water temperature as parameter slightly increased the error between observed and modeled temperature. The found correlation could be used to explain the ~10% increased annual yield of floating PV systems in Korea compared to roof-top systems.

Alshayeb et al. [23] reported on a year-long experimental study in which they compared the performance of a 4.3-kWp PV system over a green roof and a black roof. A detailed analysis of temperatures underneath the panels, in between panels and roof, and of ambient temperature helped explain the difference in energy yield between the green and black roof, which was maximally about 5% in favor of the green roof, with an average annual benefit of 1.4%.

## 3. Conclusions

The papers published as part of this Special Issue demonstrate the importance of and scientific progress in the design and performance of PV systems. The state of the art in the three fields of data analysis for optimal performance, causes for energy loss, and design and integration issues is evidenced in the content of the 21 papers from all over the world. It is clear that challenges do exist, which can be addressed in the near future. With the ever-increasing deployment of PV systems, proper design and monitoring, as well as fast and early detection of malfunctions, are prerequisites for PV to play a major role in future electricity systems.

**Author Contributions:** W.v.S. organized the Special Issue and wrote the manuscript.

**Funding:** This research was part of the IEA-PVPS Task 13 "Performance and Reliability of Photovoltaic Systems" [24] and the COST Action PEARL-PV "Performance and Reliability of Photovoltaic Systems: Evaluations of Large-Scale Monitoring Data" [25].

**Acknowledgments:** The guest editor would like to thank the authors for submitting their excellent contributions to this Special Issue. Furthermore, the present Special Issue would not have been possible without the expert reviewers that carefully evaluated the manuscripts and provided helpful comments and suggestions for improvements. A special thank you is in order for the editors and the MDPI team for their outstanding management of this Special Issue.

**Conflicts of Interest:** The author declares no conflict of interest.

## References

1. IEA. 2018: Snapshot of global photovoltaic markets: Report IEA PVPS T1-33:2018. 2018. Available online: http://www.iea-pvps.org/fileadmin/dam/public/report/statistics/IEA-PVPS_-_A_Snapshot_of_Global_PV_-_1992-2017.pdf (accessed on 22 January 2019).
2. Santiago, I.; Trillo Montero, D.; Luna Rodríguez, J.; Moreno Garcia, I.; Palacios Garcia, E. Graphical Diagnosis of Performances in Photovoltaic Systems: A Case Study in Southern Spain. *Energies* **2017**, *10*, 1964. [CrossRef]
3. Kausika, B.; Moraitis, P.; van Sark, W. Visualization of Operational Performance of Grid-Connected PV Systems in Selected European Countries. *Energies* **2018**, *11*, 1330. [CrossRef]
4. Moraitis, P.; Kausika, B.; Nortier, N.; van Sark, W. Urban Environment and Solar PV Performance: The Case of the Netherlands. *Energies* **2018**, *11*, 1333. [CrossRef]
5. Zhao, Q.; Shao, S.; Lu, L.; Liu, X.; Zhu, H. A New PV Array Fault Diagnosis Method Using Fuzzy C-Mean Clustering and Fuzzy Membership Algorithm. *Energies* **2018**, *11*, 238. [CrossRef]
6. Tsafarakis, O.; Sinapis, K.; van Sark, W. PV System Performance Evaluation by Clustering Production Data to Normal and Non-Normal Operation. *Energies* **2018**, *11*, 977. [CrossRef]
7. Vergura, S. Hypothesis Tests-Based Analysis for Anomaly Detection in Photovoltaic Systems in the Absence of Environmental Parameters. *Energies* **2018**, *11*, 485. [CrossRef]

8. Baschel, S.; Koubli, E.; Roy, J.; Gottschalg, R. Impact of Component Reliability on Large Scale Photovoltaic Systems' Performance. *Energies* **2018**, *11*, 1579. [CrossRef]

9. Kang, T.; Yao, J.; Jin, M.; Yang, S.; Duong, T. A Novel Improved Cuckoo Search Algorithm for Parameter Estimation of Photovoltaic (PV) Models. *Energies* **2018**, *11*, 1060. [CrossRef]

10. Gutiérrez Galeano, A.; Bressan, M.; Jiménez Vargas, F.; Alonso, C. Shading Ratio Impact on Photovoltaic Modules and Correlation with Shading Patterns. *Energies* **2018**, *11*, 852. [CrossRef]

11. Mirbagheri Golroodbari, S.; de Waal, A.; van Sark, W. Improvement of Shade Resilience in Photovoltaic Modules Using Buck Converters in a Smart Module Architecture. *Energies* **2018**, *11*, 250. [CrossRef]

12. Conceição, R.; Silva, H.; Mirão, J.; Collares-Pereira, M. Organic Soiling: The Role of Pollen in PV Module Performance Degradation. *Energies* **2018**, *11*, 294. [CrossRef]

13. Cotfas, P.; Cotfas, D.; Borza, P.; Sera, D.; Teodorescu, R. Solar Cell Capacitance Determination Based on an RLC Resonant Circuit. *Energies* **2018**, *11*, 672. [CrossRef]

14. Eder, G.; Voronko, Y.; Hirschl, C.; Ebner, R.; Újvári, G.; Mühleisen, W. Non-Destructive Failure Detection and Visualization of Artificially and Naturally Aged PV Modules. *Energies* **2018**, *11*, 1053. [CrossRef]

15. Kropp, T.; Schubert, M.; Werner, J. Quantitative Prediction of Power Loss for Damaged Photovoltaic Modules Using Electroluminescence. *Energies* **2018**, *11*, 1172. [CrossRef]

16. Robles Algarín, C.; Taborda Giraldo, J.; Rodríguez Álvarez, O. Fuzzy Logic Based MPPT Controller for a PV System. *Energies* **2017**, *10*, 2036. [CrossRef]

17. Brecl, K.; Topič, M. Photovoltaics (PV) System Energy Forecast on the Basis of the Local Weather Forecast: Problems, Uncertainties and Solutions. *Energies* **2018**, *11*, 1143. [CrossRef]

18. Huang, L.; Qiu, D.; Xie, F.; Chen, Y.; Zhang, B. Modeling and Stability Analysis of a Single-Phase Two-Stage Grid-Connected Photovoltaic System. *Energies* **2017**, *10*, 2176. [CrossRef]

19. PSIM power system simulation. Available online: https://powersimtech.com/products/psim/ (accessed on 5 February 2019).

20. Lourenço, L.; Monaro, R.; Salles, M.; Cardoso, J.; Quéval, L. Evaluation of the Reactive Power Support Capability and Associated Technical Costs of Photovoltaic Farms' Operation. *Energies* **2018**, *11*, 1567. [CrossRef]

21. Kim, S.; Yoon, S.; Choi, W. Design and Construction of 1 MW Class Floating PV Generation Structural System Using FRP Members. *Energies* **2017**, *10*, 1142. [CrossRef]

22. Charles Lawrence Kamuyu, W.; Lim, J.; Won, C.; Ahn, H. Prediction Model of Photovoltaic Module Temperature for Power Performance of Floating PVs. *Energies* **2018**, *11*, 447. [CrossRef]

23. Alshayeb, M.; Chang, J. Variations of PV Panel Performance Installed over a Vegetated Roof and a Conventional Black Roof. *Energies* **2018**, *11*, 1110. [CrossRef]

24. IEA-PVPS-Task 13. Available online: http://www.iea-pvps.org/index.php?id=57 (accessed on 5 February 2019).

25. COST Action PEARL-PV. Available online: https://www.pearlpv-cost.eu (accessed on 5 February 2019).

Article

# Design and Construction of 1 MW Class Floating PV Generation Structural System Using FRP Members

Sun-Hee Kim [1] , Soon-Jong Yoon [2] and Wonchang Choi [1,*]

[1]  Department of Architectural Engineering, Gachon University, Seongnam 13120, Korea;
    sunnys82@hanmail.net
[2]  Department of Civil Engineering, Hongik University, Seoul 04066, Korea; sjyoon@hongik.ac.kr
*  Correspondence: wchoi@gachon.ac.kr; Tel.: +82-31-750-5335

Received: 22 May 2017; Accepted: 1 August 2017; Published: 3 August 2017

**Abstract:** The paper investigates overview of construction process of a 1 MW class floating photovoltaic (PV) generation structural system fabricated with fiber reinforced polymer (FRP) members. The floating PV generation system consists of unit structures linked by a hinge type connection of which the effect of bending moment between the unit structures, induced by the unstable movement of the water surface, was minimized. Moreover, the unit structures were classified into three types of structures by combining the floating PV generation system and pontoon bridges, which are constructed to install the electrical equipment and a route of movement for workers. The structural safety of the connection system among the unit structures and/or the mooring system is confirmed by referring to the relevant design codes. In addition, structural analysis using the finite element method was performed to ensure the safety of the floating PV generation structure, and commercial viability evaluation was performed based on the construction cost. The FRP member shows superior performance in construction and cost effectiveness in a floating PV generation system.

**Keywords:** floating PV generation structure; fiber reinforced polymeric plastic (FRP); pultruded FRP; sheet molding compound FRP; structural design; mooring system

## 1. Introduction

Recently, environmental problems associated with the excessive use of fossil fuel have become social issues. As an alternative energy resource, the importance of renewable energy is continuously rising. Moreover, the demands for facilities to generate renewable energy are also ever increasing. In Japan, after the Fukushima Daiichi nuclear power plant disaster, a law was enacted for the development of the solar industry, wind power industry, etc. In Korea, the renewable portfolio standard (RPS), which requires electricity providers to gradually increase the amount of renewable energy sources such as wind, solar, bioenergy, and geothermal, was enacted to ensure the growth of the renewable energy market.

To satisfy such demands, a large number of photovoltaic energy generation systems are being constructed and planned. However, since these facility zones are mostly located on land, some problems, such as an increase in construction cost and environmental disruption, have occurred [1,2]. Accordingly, the floating photovoltaic (PV) energy generation system has become a desirable renewable energy alternative in Korea. Song and Choi (2016) [3] analyzed the potential of floating PV systems on a mine pit lake in Korea. Galdino and Olivieri (2017) [4] present several considerations regarding the application of floating PV plants in Brazil, whose characteristics make them different from those that have been deployed in other places, such as Japan and Korea.

Most frames that support the photovoltaic modules in the existing water levels photovoltaic power generation facilities are made of structural steel. In general, a structural steel frame is vulnerable to corrosion, which leads to increased maintenance costs and a decreased life span for the structure.

Due to the heavy unit weight of steels, the structural system needs a larger buoyant system, which increases construction cost. Therefore, in order to install a floating photovoltaic power generation structure, it is necessary to come up with a new material with light weight, good strength, and high durability that is less affected by moisture. In this study, we introduce the design and construction process of a 1 MW class floating PV generation system using fiber reinforced polymeric plastic (FRP) members. FRP has superior material properties compared with those of conventional structural materials. FRP has excellent corrosion-resistance and high specific strength and stiffness [5], both of which are highly appreciated for the design and fabrication of floating PV generation systems. In this paper, the floating PV generation system was fabricated by PFRP (pultruded FRP) and SMC (sheet molding compound) FRP; FRPs were made of polyester resin and E-glass fiber.

The 1 MW class floating PV generation complex is composed of 105 unit structures. The unit structures are connected to each other on the surface of the water. Moreover, the 1 MW class floating PV generation complex is the first commercialized system constructed at a site with such a large scale in Korea.

## 2. Design of 1 MW Class Floating PV Generation System

The floating PV generation complex is constructed by connecting unit structures using bolts or steel bars to minimize the transfer of any bending moment induced by the unstable movement of the water surface. Unit structures fabricated on the ground are connected on the water and moored using an anchor system. In this paper, the structural composition of the system, connection method for the unit structures, and mooring method of the 1 MW class floating PV generation complex considering the construction site conditions are discussed.

### 2.1. Design of Unit Structures

Unit structures consist of solar modules, structural systems to support the PV modules (which are composed of FRP structural members), a floating system, and connecting devices, as shown in Figure 1.

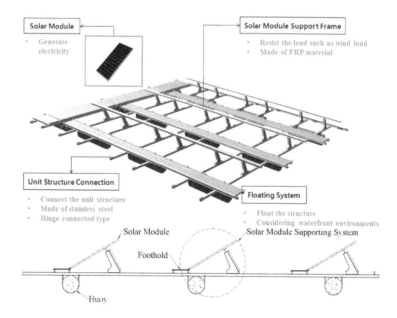

**Figure 1.** Composition of unit structure [6].

The FRP structural members are classified as vertical members, supporting members used to assemble the solar module, solar module connecting members used to assemble the solar module and foothold, foothold members to assemble the footholds and increase the horizontal stiffness of the structures, and main members to assemble the other members and the buoy. All structural members are connected using stainless steel bolts. The length of the structural members is set to 12.6 m for easy handling during fabrication of the structural system.

The 1 MW class floating PV generation complex consists of unit structures that are 12.6 m in length and 11.5 m in width. The unit structure has a 300 W generation capacity, in which 33 solar panels with dimension of 1966 mm × 1000 mm per panel are installed.

## 2.2. Complex Composition

The 1 MW class floating PV generation complex was constructed at Dangjin-city, Korea. The construction site of the 1 MW class floating PV generation complex is shown in Figure 2.

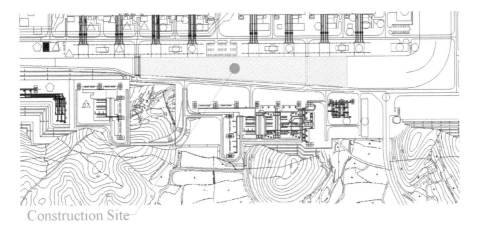

**Figure 2.** Construction site.

The 1 MW class floating PV generation complex consists of 105 unit structures, which are classified into three types: A, B, and C types. The A type is the basic model, the B type is the combination of the A-type and a pontoon bridge, and the C type is the structure used to install electrical devices such as the converter, as shown in Figure 3. The pontoon bridge is constructed to install the electrical equipment and is also used as movement route for the workers. The generation capacity of each unit structure type is given in Table 1. The total arrangement of the generation complex composition is shown in Figure 4.

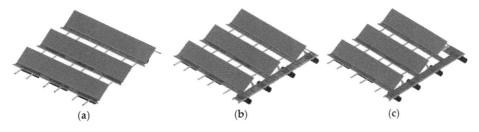

(a)          (b)          (c)

**Figure 3.** Unit structure (**a**) A type (basic mode); (**b**) B type (with foothold); (**c**) C type (to install electric device) [6].

**Table 1.** No. and generation capacity of each unit structure.

| Type | A-Type | B-Type | C-Type |
|---|---|---|---|
| No. of unit structure (EA) | 89 | 6 | 12 |
| No. of solar module per unit structure (EA) | 33 | 30 | 29 |
| Generation capacity per unit structure (kW) | 9.9 | 9.0 | 8.7 |

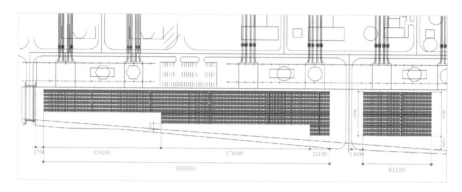

**Figure 4.** Composition of floating PV generation complex (Unit: mm).

The finite element (FE) analyses of the unit structures shown in Figure 3 are conducted using the most intelligent design and analysis system (MIDAS Civil 2012) [7]. The results of the FE analyses are used for structural design; the safety of the unit structures is estimated by the allowable stress design (ASD) according to the Structural Plastic Design Manual [8]. The mechanical properties and allowable stress of the FRP structural members are given in Tables 2 and 3, respectively [9]. In addition, the FE analysis model is shown in Figure 5.

**Table 2.** Mechanical properties.

| Mechanical Properties | PFRP | SMC FRP |
|---|---|---|
| Elastic modulus ($E$, GPa) | 33.28 | 17.33 |
| Tensile strength ($f_t$, MPa) | 402.58 | 183.85 |
| Shear strength ($f_v$, MPa) | 79.20 | 34.47 |
| Poisson's ratio ($v$, mm/mm) | 0.25 | 0.25 |
| Unit weight ($G$, kN/m$^2$) | 18.42 | 18.42 |

**Table 3.** Allowable stress.

| Allowable Stress | PFRP | SMC FRP | S.F. |
|---|---|---|---|
| Tension (MPa) | 201.29 | 91.93 | 2.0 |
| Compression (MPa) | 134.19 | 61.28 | 3.0 |
| Shear (MPa) | 26.40 | 11.49 | 3.0 |
| Flexure (MPa) | 161.03 | 73.54 | 2.5 |

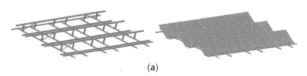

(a)

**Figure 5.** *Cont.*

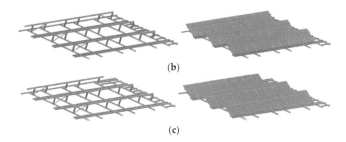

(b)

(c)

**Figure 5.** Finite element (FE) analysis model; (**a**) A type; (**b**) B type; (**c**) C type.

Wind load for the design of the unit structures is determined based on the wind speed of 30 m/sec according to the requirements in the Korean Building Code and Commentary [10]. The design wind load at the construction site (Dangjin-city, Korea) is suggested in the Korean Building Code and Commentary [10]. In addition, snow load (0.4 kN/m$^2$) is applied on the PV module and foothold. Crowd load (4.9 kN/m$^2$) is additionally applied on the foothold. Snow load and crowd load are determined according to the Korean Building Code and Commentary [10] and the Korean Bridge Design Specification [11].

From the FE analysis results, it was found that the ratio of the maximum stress to the allowable stress occurring on the A-type structure is 47.73%; on the B-type structure this value is 45.68%; and on the C-type structure this value is 46.02% as shown in Table 4. Shear stress on the PFRP member in each type of structure was the critical stress governing the structural behavior. Therefore, according to the FE analysis results, it was also found that the strength and stiffness of the PFRP and SMC FRP structural members are acceptable for the design and fabrication of the system.

**Table 4.** FE analysis results.

| Stress | | A-Type (MPa) | B-Type (MPa) | C-Type (MPa) | Allowable Stress (MPa) |
|---|---|---|---|---|---|
| Tension | PFRP | 1.32 | 2.89 | 2.89 | 2.89 |
| | SMC FRP | 0.00 | 0.00 | 0.00 | 0.00 |
| Compression | PFRP | 1.79 | 6.90 | 6.94 | 6.94 |
| | SMC FRP | 1.52 | 1.42 | 1.41 | 1.41 |
| Shear | PFRP | 12.60 | 12.06 | 12.15 | 12.15 |
| | SMC FRP | 0.36 | 1.53 | 2.17 | 2.17 |
| Flexure | PFRP | 25.86 | 24.36 | 24.59 | 24.59 |
| | SMC FRP | 2.51 | 8.43 | 11.91 | 11.91 |
| Remark | | OK | OK | OK | - |

### 2.3. Design of Connection System

The 1 MW class floating PV generation complex is composed of unit structures that are connected using the connection system shown in Figure 6. The connection system is devised to minimize welding and increase the fatigue resistance. Moreover, a C-shape device made of stainless steel (STS304), which has high corrosion resistance, is connected using stainless steel bolts. The C-shape device can be made by cutting a stainless steel plate. This connection system behaves as a bending moment free system on the fluctuating water surface.

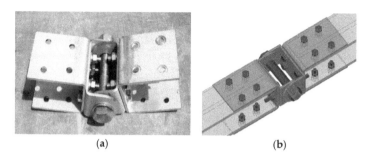

(a)                                                    (b)

**Figure 6.** Connection system between unit structures; (**a**) C-shape device; (**b**) Member connected with C-shape device.

Structural design using the connection system suggested is conducted according to the Pre-Standard for Load and Resistance Factor Design of Pultruded Fiber Reinforced Polymer Structures [12]. American Society of Civil Engineers (ASCE) [12] suggests that the connection system may use the minimum requirements for bolted connection geometries, as shown in Table 5. In Table 5, end distance, edge distance, stagger distance, pitch, and gage are given; notations are illustrated in Figure 7. In addition, bolt shear strength, pin-bearing strength, net tension strength, shear-out strength, and block shear strength must also be considered. The final results of the design of connection system are given in Table 6.

**Table 5.** Minimum requirements of bolted connection geometries.

| Notation | Definition | Minimum Required Spacing (mm) | | Distance of Suggested Connection (mm) | Remark |
|---|---|---|---|---|---|
| $e_{1,min}$ | End distance of all connections | $2.0d$ | 20 | 40 | OK |
| $e_{2,min}$ | Edge distance | $1.5d$ | 15 | 20 | OK |
| $s_{min}$ | Pitch distance | $4.0d$ | 40 | 50 | OK |
| $g_{min}$ | Gage distance | $4.0d$ | 40 | 50 | OK |
| $g_{s,min}$ | Gage spacing with staggered bolts | $2.0d$ | 20 | 50 | OK |
| $l_{s,min}$ | Stagger distance | $2.8d$ | 28 | 74 | OK |

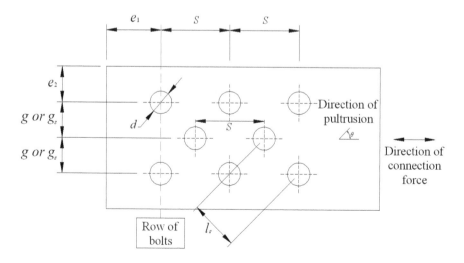

**Figure 7.** Connection geometry and definition for a row of bolts [12].

**Table 6.** Result of design on the connection system.

| Step | Consideration | Load | Note |
|---|---|---|---|
| | Bolt failure (kN) | 65.16 | N/A |
| | Pin-bearing failure (kN) | 128.83 | N/A |
| Checking the failure mode | Net tension failure (kN) | 58.24 | N/A |
| | Shear-out failure (kN) | 101.21 | N/A |
| | Block shear failure (kN) | 57.81 | Failure mode |
| | Critical load (kN) | 57.24 | N/A |
| | Design load (kN) | 46.25 | A |
| Connection design | Applied load (kN) | 29.82 | B |
| | Remark | OK | - |
| | A/B | 1.55 | - |

*2.4. Design of Mooring System*

The mooring systems of the floating structures that are installed at the dam or reservoir may be classified into gravity type, anchor-tension type, semi-rigid type, tension type, and modified type [13]. In the mooring systems, the anchor-tension type using an anchor is generally applied. However, the fluctuation of the water level at the construction site of the 1 MW class floating PV generation complex is less severe than that of a dam or reservoir. The tension type mooring system is not applicable to the present construction site. Therefore, for the safe and commercial design of the mooring system, the anchor-tension type, which is fixed with a chemical anchor and cable to the wall of the waterway in the cooling water intake channel, is adopted.

The anchor-tension type mooring system for the 1 MW class floating PV generation complex is shown in Figure 8. In Figure 8, the level of the chemical anchor is similar to the level of the PFRP member; therefore, cable (wire rope) is attached to maintain a constant tension force in the cable, as shown in Figure 8. The level of the PFRP member is determined by referring to the observation source, which indicates the average water level of the site.

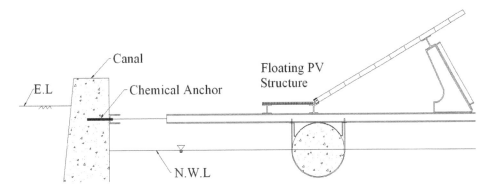

**Figure 8.** Anchor mooring system.

Connection methods between the chemical anchor or the PFRP member and the wire rope are shown in Figure 9. All connection devices are made of stainless steel (STS304), which is cut and bent from stainless steel plate.

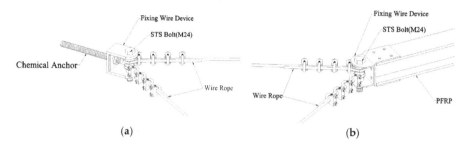

Figure 9. Connection of mooring system; (a) Between chemical anchor and cable; (b) Between PFRP member and cable.

According to the Ministry of Land, Infrastructure, and Transport, Korea (2008), the criteria concerning structures, facilities, etc., of a floating maritime type is specified by the weight of the anchor and the type of cable. The weight of the anchor and the type of rope for floating type maritime structures are determined according to the flowchart shown in Figure 10. In Figure 10, the total resistance ($R$) is determined by combining the wind resistance ($R_a$), flow resistance ($R_w$), and effect of the shape of the structure ($R_v$) as given in Equation (1).

$$R = \sqrt{R_a{}^2 + (R_w + R_v)^2} \tag{1}$$

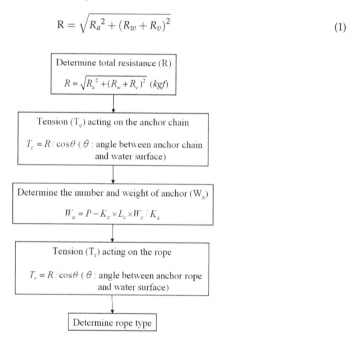

Figure 10. Flowchart for mooring system design [2].

In Figure 2, the construction site can be seen to be separated from an existing bridge. In the design of the system, the effect of the horizontal and vertical directions of the water flow should be considered. The result of the mooring system design is given in Table 7. In Table 7, the diameter of the cable is 10 mm; the safety factor (S.F.) of the cable is 2.

**Table 7.** Result of mooring system design.

| Location | Left | | Right | |
|---|---|---|---|---|
| Direction | Horizontal | Vertical | Horizontal | Vertical |
| Total resistance (kN) | 556.15 | 13.92 | 136.61 | 13.82 |
| Design strength of cable (kN) | 20.97 | 9.31 | 20.97 | 9.31 |
| No. of cable (EA) | | 38 | | 12 |
| Mooring load (kN) | 796.74 | 363.09 | 251.66 | 111.72 |
| Remark | OK | OK | OK | OK |

The total number of cables installed is 60; this is more than specified by the design results (50). The plan view of the mooring system is given in Figure 11. At each mooring point, the number of connected cables at the anchor is between one and three, and the length of the cable is determined by the distance between the anchor and the PFRP member.

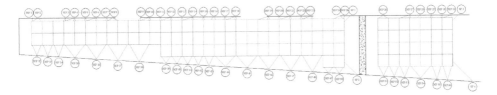

**Figure 11.** Mooring system of 1 MW class floating PV generation system.

## 3. Construction of 1 MW Class Floating PV Generation System

The 1 MW class floating PV generation complex consists of 105 unit structures, as mentioned earlier. The construction process of the 1 MW class floating PV generation complex is classified into the following steps: "fabrication," in which each unit structure is fabricated on the ground, "lifting and launching," in which the fabricated unit structure is lifted and launched onto the water surface, and "mooring," in which each unit structure and moor is connected using anchors and cables.

In the fabrication process, unit structures are fabricated using FRP members, which are cut at the manufacturing facility. At first, plane frames are fabricated on the buoys. SMC FRP vertical members, PV modules, footholds, etc., are assembled in sequence. The fabrication process (A-type) is shown in Figure 12.

| (a) | (b) |
|---|---|

**Figure 12.** *Cont.*

**Figure 12.** Fabrication process; (**a**) Fabricating plane frame; (**b**) Assembling buoy; (**c**) Assembling foothold; (**d**) Assembling SMC FRP member; (**e**) Assembling module support frame; (**f**) Assembling solar module.

In the lifting and launching process, a crane is used to lift and launch the unit structures. To prevent unexpected accidents during the tilting and lifting operation, lifting points in the structure are selected by referring to the FE analysis results. The lifting and launching processes are shown in Figure 13.

**Figure 13.** Lifting and launching process; (**a**) Lifting; (**b**) Launching.

In the mooring process, all unit structures are connected using the connection system shown in Figure 6; chemical anchors are fixed on the wall of the waterway. Chemical anchors installed on the wall are connected to the cables as shown in Figure 9. The mooring process is also shown in Figure 14. Finally, the 1 MW class floating PV generation complex constructed at the site is shown in Figure 15.

(a)

(b)

(c)

**Figure 14.** Mooring process; (**a**) Connecting; (**b**) Anchoring; (**c**) Mooring.

(a)

(b)

**Figure 15.** 1 MW class floating PV generation complex; (**a**) Side view; (**b**) Top view.

## 4. Commercial Viability Evaluation of 1 MW Class Floating PV Generation System

In Korea, the electricity generated in power plants has to be sold to the Korea Electric Power Corporation (KEPCO) at a cost determined by the government. Therefore, the commercial viability evaluation of the 1 MW class floating PV generation complex through the generated electricity is not appropriate.

In this paper, structural analyses with various construction materials using the finite element method were performed to ascertain the safety of the floating PV generation structure, and a commercial viability evaluation was conducted based on the cost of materials. The construction materials used in the fabrication of floating PV generation structures are steel, aluminum, FRP, etc.

The material properties of steel, aluminum, and FRP are summarized in Table 8. The allowable stress of steel, aluminum, and FRP are also given in Table 9.

**Table 8.** Mechanical properties of steel, aluminum, and fiber reinforced polymer (FRP).

| Material | | Yield Strength (MPa) | Elastic Modulus (E, GPa) | Unit Weight (G, kN/m²) |
|---|---|---|---|---|
| Steel | SS400 | 280.00 | 200.00 | 76.91 |
| Aluminum | 6063-T6 | 170.00 | 69.60 | 26.49 |
| FRP | GFRP | 402.58 | 33.28 | 18.42 |

**Table 9.** Allowable stress of steel, aluminum, and FRP.

| Allowable Stress | | Tension (MPa) | Compression (MPa) | Shear (MPa) | Flexure (MPa) |
|---|---|---|---|---|---|
| Steel | SS400 | 140.00 | 140.00 | 80.00 | 140.00 |
| Aluminum | 6063-T6 | 103.03 | 103.03 | 61.82 | 103.03 |
| FRP | GFRP | 201.29 | 134.19 | 26.40 | 161.03 |

From the structural analyses, the cross-sectional dimensions of the member made of each material are determined; H-125 × 125 × 7 × 9 (in mm scale) for steel, H-100 × 100 × 10 × 10 (in mm scale) for FRP, and H-130 × 120 × 10 × 10 (in mm scale) for aluminum, respectively. The results of structural analyses are summarized in Table 10.

**Table 10.** FE analysis results of steel, aluminum, and FRP.

| Material | Tension (MPa) | Compression (MPa) | Shear (MPa) | Flexure (MPa) | Remark |
|---|---|---|---|---|---|
| Steel | 30.02 | 28.94 | 30.38 | 116.24 | OK |
| Aluminum | 15.43 | 15.24 | 13.20 | 89.04 | OK |
| FRP | 1.32 | 1.79 | 12.60 | 25.86 | OK |

Buoys are made of high density polyethylene (HDPE) that are 590 mm in diameter and 2300 mm in length. Floating design is conducted to determine the number of buoys for the structural system made of each material. The results of the floating design are given in Table 11.

**Table 11.** Results of the floating design.

| Material | Length of Buoys (mm) | No. of Buoys (EA) | No. of Structures (EA) | Total No. of Buoys (EA) | Total Volume (m³) |
|---|---|---|---|---|---|
| Steel | | 39 | | 4095 | 2574.99 |
| Aluminum | 2300 | 14 | 105 | 1470 | 924.36 |
| FRP | | 9 | | 945 | 594.23 |

The commercial viability evaluation of floating PV generation complex is conducted by comparing the cost of materials (structural members and buoys). In the evaluation process, common items such as solar modules and mooring systems are excluded. In addition, in the evaluation, the durability of the material, their ease of handling due to light-weight of the material, etc., are not included.

From the results of the viability evaluation, the unit material cost of aluminum is higher than the FRP, and the unit cost of FRP is higher than the steel. However, the cost of aluminum members is higher than the steel members, and the cost of steel members is higher than the FRP members because of the difference of the unit weight (or specific gravity) of materials.

When a 1 MW class floating PV generation complex is constructed using FRP members, the total cost for structural system is 2.49 times lower than the steel members and 1.77 times lower than the aluminum members, mainly because of low specific gravity. Therefore, when a 1 MW class floating PV generation complex is constructed, it is estimated that FRP is the most cost effective material. The results of the commercial viability evaluation are summarized in Table 12.

**Table 12.** Commercial viability evaluation of floating PV generation complex.

| Material | Steel | Aluminum | FRP |
|---|---|---|---|
| Self weight of structure (kN) | 48.84 | 17.60 | 11.48 |
| Unit cost of member (US$/N) | 0.16 | 0.54 | 0.44 |
| No. of structures (EA) | | 105 | |
| Cost of members (US$) | 820,512 | 997,920 | 530,376 |
| Unit cost of buoy (US$) | | 285 | |
| Total No. of buoys (EA) | 4095 | 1470 | 945 |
| Cost of buoys (US$) | 1,167,075 | 418,950 | 269,325 |
| Total Cost (US$) | 1,987,587 | 1,416,870 | 799,701 |

## 5. Conclusions

In this paper, we present the design and construction process of a floating PV generation system with details of its actual construction. Moreover, we suggest the composition of the unit structure and the process for the construction of the large-scale floating PV generation complex.

The 1 MW class floating PV generation complex is not constructed at a dam or reservoir, but is constructed at the waterway of the cooling water intake channel in a thermoelectric power plant. Considering the site conditions, anchors of the tension type that can be fixed to the wall of the waterway using a chemical anchor and cable are adopted. The 1 MW class floating PV generation complex is designed to consider the conditions of the construction site and to promote the commercialization of the floating PV system.

A brief commercial viability analysis was performed based on the material costs of the structural system fabricated with different materials. It found that the structural system fabricated with FRP is the most cost effective due to the light weight of the material.

For the commercialization of a large-scale floating PV generation systems using FRP members, it may be necessary to develop appropriate elemental techniques, construction skills, mooring systems, etc. Therefore, the design and construction techniques developed in the 1 MW class floating PV generation complex may be positive examples for the construction of large-scale floating PV generation complexes in the future.

**Acknowledgments:** This work is supported by the Korea Agency for Infrastructure Technology Advancement (KAIA) grant funded by the Ministry of Land, Infrastructure and Transport (Grant 17AUDP-B099686-03).

**Author Contributions:** S.J. Yoon contributed to conception and design and acquisition of data; S.H. Kim wrote the paper; W. Choi critically revised the paper.

**Conflicts of Interest:** The authors declare no conflict of interest.

## References

1. Choi, H.; Joo, H.J.; Nam, J.H.; Kim, K.S.; Yoon, S.J. Structural Design for the Development of the Floating Type Photovoltaic Energy Generation System. *Mater. Sci. Forum* **2010**, *654–656*, 2803–2806. [CrossRef]

2. Nam, J.H. Development of Floating Type Photovoltaic Energy Generation System Using the Pultruded Structural Members. Ph.D. Thesis, Hongik University, Seoul, Korea, 2010.

3. *Korean Bridge. Design Specification*; Korea Road and Transportation Association: Seongnam, Korea, 2012.

4. Galdino, M.A.E.; Olivieri, M.M.A. Some Remarks about the Deployment of Floating PV Systems in Brazil. *J. Electr. Eng.* **2017**, *1*, 10–19. [CrossRef]

5. Jones, R.M. *Mechanics of Composite Materials*, 2nd ed.; Taylor & Francis: Philadelphia, PA, USA, 1998.

6. Yoon, S.J. *Estimation of Structural Behavior for the Floating PV Generation System with High. Durability, Technical Report*; Shinhwa E&E and ISIS E&C: Seoul, Korea, 2013. (In Korean)

7. *Analysis Reference*; MIDAS IT: Seoul, Korea, 2012. (In Korean)

8. *Structural Plastic Design Manual-ASCE Manual and Reports on Engineering Practice No. 63*; American Society of Civil Engineers (ASCE): Reston, VA, USA, 1984.

9. Choi, J.W.; Joo, H.J.; Nam, J.H.; Hwang, S.T.; Yoon, S.J. Performance Enhancement of Floating PV Generation Structure Using FRP. *Compos. Res.* **2012**, *26*, 105–110. (In Korean) [CrossRef]

10. *Korean Building Code and Commentary*; Architectural Institute of Korea: Seoul, Korea, 2009.

11. *Pre-Standard for Load and Resistance Factor Design of Pultruded. Fiber Reinforced Polymer Structures*; American Society of Civil Engineers (ASCE): Reston, VA, USA, 2010.

12. Kim, J.H.; Kim, H.J.; Hong, S. Experiment and Analysis of Mooring System for Floating Fish Cage. *J. Korean Soc. Fish. Aquat. Sci.* **2001**, *34*, 661–665.

13. Song, J.Y.; Choi, Y.S. Analysis of the Potential for Use of Floating Photovoltaic Systems on Mine Pit Lakes: Case Study at the Ssangyong Open-Pit Limestone Mine in Korea. *Energies* **2016**, *9*, 102. [CrossRef]

*Article*

# Graphical Diagnosis of Performances in Photovoltaic Systems: A Case Study in Southern Spain

Isabel Santiago *, David Trillo Montero, Juan J. Luna Rodríguez, Isabel M. Moreno Garcia and Emilio J. Palacios Garcia

Departamento de Arquitectura de Computadores, Electrónica y Tecnología Electrónica,
EscuelaPolitécnica Superior, Universidad de Córdoba, Campus de Rabanales, Edificio Leonardo da Vinci,
E-14071 Córdoba, Spain; ma2trmod@uco.es (D.T.M.); el1luroj@uco.es (J.J.L.R.); p92mogai@uco.es (I.M.M.G.);
p92pagae@uco.es (E.J.P.G.)
* Correspondence: el1sachi@uco.es; Tel.: +34-957-218-699

Received: 3 November 2017; Accepted: 22 November 2017; Published: 25 November 2017

**Abstract:** The starting point of the operation and maintenance tasks in photovoltaic plants is the continuous monitoring and supervision of its components. The great amount of registered data requires a major improvement in the ways this information is processed and analyzed to rapidly detect any potential fault, without incurring additional costs. In this paper, a procedure to perform a detailed graphical supported analysis of the operation of photovoltaic installations, based on inverter data, and using a self-developed application, is presented. The program carries out the automated processing of the registered data, providing their access and visualization by means of color maps. These graphs allow a large volume of data set to be simultaneously represented in a readable way, enabling operation and maintenance operators to quickly detect patterns that would require any type of intervention. As a case study, the operation of a grid-connected photovoltaic plant located in southern Spain was studied during a period of three years. The average daily efficiency values of the PV modules and inverters were in the range of 7.6–14.6%, and 73.5–94% respectively. Moreover, the presence of shadings, as well as the hours and days mainly affected by this issue, was easily detected.

**Keywords:** photovoltaic plants; software development; performance analysis; loss analysis; graphical malfunction detection

---

## 1. Introduction

According to Solar Power Europe's report entitled Global Market Outlook for Solar Power 2017–2021 [1], in 2016, there was a record of solar growth as 76.6 GW of solar plants was installed and connected to the grid. That is the largest amount of solar power installed in a year so far, which supposed a 50% year-on-year growth over the 51.2 GW added in 2015. In just a decade, the world's cumulative solar capacity increased by over 4500%, from 6.6 GW in 2006. The 306.5 GW of total grid-connected PV capacity installed at the end of 2016 generated around 2% of the world's electricity demand. From today's perspective, it is expected that the total global installed PV capacity will exceed 400 GW in 2018, 500 GW in 2019, 600 GW in 2020 and 700 GW in 2021 [1].

Although all these solar growth numbers may sound very impressive, solar energy still has a long way to go before it can achieve its full potential, having many obstacles and technical challenges to overcome in order to improve its profitability and to be fully integrated as a dispatchable energy resource in the electricity market [2–6]. The 1.5 °C Paris goal will require gigantic efforts, and, among others, it will need to direct much more money into renewable resources, and new business opportunities may be open in this sector in the near future [1].

Improving the profitability of photovoltaic (PV) installations requires optimizing their production, thus reducing their costs and increasing the plant's useful life [5,7,8]. To achieve this while still using the currently available technology, a correct design of the installations and an adequate intervention of the services of operation and maintenance (O&M) are crucial [9]. These services are playing an increasing role, and are being recognized for their determining function in ensuring long-term revenues, which is very important when taking into account the current tendency to install, mainly in emergent markets, megawatt-scale PV parks [1,2,7,9]. O&M has even charted its own course to become a standalone business segment and a critical component of the solar energy value chain [1]. Although O&M markets will continue their consolidation in Europe, in emerging markets, ambitious national solar programs are being carried out with few experienced O&M service providers, thus improving these services is one of the challenges for the next few years.

For an adequate intervention of all these O&M activities, the starting point is the continuous monitoring and supervision of both the PV power plant conditions and the performance of all its components [10,11]. An important requirement in this sector is to improve the ways of effectively processing and analyzing the large amount of data registered. In this way, they can be converted into useful information, for faster identification of behavior changes that might compromise the systems' performance and for faster intervention at the plant if necessary [7,12]. In addition to technical activities, the financial aspects of these tasks are crucial in order to identify and accomplish the appropriate actions in an optimal mode, based on a cost-benefit analysis, with the objective of these services being profitable for both the companies carrying O&M services out and for the owners of the PV installations [12]. Optimal operations must strike a balance between maximizing production and minimizing cost [7].

In this context, the main objective of this paper was to implement a procedure to perform a detailed graphical supported analysis of the operation of PV installations, using a self-developed application named *S·lar*2. This software enables us to carry out a detailed performance study of all PV plant's components in a fully automated way. To do this, it calculates the behavior indexes and loss rates of the PV system components [13–17] using data recorded by the inverters, having previously provided all technical specifications of the plant. All parameters analyzed were graphically represented in the form of color maps, boosting the application of advanced data visualization (ADV) techniques [18–23] and improved interfaces and decision support (IIDS) tools [24] in the analysis and management of energy production in PV plants. With this type of data visualization, very easy and intuitive information about the behavior of panels and inverters, as well as possible deviations from normal operation or anomalous situations, can be obtained at a glance. The instants of time at which such events take place can be quickly detected. Using this procedure, the operation of a grid-connected PV plant located in southern Spain was analyzed in detail during a period of three years.

The diagnosis of the operating mode and fault detection in PV plants is increasingly attracting the interest of researchers [25–31], and numerous papers have been published in the last few years dedicated to analyzing the operation and performance of the components of working PV plants [4,32–53]. Compared to most previous publications, where usually only the daily and/or monthly values of yield and loss indexes are represented, layering the data on a color map allows simultaneously displaying all instantaneous data registered throughout the monitoring period, even if it comprises several years, easing the comprehension and interpretation of large amounts of data. Large-scale visualizations would enable O&M operators to quickly detect patterns in data that would otherwise go unnoticed, helping to close the gap between information and insights, and facilitating diagnostics or alarm management with a minimal effort. This would allow decision-makers to act as quick as possible. Moreover, this procedure does not include additional economic costs since data recorded by the inverters themselves are used.

The structure of the paper is organized as follows. Section 2 describes the magnitudes to be analyzed and the procedure carried out by the developed software. Section 3 describes the characteristics of the grid-connected PV plant analyzed as a case study. Section 4 presents and discusses the results, and, finally, the conclusions obtained in this paper are exposed in Section 5.

## 2. Functionalities Included in the Software *S·lar2*

In this section, the parameters involved in the study are described, as well as the most important characteristics of the software used and the procedure carried out to process and visualize all data.

### 2.1. Description of Monitored Data and Calculated Parameters

Data monitored by the inverters of a PV plant connected to the grid are shown in Table 1. The values of the irradiance in the plane of the PV panels, $G$, the ambient temperature, $T_a$, and the temperature measured on the back surface of the PV modules, $T_m$, were also recorded.

**Table 1.** Analyzed parameters measured by inverters.

| Parameters | Notation |
|---|---|
| DC current from PV modules | $I_{DC}$ |
| Inverter input DC voltage from PV modules | $V_{DC}$ |
| Inverter output AC power | $P_{AC}$ |
| Inverter output total energy | $E_{AC}$ |
| AC current injected into the grid | $I_{AC}$ |
| Grid current | $I_{grid}$ |
| Grid AC phase voltage | $V_{AC}$ |
| Grid frequency | $f_{grid}$ |
| Inverter operating status | *Status* |
| Inverter error code | *Error* |

With these measured data, while also using technical information of the plant, a series of standardized behavior magnitudes and rates must be calculated in order to analyze the performance and losses of each of the installation components [13–17,54].

On the one hand, the so-called reference, $Y_r$, temperature corrected, $Y_T$, array, $Y_A$, and final yields, $Y_F$, were determined. The reference yield is defined as

$$Y_r = E_G/G_{STC} \tag{1}$$

where $E_G$ is the tilted irradiation, determined by multiplying the value of the in-plane irradiance, $G$ (W/m²), by the monitoring time interval expressed in hours, $t$ (h). $G_{STC} = 1000$ W/m² is the irradiance in STC standard conditions (AM1.5, 1kW/m², 25°C). The reference yield represents the time that radiation should be received by the PV array with a value of $G_{STC} = 1000$ W/m² in order to generate $E_G$ [13,17,33,54]. Its value depends on the location, orientation and inclination of the PV system as well as on the climate and weather conditions.

The reference yield corrected by the effect of temperature is determined by means of the expression

$$Y_T = Y_r \cdot [1 - c_T(T_c - T_{STC})] \tag{2}$$

where $T_c$ is the PV cell temperature (°C); $T_{STC}$ is the temperature at STC, 25°C; and $c_T$ is the temperature coefficient of the PV modules (%·°C$^{-1}$), given by the manufacturer in its specifications. This expression reflects the decrease of the production due to the modules' PV cells work with a temperature above 25 °C. The values of $T_c$ were determined using the expression

$$T_c = T_a + (G/800) \cdot (T_{INOCT} - 800) \tag{3}$$

This is the most frequently used NOTC model to determine the value of $T_c$ [4,29,37,47,53–59], but, in this case, the parameter $T_{INOCT}$ is used instead of the normally used $T_{NOCT}$. This latter value is defined as the Normal Operating Cell Temperature (NOCT) and represents the mean PV cell junction temperature in the so-called Nominal Terrestrial Environment (NTE) conditions (ambient temperature $T_{a,NTE} = 20\,°C$, wind speed $v_{NTE} = 1$ m/s on the front and back sides of the PV module and irradiance $G_{NTE} = 800$ W/m$^2$) in an open-rack mounted, open-circuit module tilted perpendicular to the solar noon. The $T_{NOCT}$ should characterize the module's temperature dependence and should allow the estimation of the module's performance and energy yield over time [60]. The value of $T_{NOCT}$ is provided for standard conditions by the module manufacturer in its specifications, while the parameter $T_{INOCT}$ is defined taking into account the particular shape in which the panels have been mounted and oriented, which may not be the same used by the manufacturer for $T_{NOCT}$ measurements in the laboratory. The parameter $T_{INOCT}$ depends on the value of $T_{NOCT}$ and the configuration used in the installation to mount the PV panels [61]. For the case of open-rack mounting, which is the case of the installation analyzed in this work, $T_{INOCT}$ is given by the expression [62,63]

$$T_{NOCT} = T_{INOCT} - 3\ (°C) \tag{4}$$

Greater differences between both temperatures are found in the case of panels integrated directly in façades or building roofs, where the variation between both can be up to 18 °C [62–64].

The array yield is defined as

$$Y_A = E_{DC}/P_{STC} \tag{5}$$

where $E_{DC}$ is the DC energy from the PV modules, determined from the DC current and voltage from PV panels multiplied by the recording time interval expressed in hours, $t$, and $P_{STC}$ is the PV nominal power under standard conditions measured in W. This quantity represents the time in which the PV generator must be operating with nominal power $P_{STC}$ to generate $E_{DC}$ energy [13].

The final yield is determined by

$$Y_f = E_{AC}/P_{STC} \tag{6}$$

where $E_{AC}$ is the AC energy from inverter, determined from the monitored inverter output AC power, $P_{AC}$, and the recording time interval expressed in hours, $t$ (h). This quantity represents the time in which PV generator plus the inverter must be operating with nominal power $P_{STC}$ to generate $E_{AC}$ energy [13]. It depends on the location and type of installation, so it enables one to compare the production of similar PV installations with different sizes but located in a specific geographic region.

Calculating the differences between the previous parameters, the losses of the different components may be determined. The array capture losses may be calculated as the difference between the reference and the array yields as

$$L_c = Y_r - Y_A \tag{7}$$

Capture losses are given by the sum of thermal losses, $L_{ct}$, and miscellaneous losses, $L_{cm}$. The thermal capture losses are determined by the difference between the reference yield and the temperature corrected reference yield

$$L_{ct} = Y_r - Y_T \tag{8}$$

and represent the losses due to PV modules that are operating above 25 °C [62]. The miscellaneous capture losses are determined by means of the difference between the temperature corrected reference yield and the array yield, and it is given by

$$L_{cm} = Y_T - Y_A \tag{9}$$

It represents the rest of the losses in the PV system, which may be associated with multiple causes such as the Joule effect in the wiring, diodes losses, shading effects, inhomogeneous or low irradiance, snow, contamination or dirt accumulation, mismatch, maximum power tracking error, etc. An incorrect operation of a PV grid-connected installation leads to a significant increase in $L_{cm}$ value, being this magnitude an indicator of the existence of problems in the system. Finally, the difference between the array and the final yields is given by

$$L_s = Y_A - Y_f \qquad (10)$$

which corresponds to the losses produced in the inverter.

The performance ratio, expressed by the ratio of the final and reference yield

$$PR = Y_f/Y_r \qquad (11)$$

is a dimensionless index mainly used to characterize the performance of PV installations, which measures the degree of utilization of an entire PV system [65,66]. It is the ratio of useful generated energy versus the energy that should be generated by a lossless ideal PV plant at 25 °C and receiving the same irradiation. This parameter was defined in the recently revoked standard IEC 61724:1998 [16], and it is considered with more detail in the recently published standard IEC 61724-1:2017 [17]. It takes into account all the effects that involve the PV system performance and indicates the overall effect of losses on the PV system's rated output due to array temperature, incomplete utilization of the irradiation and system component inefficiencies or failures [16,33,65].

For determining the global value of *PR* for a PV installation, $Y_f$ must be determined using the $E_{AC}$ values measured at the PCC by the meter installed by the electric company [65]. However, in this work, the objective is to be able to compare the operation of the three inverters available in the PV installation. The *PR* value of the ensemble formed by each inverter of the installation, together with the PV array connected to it, was determined, and, for this, the value of $E_{AC}$ registered by the inverters was taken into account for determining the value of $Y_f$ for each inverter. In this case, the losses that would take place between the inverter and the meter were not considered in this work.

The temperature-corrected performance ratio, $PR_T$, was defined by the IEC 61724-1:2017 [17]. In this case, instead of using the value of the final yield previously defined, the value of a corrected temperature final yield, $TC\_Y_f$, defined as follows, will be used

$$TC\_Y_f = E_{AC}/(P_{STC} \cdot [1 - c_T(T_m - T_{STC})]) \qquad (12)$$

Instead of using a constant value of $P_{STC}$, the reference power is calculated at each recording interval to compensate for the differences between the actual module temperature ($T_m$) and the STC reference temperature of 25 °C. The standard IEC 61724-1:2017 [17] indicates that, in this case, the measured module temperature $T_m$ may be used.

The efficiencies with which the components of the installation are working were also determined. The efficiencies of the PV modules were calculated by means of the expression

$$\eta_G = (E_{DC}/E_r) \times 100 \qquad (13)$$

where $E_r$ is the reference energy calculated by the product of the tilted irradiation, $E_G$, by the PV module area associated to each inverter, *A* [54]:

$$E_r = E_G \cdot A \qquad (14)$$

This reference energy represents the available irradiation on the total PV module surface. The efficiency of the inverter was determined by the ratio of the AC energy from inverter versus the DC energy from PV modules:

$$\eta_{inv} = (E_{AC}/E_{DC}) \times 100 \qquad (15)$$

It depends on the input voltage to the inverter (some inverters operate more efficiently in the upper area of the MPP voltage window, while other manufacturers prefer the lower, and even some choose the intermediate zone). This should be taken into account when choosing the number of modules in the series associated to each inverter [67].

Finally, the system efficiency, given by the product of the inverter efficiency and the efficiency of the PV modules associated with it,

$$\eta_s = (\eta_G \cdot \eta_{inv})/100 \tag{16}$$

provides the efficiency of the system composed by both of them. In the case of installations with more than one inverter, these magnitudes can be determined for the systems comprised by each inverter and its associated PV modules and/or for the complete installation, adding or multiplying the results for each inverter, as appropriate.

## 2.2. Description of Data Processing and Visualizing Procedure

The management and processing of the monitored data were carried out using our self-developed software *S·lar2*. This software was specifically designed for the automatic reading, management, treatment and storage of data measured in PV installations. A previous version of this software was the objective of the study in a previously published work [54], where it was described in detail. In the new version, this software was modernized to a web application, more flexible to be installed on the customer's equipment, and with shorter processing and consulting times. It was developed using Python [68], HTML [69], CSS [70] and Typescript [71].

Figure 1 shows a schematic diagram where the steps that are carried out in the developed program for processing and visualizing the information are reflected. The monitored data of the PV installation, as provided by the company which are the owner of the plant, are distributed in one file per day and organized in directories by months and years. *S·lar2* enables monitoring data to be automatically entered into different tables in a Relational Database Management System (RDBMS), specifically designed to contain all of this information. The application has a graphical user interface (GUI) that facilitates data migration and all their processing. Once all recorded data have been stored, the software allows for performing a series of mathematical operations on the monitored parameters for the calculation of Equations (1)–(16) for each inverter. Therefore, each of the previously shown standard magnitudes and performance rates of each individual component of the PV installations can be calculated and stored.

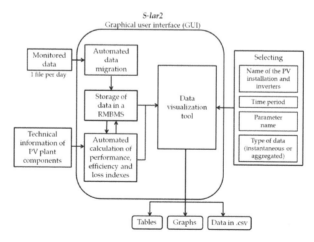

**Figure 1.** Esquematic diagram of the procedure used by *S·lar2* for the automated processing of monitored data in PV installations.

Moreover, the application determines both the "instantaneous values" of all parameters and calculated magnitudes, which will have the same temporal frequency as the recording parameters, and all of their average and/or accumulated or aggregated values, which are also automatically calculated. These last values must be calculated to produce an hourly scale (the sum or average of the instantaneous values recorded or calculated for each hour), a daily scale (sum or average of the 24-hourly values for each day), and a monthly scale (sum or average of the daily values for each month).

Once data are calculated, the *S·lar2*'s graphical user interface (GUI) also enables the user to easily access and select all data stored in the databases, by means of a series of queries via its data visualization tool. A screenshot of this data visualization tool can be observed in Figure 2a. By means of a series of buttons and menus, selected data can be directly visualized either graphically (Figure 2b) or numerically, in tabular form, or even being exported as .csv files with the aim of using them with other third-party software for its analysis.

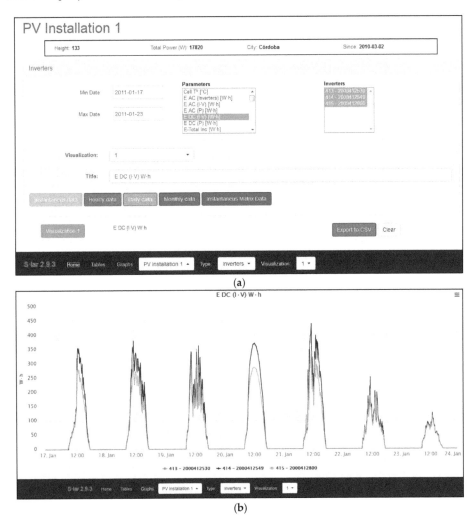

(a)

(b)

**Figure 2.** *S·lar2*'s graphical user interface (GUI): (a) selection of data; and (b) graphical visualization of selected parameter ($E_{DC}$) for the three inverters.

The PV installation was analyzed as a case study during a period of three years (from 1 January 2011 to 31 December 2013), for which the monitored data of the installation were available. The 1096 files, which contain the parameters recorded by the three inverters as well as the irradiance and temperature data, were provided by the company that owns the facility. These files were then uploaded using the software and subsequently calculating all the yield and loss indexes. By using the developed application, the same parameters for the three inverters were queried.

For the analysis, all the instantaneous data of each parameter were represented in the form of color maps. With this type of data visualization, a large volume data set can be represented in a readable way, easing its interpretation, and with the aim of assisting in the decision-making process of PV plant maintenance tasks. To do this type of representation, in *S·lar*2, the option has been added to request for any of the parameter's data in a matrix form. The application will supply us data in an array that will have as many columns as the requested days corresponding to the monitoring period, and as many rows as records have been made throughout each day. The days in which, due to technical problems, there are no monitoring data, the program provides the columns corresponding to those days with null or empty values, according to interest. Since the PV installation has been analyzed in this work during a period of three years, there were 1096 columns in total in these matrices, but obviously the matrix's number of columns can be adjusted to the number of days in which this analysis of the PV installation is desired to be done. On the other hand, for a recording interval of 5 min, as is the case of the PV installation analyzed in this work, there are 288 data recorded per day for each parameter, so this will be the number of rows that the matrix will have: the first data or row being the one corresponding to the registration at 00:00, and the last one corresponding to the registration at 23:55. Then, the data of these matrices will be represented by color diagrams. In the *Y*-axis, the 288 registers or rows are represented, while the *X*-axis shows the monitored days, or the different columns that the matrix have, which is the same. The magnitude of the values will be given in such graphical representations by a color scale.

Graphs with daily data, which will correspond to the sum or average of the values of each of the matrix's columns, will be presented after the color maps, such that the union of both graphs will provide complete information of each parameter analyzed.

Although *S·lar*2 carries out the graphical representations of the data, in this work the .csv files provided by the software was exported and loaded into MATLAB (version 2013, MathWorks, Natick, MA, USA), with the aim to provide high-quality figures in the paper. By means of a routine, instantaneous data matrices and daily data vectors were uploaded, and all figures were represented in an automated way.

## 3. Specifications of the PV Plant Analyzed for the Case Study

In this paper, a grid-connected PV system currently operating in Andalusia, Spain, was analyzed. This installation belongs to the company Solar del Valle SL, which provided both the installation's technical specifications and the data obtained from monitoring [72]. This facility is located in the city of Córdoba (Cordova), with a nominal power of 17.8 kW. It comprises three inverters, each one with 36 modules connected to it (three parallel strings with 12 modules in series each one). The parameters of its modules, mounted in an open rack on a rooftop (tilt: 30°, azimuth: 18°), are shown in Table 2, whereas inverter specifications are listed in Table 3. The three inverters of the PV installation has been named inverter 413, inverter 414 and inverter 415, which are labels used by the owners of the plant.

Monitoring was carried out in the PV installation with the following equipment. In-plane irradiance and temperatures were recorded by using a device called Sunny SensorBox, from the SMA company [73], whose sensors and specifications are reflected in Table 4. The other measurement sensors and the data acquisition system were located in the inverters. The Sensor Box and inverters were connected via RS485 to a Sunny WebBox, also from SMA. This multifunctional device, with a low-consumption data logger, is responsible for collecting data continuously from inverters and the SensorBox. Measurement data are transmitted through a GSM modem from remote locations where a telephone or ADSL connection is available. In the installation, monitoring data, recorded at 5-min intervals in the WebBox, were extracted via an SD card.

**Table 2.** PV module technical specifications.

| PV Module | Specifications |
|---|---|
| Model | BP-3165 |
| Type | Polycrystalline Silicon |
| Number of cells | 72 (6 × 12) |
| Nominal power ($P_{STC}$) | 165 W |
| Power tolerance | ±3% |
| Module efficiency | 13.1% |
| Maximum power current ($I_{pm}$) | 4.7 A |
| Maximum power voltage ($V_{mpp}$) | 35.2 V |
| Short circuit current ($I_{sc}$) | 5.1 A |
| Open circuit voltage ($V_{oc}$) | 44.2 V |
| NOTC | 47 ± 2 °C |
| Module area | 1593 × 790 × 50 mm |
| Weight | 15.4 kg |
| Top side | Tempered glass |
| Encapsulating material | EVA |
| Back side | White polyester |

**Table 3.** Inverter specifications.

| Inverter | Specifications |
|---|---|
| Model | SMA SMC-5000 [73] |
| **Input** | |
| Recommended Maximum DC power | 6.35 kWp |
| Maximum DC voltage | 600 V |
| Maximum DC current | 26 A |
| Nominal DC voltage | 270 V |
| **Output** | |
| Nominal AC power | 5 kW |
| Maximum output current | 26 A |
| THD of grid current | <4% |
| Number of phases | 1 |
| Maximum efficiency | 96.0% |
| Euro-eta | 95.1% |
| Weight | 63 kg |

**Table 4.** Irradiance and temperature sensors.

| Sensor | Measurement Range | Accuracy | Resolution | Magnitude Measured |
|---|---|---|---|---|
| Calibrated ASI amorphous PV cell | [0,1500] W/m$^2$ | ±8% | 1 W/m$^2$ | In-plane total irradiance on the PV modules (it is located close to one of them, with the same inclination and orientation) |
| PT-100M | [−20,+110] °C | ±0.5 °C | 0.1 °C | Module temperature |
| PT-100M-NR | [−20,+110] °C | ±0.7 °C | 0.1 °C | Ambient temperature |

## 4. Results

### 4.1. Irradiance, Ambient and Module Temperature

Firstly, the external parameters of irradiance and temperature to which the PV modules were subjected to are shown in order to have them as a reference to the dependence of the production indices and losses with these parameters. In Figure 3a the tilted irradiance registered during the monitored period of three years is represented. The first day corresponds to 1 January 2011, whereas the 1096th day corresponds to 31 December 2013. Firstly, it is notable that the set of 315,648 records over the three years (288 records per day for 1096 days) is represented on the same graph, and the set of all data is easily interpretable at a glance. The Y-axis shows the hours of the day in which the registers take place, from 00:00 to 23:55, which correspond to the first and last data measured each day. It is possible to observe the seasonality of the irradiance values for the geographic location as well as for the inclination and orientation of the PV panels of the studied installation. In dark blue color, the null values and the instants of time in which there is no record of irradiance, either because it corresponds to night hours or because of the absence of data for some type of technical problem, can be seen. The dark red color corresponds to the maximum values of irradiance. The maximum recorded value was 1141.6 W/m², which took place on 14 April 2012 at 13:50. It is observed in the graph that the months of January and December are the period of the year in which days have the lowest number of hours with non-zero values of irradiance. The hour daily range with solar radiation increases progressively until the summer months, where the number of hours with sun and therefore electricity production is the highest. It can also be observed that the greatest irradiance records take place during the central hours of the day, between 12:00 and 15:00. The two time changes that take place throughout each year in Spain, in the passage from winter to spring (around the end of March), in which there is an advance of 1 h (from Greenwich Mean Time (GMT)+1 to GMT+2), and the transition from summer to autumn (at the end of October), where there is a delay of 1 h (From GMT+2 to GMT+1), are also reflected in the image.

It is observed in Figure 3a, as well as in many others of the work, the range of days, between 14 November 2012 and 1 January 2013 (Days 684–731), in which there was no data in the PV plant's monitoring system due to technical problems, as well as 26–28 July 2011 (Days 207–209), in which there was also fault in the data.

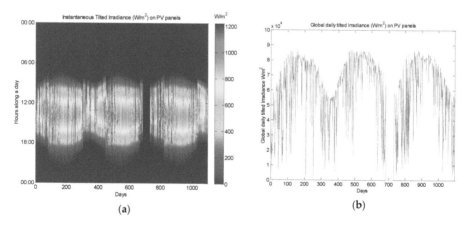

**Figure 3.** Tilted irradiance registered from 1 January 2011 to 31 December 2013: (**a**) instantaneous values, during 24 h a day; and (**b**) total daily values.

In Figure 3a, it is possible to observe the difference between the irradiance profiles that occur on clear cloudless days, in which the variation experienced by irradiance values throughout the day is exclusively due to variations in the solar path, with respect to the days with clouds passing, in which, from some registers to others, there were great variations in the values of the irradiance. It can also be observed in those covered days that the values of irradiance were very low throughout the day.

The correspondence between the data of Figure 3a and those reflected in Figure 3b, where the total daily values of irradiance are represented, can be observed. In Figure 3b it is shown, in addition to the annual periodicity that takes place due to the seasonality, already reflected in Figure 3a, the great variation that can take place from day to day in the total daily values of irradiance, present in general throughout all the seasons, although this variation is inferior in summer due to a greater predominance of clear days. This fact results in a great variability and randomness in the production of energy in this type of renewable installations, and therefore in their difficulty to be considered as manageable energy generation sources in the electricity markets. The higher daily total values, above 80 kW/m$^2$ per day, were concentrated between the months of April and July. The highest total daily value corresponded to 29 May 2013, with a total of 86.89 kW/m$^2$. In contrast, in the months from January to December, the total daily irradiance values were generally the lowest. In these months, the highest total irradiance did not exceed 55 kW/m$^2$ per day. Although, as can be observed, it was possible in summer to find days with a total irradiance collected by the panels of less than 55 kW/m$^2$, this may be less than 10 kW/m$^2$ in winter days.

The ambient, the measured at the back of the modules, as well as the estimated PV cell temperatures are reflected in Figure 4a. The color maps allow an intuitive visualization of the temperature distribution over the three-year period. The average daily values of these three temperatures are shown in Figure 4b. These figures show the seasonality of these three magnitudes, with values being very high in summer, greater between 12:00 and 18:00, and lower in winter, mainly in the months of January and December. The highest registered ambient temperature value, 48.23 °C, took place on 10 August 2012 at 15:15, while the lowest value, −1.17 °C, was recorded on 5 February 2012 at 8:05. The highest values of module and cell temperatures obtained, 73.15 °C and 75.38 °C, respectively, also took place on 10 August 2012 at 15:05, while the lowest values, −5.25 °C and −5.23 °C, respectively, were recorded on 12 February 2012 at 8:15.

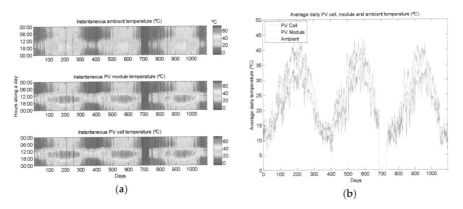

**Figure 4.** Registered ambient and module temperature, and calculated cell temperature registered from 1 January 2011 to 31 December 2013: (**a**) instantaneous values, during 24 h a day; and (**b**) average daily values.

The high temperature of the PV modules, being an average of 4.44 °C above the ambient temperature is worth noting. The maximum increase of the module temperature compared to the registered ambient temperature was of 35.96 °C, which was registered on 11 April 2012 at 13:40, while the maximum variation in which the temperature of the module was lower than ambient, 6.6 °C, took place on 13 June 2012 at 7:35. The mean difference found between $T_c$ and $T_m$ is

0.57 °C, although the maximum temperature difference obtained between the both values was 3.43 °C, also recorded on 11 April 2012 at 13:40. The average ambient temperature values range from 5 °C to 36 °C, between 6 and 43 °C for the module temperature and between 6.5 °C and 43.7 °C for the cell temperature.

*4.2. DC Energy Produced by PV Panels*

Once the conditions to which the components of the PV system were exposed to have been shown, in Figure 5a the instantaneous DC energy generated in the PV arrays connected to the three inverters of the installation is represented. The total daily values of this energy are shown in Figure 5b. It can be noticed how the values of this parameter are directly related to the irradiance values received by the panels previously shown in Figure 3a. The seasonality in production over the three years, the highest values of production in the mid hours, the highest number of hours of production in spring and summer, and the differences of the daily DC production profiles on clear days, with small transitions, compared to that of the days with the passage of clouds, with a greater number of transitions, can be observed, all aspects previously commented on when analyzing the color map of irradiance values.

Since the nominal power of panels associated with each inverter is the same, the energy produced in all three should be practically the same. However, if one observes Figure 5a, it can be clearly seen that, in the winter months, during the period from September to March, the electricity output in the arrays connected to inverter 415 is lower than that of the other two inverters. This fact is also shown in Figure 5b, where it is observed that the daily output of the panels corresponding to inverter 415 is lower during these months. The maximum difference occurred on 2 February 2013, in which the daily output of inverter 415 panels was lower by 7.56 Wh than that occurring in inverter 414 panels, which means a daily production that is 24.37% lower. However, the color map allows us to discern, throughout the year, the hours of the day when the production of the third inverter is lower, an aspect that obviously is not evident if only the daily total data are represented.

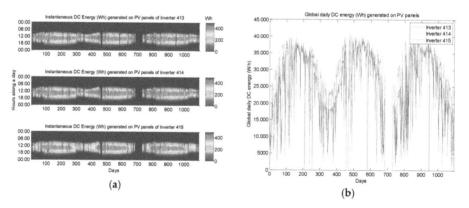

**Figure 5.** Instantaneous DC energy generated in the PV arrays connected to inverters registered from 1 January 2011 to 31 December 2013: (**a**) instantaneous values, during 24 h a day; and (**b**) total daily values.

Figure 5a also shows, during the period from 2 to 12 April 2012 (Days 458–468), a stop in the production of the three inverters, as well as a series of stops that took place only in inverter 413, and that corresponded to 26 June 2011 (Day 238), from 3 to 6 August 2012 (Days 581–584), 17 April 2013 (Day 838) and from 3 to 5 August 2013 (Days 946–948). These stops are also reflected in both Figure 5b and in the enlargement shown in Figure 6. This last figure represents the daily DC energy generated from 23 June 2012 (Day 540) to 15 September 2012 (Day 624). The stop in the production of panels can be clearly seen, which corresponds to inverter 413 from 3 to 6 August 2012 (Days 581–584).

It can also be seen that the total daily output of inverter 413's panels is slightly lower than that of the inverter with the highest production, which is 414, and this difference remains practically constant over the three monitored years. These smaller differences are more difficult to visualize in the color maps, so it is recommended that these maps be accompanied by graphical representations of the daily values in order to complement the information of the first ones. In total, the production of panels associated with inverters 413, 414 and 415, during the three monitoring years, was 28.2, 28.9 and 26.4 MWh, respectively. The output of the 415 inverter was 2.5 MWh lower than that of inverter 414, which is 8.6% lower. This means an average difference between both inverters of 0.83 MWh/year. The difference between inverters 413 and 414 is smaller, 0.7 MWh, which is 2.5% lower. This implies an average difference of 0.24 MWh/year. These differences in the production of the PV panels were reflected in a smaller injection of energy into the grid (0.78 and 0.22 MWh/year respectively for 415 and 413 compared to inverter 414). Considering the remuneration tariff for this type of installations (Type I.1) in 2011 [74], this meant a loss of 245 and 70€ per year respectively for inverters 415 and 413, which despite variations in tariffs, over the plant service life may lead to a considerable decrease in revenues. In the case of inverter 415, this deviation is mainly due to the presence of shading effects.

### 4.3. Maximum AC Energy Produced by Inverters

In this study, the behavior pattern of the instantaneous and total daily energy $E_{AC}$ measured in the inverters' outputs are really similar to those of the $E_{DC}$ represented in Figure 5a,b respectively. Thus, instead, in Figure 7 there is a representation in which, for each instant of time, the inverter with the maximum $E_{AC}$ energy production is indicated. A code is used, such as 0 (dark blue) means there is no electricity production; 1 (intermediate blue) represents that inverter 413 is producing the maximum value of $E_{AC}$ in that instant of time; 10 (light blue) represents that inverter 414 is producing the maximum value of $E_{AC}$; and 20 (yellow) represents that inverter 415 is producing, in this case, the maximum value of $E_{AC}$. The values 15 (green), 25 (orange) and 30 (dark red) would represent the instants of time in which two of the three inverters (413–414, 413–415 and414–415, respectively) are simultaneously producing the maximum amount of $E_{AC}$, although these situations are infrequent during these three years. The situation in which the three inverters are producing the same quantity of $E_{AC}$, which would correspond to number 35, according to the code used, does not occur during the entire monitoring period. It can be seen that the inverter that is more often producing more energy than the others is inverter 414, which results in the fact that its daily production, as already seen above, is superior to that of the other two inverters in the greater part of the monitoring period. However, it can be observed that the behavior reflected in the figure presents an annual periodicity. The 414 inverter has maximum production during the central hours of the day in the spring and summer months. Inverter 413 produces the most during the first hours of the day, especially during the summer months, and during the last hours of the day, especially during the months corresponding to autumn and winter. In addition, throughout autumn and winter, the maximum production oscillates between inverters 413 and 414. Inverter 415, in a smaller number of occasions, presents the maximum production. However, this occurs during the first and mainly during the last daily production records, but this situation is much less frequent during the central hours of the day.

In accordance with this figure are the data represented in Figure 8. In it, for each of the registers over the three years, the inverter that is having the maximum output has also been determined. Considering that this should be the production expected for the other two inverters, the difference between the output power in each inverter, $P_{AC}$, and the maximum output found in the three are calculated for each of them. This difference, in percent, with respect to the value of the nominal power of PV panels associated to each inverter, $P_{STC}$, is represented. Figure 8b shows the same values, but on 25 February 2011, as an example of one day in which there are more differences in the production of inverters (around 20%), and on 17 June 2011, as an example of a day in which the difference among the output of the different inverters is the smallest recorded (around 2%). As shown in Figure 7, inverter 414 had the highest production on a greater number of occasions (for that reason, the value corresponding to it in Figure 8b in the central hours of the day was zero), except in some instants of time in the first and last hours of the day, in which the production of this inverter was inferior, thus making inverter 413 the maximum producer in those instants of time.

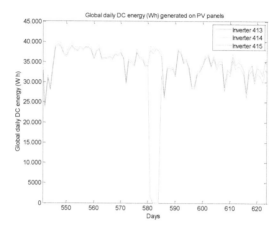

**Figure 6.** Total daily DC energy generated in the PV arrays connected to inverters, from 23 June 2012 to 15 September 2012.

The influence of some aspects that show a periodicity throughout the year is clearly manifest in Figures 7 and 8, indicating, because the PV installation is located on the roof of a building, the presence of shadows caused by the structure of the building and/or those surrounding it, which is affecting to a greater or lesser extent the production of panels associated with each of these inverters throughout the year. This reduction in production can be improved by rectifying, as far as possible, the design of the PV plant or, at least, by taking this aspect into account in other future installations in which similar circumstances may occur. If rectification is not possible, at least they are foreseeable variations that can be considered in a possible management of the production of the analyzed PV plant.

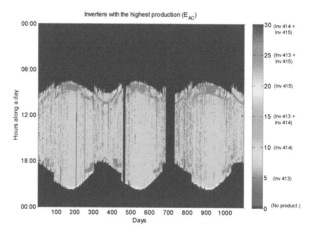

**Figure 7.** Inverters that are producing the maximum amount of energy $E_{AC}$ in each instant of time from 1 January 2011 to 31 December 2013.

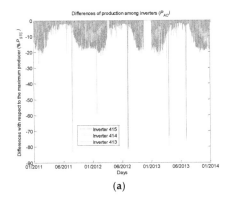
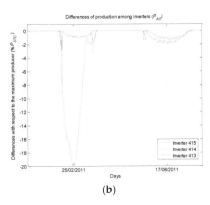

(a)        (b)

**Figure 8.** Instantaneous differences, in percentage of nominal power, of the production of each inverter with regard to that of the maximum production, in each instant of time: (**a**) from 1 January 2011 to 31 December 2013; and (**b**) on 25 February 2011 and on 17 June 2011.

On the other hand, the presence of faults, for example, in inverter 413, is clearly shown in Figure 8a with differences in the output, which may be even greater than 80% of the nominal power of the array. There are numerous authors who model the production of the PV plant and compare the actual production with that obtained by the model in order to verify the presence of faults in the installation [25–31]. Another option is, as presented in this work, to compare the production of arrays that are located together and exposed to the same conditions. The deviation in the production of one of them with respect to the others is a sign of the presence of some type of disturbance that is affecting the production. Obviously, this method does not allow for detecting deviations that are simultaneously affecting the output of all inverters, but it avoids having to use a model of both the array and the inverter, whose results will also present errors with respect to the actual production, which is higher the more simplified the model used is. Once the behavior of each inverter has been characterized, alarms can be set for deviations higher than those normally detected for each time of the year in order to control possible anomalous deviations.

*4.4. Performance Ratio*

In Figure 9a, the *PR* value corresponding to the output in each of the three inverters is represented during the entire monitoring period. In Figure 9b an enlargement corresponding to values of inverter 413 is represented. In agreement with the aspects previously seen, it is observed that inverter 415 presents a smaller value of this parameter, and more significantly during the months of autumn and winter. From November to January, the value of *PR* in this inverter is lower than that of the other two throughout the day, while, during the months of September, October, February and March, these lower values are only concentrated in the first hours of the day. This behavior is repeated during the three years in which this installation is analyzed, which is showing, as already seen, the presence of shadows in the panels corresponding to this inverter during these months of the year, which do not affect the panels of the two remaining inverters or do so at very specific times of the day. This behavior is easily revealed thanks to the visualization of this parameter using color maps. On the other hand, the daily profiles corresponding to *PR* values are subject to irradiance and temperature conditions. When days are clear, without clouds, the daily profiles of this parameter, especially in inverters 413 and 414, present a behavior that is being repeated throughout the year. In the mornings, after the grid connection, when the inverter starts to generate electricity, the value of *PR* goes from a null value to a value close to 1 in a time of about one hour or even less. In the central hours of the day, the value of *PR* decreases, with a value between 0.8 and 0.9, depending on the time of year. This decrease could be due to the higher temperature losses that take place during this time and more significantly in the summer months. Subsequently, a new increase in the value of *PR* occurs, reaching from 18:00 the highest values of the day, even higher than 1 (red color in the figure), and then decreasing its value in the last moments of production. The shape of this curve is also observed by other authors, who call it a "bath tube" [12]. These authors indicate that the high temperature in the central hours of the day contributes to the reduction of the *PR* value.

In theory, the value of *PR* can be effectively higher than 1 because the panel nominal power is measured at STC, so it can be higher than 1 under more favorable conditions, such as higher irradiation or lower temperature than STC [12]. During the hours in which the *PR* values above 1 were obtained in this work, the irradiance did not present values superior to 1000 W/m$^2$ (Figure 3a). At these instants of time in the winter, the temperature of the module decreased below 25 °C. In summer, at these instants of time, the temperatures had already begun to fall, but they were still high, above 25 °C, and the *PR* peak with values greater than 1 occurred even with more incidence. This apparent increase in *PR* value could be influenced by a lack of synchronization in the data record, by a sub-estimation of the nominal power of the plant used for normalization, $P_{STC}$, or by an overestimation of the $E_{AC}$ values recorded by a calibration error, although all these reasons would have the same effect during all the hours of the day and would not have a greater incidence at specific hours of the day. One might think this is due to an underestimation of the measured reference energy values, $E_r$, due to the presence of shading in the irradiance sensor, which did not affect the PV panels in the same way. There may also be affecting a different response of the PV panels, which are of polycrystalline silicon, compared to those of the reference cells used to record the irradiance, which is of amorphous silicon, since both materials have a different response to the spectrum of solar radiation [75].

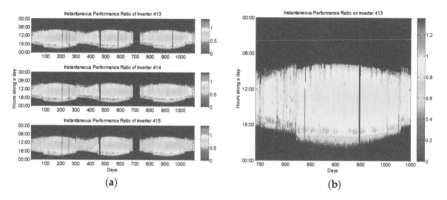

**Figure 9.** (**a**) Instantaneous values of performance ratio of the three inverters, during 24 h a day, from 1 January 2011 to 31 December 2013; and (**b**) enlargement of performance ratio values of inverter 413.

The average daily values of *PR* are shown in Figure 10a. In this case, the *PR* values have been determined by only taking into account the instants of the monitoring period in which each inverter was available and producing electricity, not counting the instants in which there have been stops in the PV production. With the data of this figure, several conclusions can be obtained. First, the seasonality in the annual variation of these daily average values can be seen, mainly in the case of inverter 415. The months in which the *PR* values are higher correspond generally to the months of April and May. During these months, the values of irradiance have high values, but the higher temperatures that occur later during the summer months, which increase the losses in the capture system due to the temperatures, are not yet recorded. This is the reason the *PR* highest values do not take place in the summer. The highest *PR* values are those corresponding to inverter 414, being slightly lower, although with the same behavior throughout the year, those corresponding to inverter 413.

The lower values correspond to inverter 415. The maximum values of the daily *PR* values found during the monitoring period were 0.92 (22 March 2012), 0.93 (22 March 2012) and 0.86 (27 April 2011), corresponding to inverters 413, 414 and 415, respectively, being the minimum values 0.40%, 0.42% and 0.39%, all registered on 19 December 2013.

The values of the temperature-corrected performance ratio, *TC_PR* (Figure 10b), are higher than those of *PR* because the effects of temperature losses are removed. The maximum values were 0.98, 1.02 and 0.97 for inverters 413, 414 and 415, respectively, whereas the minimum values were 0.37, 0.39 and 0.37, respectively. It can be observed that the seasonal variation is not eliminated at all due to the presence of other factors that are causing seasonal variations, with a high impact in this PV installation, such as seasonally dependent shading and/or spectral effects.

To compare these results with those of other authors, it should be noted that the *PR* value has been determined with tilted irradiance instead of horizontal values. Moreover, it must be considered that, on an annual basis, crystalline silicon sensors measure less irradiation than pyranometers, although the difference between the two types of sensors depends very much on the sensor and the location (in Germany, the difference is around 2–4%). Therefore, the annual *PR* of a PV plant that is calculated on the basis of crystalline silicon sensors may be higher than *PR* based on pyranometer measurements [12,65,66]. In that regard, no information was found in the case of amorphous silicon sensors, which was the sensor used in this work.

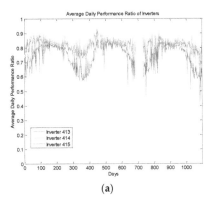
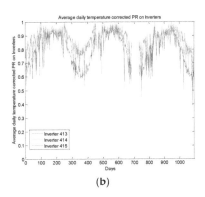

(a)                                                   (b)

**Figure 10.** (a) Average daily performance ratio; and (b) average daily corrected temperature performance ratio of the production of the three inverters, from 1 January 2011 to 31 December 2013

### 4.5. PV Panel Efficiency

In Figure 11a, the instantaneous values of the efficiency of the PV panels during the whole monitoring period are depicted. It is observed that the behavior presented by this parameter during the day was very similar to the one that has the *PR* of the production associated to each one of the inverters during the whole period of monitoring (Figure 9a). The values of the modules' efficiency were slightly higher in the first and last hours of the day, with values higher than 15% and even 20% (which is a value above the 13.1% given by the manufacturer), while the value was lower in the central hours of the day, around 12–13% (depending on the time of year). The upper values in the first and last hours of the day also presented a certain periodicity such that these values were also slightly higher in the spring and summer months. If Figure 11b is observed, in which the average daily values of the efficiency of the panels are represented, the same behavior as that of the daily mean values of *PR* can be noticed. The lowest values are presented by inverter 415, mainly during the months from September to March. The maximum values of the average daily module efficiency were 14.36% (19 and 22 April 2012), 14.66% (22 March 2012) and 13.50% (27 April 2011 and 2 May 2012), respectively, for modules connected to inverters 413, 414 and 415, in the same range of values found by other authors for a near geographical location [76], being the minimum values 7.76%, 8.14% and 7.65% (19 December 2013).

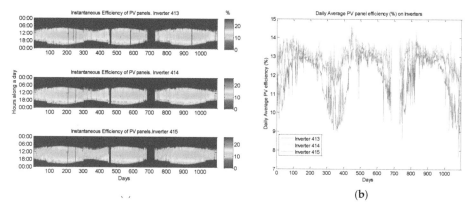

**Figure 11.** PV panels' efficiency from 1 January 2011 to 31 December 2013: (**a**) instantaneous values, during 24 h a day; and (**b**) total daily values.

*4.6. Inverter Efficiency*

Figure 12a shows the instantaneous values of the efficiency of the three inverters, observing that, in this case, the profiles throughout the day were different from those presented by the efficiency of the panels. Although the efficiency was lower in the first and last hours of production, it rapidly increased to reach, during almost all the hours of production, values higher than 90%, which remained practically constant. In this case, the profiles and values found were very similar for the three inverters, as is also shown in Figure 12b, where it is observed that the average daily values of the inverters' efficiency were practically the same in all three. The effect of the seasonality, although present, was lower in this parameter. These average values were above 90%, being slightly lower in the winter months and with greater variations from one day to another, between 75% and 92%. The highest values found were 94.00% (25 August 2011, Day 237), 93.37% (1 September 2012, Day 610) and 93.21% (6 October 2011, Day 279), respectively, for inverters 413, 414 and 415, being the minimum values equal to 75.02%, 73.57% and 73.92%, which took place on 19 December 2013 (Day 1084).

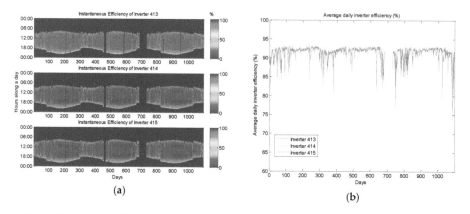

**Figure 12.** Inverters' efficiency from 1 January 2011 to 31 December 2013: (**a**) instantaneous values, during 24 h a day; and (**b**) total daily values.

*4.7. Capture Losses*

Figure 13a shows at a glance the distribution of losses occurring in the capture systems connected to the three inverters throughout the whole monitored period. One the one hand, the losses occurring in the central hours of the day can be distinguished. They are mainly caused by temperature losses (which present the same daily profile and the same seasonality as the temperature values reflected in Figure 4a,b), which occurred with greater incidence in summer months and equally affect the panels of the three inverters. The wiring losses will be proportional to the production, so they will also be higher in the central hours of the day. On the other hand, the presence of losses that occur in the first and last hours of the day can be observed, which present a seasonal behavior, affecting to a greater extent the PV panels corresponding to inverter 415. These losses had a higher incidence in the months of autumn and winter, being higher the time interval in which inverter 415 is affected.

The losses by dirt and dust accumulated on the solar module surface, blocking some of the sunlight and reducing the output production, also affected the reference cell used to measure the radiation received by the panels [12]. Although dirt and dust are cleaned off during every rain event, it is necessary to clean them during dry periods. This type of intervention was not carried out in the installation, so this type of loss could not be evaluated. To do this, it is necessary to compare the performance, in situations where only the sensor is cleaned, with those in which both sensors and PV modules were cleaned [77].

Figure 13a also shows losses that occurred due to the presence of stops in production, already noticed in Figure 5a,b and Figure 6. On one occasion, this shutdown affected the three inverters, while it only affected inverter 413 on four other occasions.

In accordance with these results are those reflected in Figure 13b, in which the total daily values of capture losses are represented. For a typical day of July, total daily losses have been quantified at 0.72, 0.60 and 0.75 h/day for the arrays corresponding to the panels of inverters 413, 414 and 415, respectively, whereas, for a clear day in January, these losses may be 0.65, 0.60 and 1.63 h/day. It becomes clear that, although the temperature losses are higher in summer, they are more affected by shading losses in winter, influencing inverter 415 to a greater extent. The lowest losses were recorded in the months of April and May, the same ones in which the *PR* values were improved. This information can be taken into account when making stops at the facility for maintenance issues in order to carry them out in the periods of lower electricity production. It would be recommended that these two graphs be updated daily and monitored by O&M operators to quickly detect any deviations in production or abnormal operation.

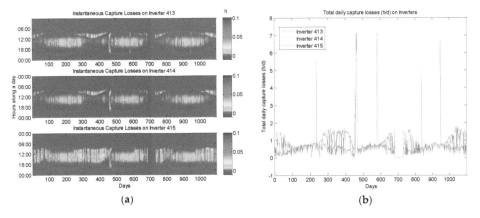

**Figure 13.** Capture losses from 1 January 2011 to 31 December 2013: (**a**) instantaneous values, during 24 h a day; and (**b**) total daily values.

*4.8. Inverter Losses*

With respect to inverter losses, their instantaneous and daily total values are reflected in Figure 14a,b, respectively. As shown in Figure 14a, these losses were higher in the central hours of the day in which the production was higher. Since inverter 415 has a lower production in the autumn-winter months, it also has fewer losses during those months than the other two inverters, which is a fact that can be observed in both figures.

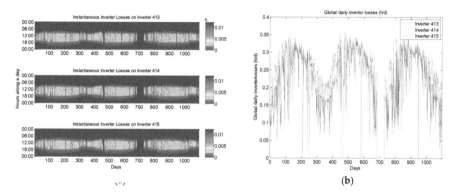

**Figure 14.** Inverter losses, from 1 January 2011 to 31 December 2013: (**a**) instantaneous values, during 24 h a day; and (**b**) total daily values.

The daily total losses in the inverters were lower than those that take place in the PV modules, in line with the greater efficiency with which the inverters worked. The maximum values determined were 0.36, 0.36 and 0.35 h/day for inverters 413, 414 and 415, respectively, which took place on 16 April 2012 (Day 472). In general, although there were not any large variations in this type of losses from one day to the other throughout the year, the periods in which the highest losses in inverters were recorded were the months of April, May and June.

*4.9. Inverter States*

To analyze the behavior of inverters in Figure 15a their states have been represented during the entire monitoring period, and are codified with a series of numbers. An enlargement corresponding to inverter 413's states is represented in Figure 15b.

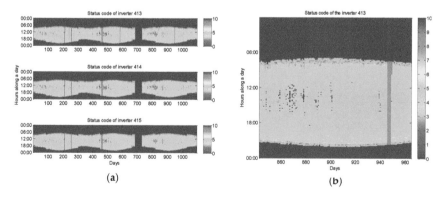

**Figure 15.** (**a**) instantaneous values of performance ratio of the three inverters, during 24 h a day, from 1 January 2011 to 31 December 2013: and (**b**) enlargement of status values of inverter 413.

When inverters are not working, the null value has been assigned (dark blue color). In the first and last hours of production, in which the inverter is in the start-up and shut-down phases, the inverters fundamentally monitored States 2 and 3 [13] (light blue color), which correspond, respectively, to a state in which inverters are monitoring the grid (which takes place just before connecting and disconnecting from the grid) and to a standby state, in which the requirements for the connection to the inverter to the grid are not yet fulfilled and the injection of current is waiting. Code 1 was also usually registered at start-up times, which is a state in which the inverter indicates an interruption after an error.

Once inverters are connected and in operation and are already injecting current into the grid, in the first few hours of operation or also in the later hours of the afternoon, the state that may be monitored was 6 (yellow color), in which the MPP (maximum power point) is being calculated. State 7 (orange color) is registered when the inverter is working in MPP. It was the most frequent state in which the inverters can be found. In it, the inverters ensure that the modules are in their optimum operating point for each external conditions, thus trying to extract the maximum power output of PV systems [39,78,79]. It can be observed that, except in the initial and final hours of the day, inverters were usually working in this state. Depending on the algorithm they use for the calculation, especially in the presence of shading, they will be getting more or less power from the PV array [80,81]. State 10 (brown color), named MPP peak, corresponds to that in which the inverter is operating in MPP mode, but above its rated capacity. This state occurred at the time of the year in which the higher values of irradiance were recorded, which took place in the months of February to June for the geographic location and the inclination and orientation of the PV panels, registering the highest values during the month of April. On clear days, inverters worked in this mode in the central hours of the day, around zenith. During these instants of time, inverters operated above their nominal power. The days of this period of the year with the passage of clouds cause the inverters to alternate their operation mode between MPP and MPP peak (compare Figures 15a and 3a). For these instants in which inverters were worked in state 10 or MPP peak, the size of the PV array was overestimated, and the DC/AC inverter load ratio was higher than 1. State 10 appeared less frequently in inverter 415 than in the other two, since the output in the latter was lower. When inverters work above their nominal power, inverter heating increases. If this takes place for longer periods, its lifetime may be affected, so the cost of this type of degradation should be assessed against the initial savings that the underestimation of the inverter may involve.

On the other hand, State 8 (red color) corresponds to instants of time in which the inverter is recording information, but it is in a state of alert or warning, and it is not injecting electricity to the grid during those instants of time as a result of some kind of failure. The presence of this state in the graphs coincides with the instants in which there have been interruptions in the production, previously seen in the representation of both the energy generated in the PV arrays and in the representation of losses in the capture system. One or several daily supervisions of graphs such as the one presented in Figure 15 allow the O&M operators to quickly detect, simply by observing the color variations, deviations that would require an intervention in the plant.

## 5. Conclusions

The procedure proposed in this work has enabled us to effectively manage the large amount of data recorded in PV plants. The developed software accomplishes an automated migration of the registered data, the calculation of a series of performance and loss indexes of the components in the installation, and its storage in a RDBMS. Furthermore, it allows an easy and orderly access to the data the user is interested in. Specifically, the application provides the values of each parameter in the form of matrices, which have been represented in the form of color maps for their further study.

Even with the great amount of data generated, color maps enable representing each of these parameters during all the monitoring period in a readable way. This type of data visualization, with a broader scope than the daily or monthly values, provides a visual tool with which much information about the components' behavior can be easily revealed in any instant of time. Its visual inspection enables one to detect the normal and abnormal modes of operation and to compare the behavior of the different components, helping in decision making related to O&M tasks. The analysis of these figures may serve to identify and interpret common design flows and operational problems or to simply document the operational mode of the installations in real conditions.

As a case study the operation of a grid-connected PV plant located in a city in southern Spain was studied using this type of graphs. All parameters were analyzed over a monitored period of three years to carry out a detailed study. With the color maps, it was possible to simultaneously represent the large volume data set obtained during the three years, and information about the behavior of panels and inverters was obtained. The average daily module efficiency values were in the range of 7.6–14.6%, whereas, for inverters, their average daily efficiency was in the range of 73.5–94%. It was possible to emphasize the presence of shadings in PV panels, with a greater impact on inverter 415. The hours and days of the year in which they take place were easily detected through this type of graphics. This has meant that for this inverter the daily values of *PR* range between 0.6 and 0.8, lower than the values of the other two inverters under normal operating conditions, between 0.7 and 0.9. The presence of stops or failures in the production was also easily detected by the visualization of the color maps corresponding to the loss indexes and to the inverter status.

Therefore, the developed application and this type of data visualization enable us to easily analyze the performance of PV systems and provide an affordable tool that may facilitate the corrective, preventive and predictive interventions of the O&M activities, generating information that can be considered in the design of future PV plants.

In addition to being easy to interpret, their use does not require an additional economic cost, since the data basically recorded by inverters were used. As it is indicated in the report of the International Energy Agency (IEA) [12], the inverter integrated measurements are usually not sufficiently precise and, in absolute terms, a more rigorous monitoring system would be necessary. Nevertheless, inverter data may be useful enough to know the behavior of the components of the PV installation and for identifying relative changes over time or deviations from their normal operation. However, it is important to know the limitations of this type of measure to correctly interpret the results. The use of an advanced monitoring system compared to the simple inverter monitoring depends very much on the individual project, and only if its use provides economic benefits will its implementation be justified.

The next objective is to apply the use of these graphs to the analysis of a megawatt PV park to manage the information corresponding to the monitoring of its components and to facilitate its maintenance, which will become the object of a future paper.

**Acknowledgments:** This work is supported by the Spanish Ministry of Economy and Competitiveness under Research Project SCEMS TEC2013–47316–C3–1–P. It is also supported by an agreement with Solar del Valle SL company.

**Author Contributions:** D.T.M. implemented and developed the software for the automated processing of data. D.T.M., J.J.L.R. and I.S. conceived of the architecture and the management strategies for the processing of data. D.T.M., J.J.L.R. and I.S. performed the literature review. D.T.M., E.J.P.G., I.M.M.G. and I.S. provided the graphical representations and interpretation of results. D.T.M. and I.S. coordinated and wrote the paper. All authors supervised and approved the final version of the manuscript.

**Conflicts of Interest:** The authors declare no conflict of interest.

## References

1.  Schmela, M. Global Market Outlook for Solar Power 2017–2021. Available online: http: //www.solarpowereurope.org/fileadmin/user_upload/documents/WEBINAR/Free_SolarPower_ Webinar_Global_Market_Outlook_2017-2021.pdf (accessed on 24 November 2017).
2.  Obi, M.; Bass, R. Trends and challenges of grid-connected photovoltaic systems—A review. *Renew. Sustain. Energy Rev.* **2016**, *58*, 1082–1094. [CrossRef]
3.  Polman, A.; Knight, M.; Garnett, E.C.; Ehrler, B.; Sinke, W.C. Photovoltaic materials: Present efficiencies and future challenges. *Science* **2016**, *352*, aad4424. [CrossRef] [PubMed]

4.  Bhakta, S.; Mukherjee, V. Solar potential assessment and performance indices analysis of photovoltaic generator for isolated Lakshadweep island of India. *Sustain. Energy Technol. Assess.* **2016**, *17*, 1–10. [CrossRef]

5.  Nižetić, S.; Papadopoulos, A.M.; Giama, E. Comprehensive analysis and general economic-environmental evaluation of cooling techniques for photovoltaic panels, Part I: Passive cooling techniques. *Energy Convers. Manag.* **2017**, *149*, 334–354. [CrossRef]

6.  Sampaio, P.G.V.; González, M.O.A. Photovoltaic solar energy: Conceptual framework. *Renew. Sustain. Energy Rev.* **2017**, *74*, 590–601. [CrossRef]

7.  *Utility-Scale Solar Photovoltaic Power Plants: A Project Developer's Guide*; International Finance Corporation: Washington, DC, USA, 2015.

8.  Honrubia-Escribano, A.; Ramirez, F.J.; Gómez-Lázaro, E.; Garcia-Villaverde, P.M.; Ruiz-Ortega, M.J.; Parra-Requena, G. Influence of solar technology in the economic performance of PV power plants in Europe. A comprehensive analysis. *Renew. Sustain. Energy Rev.* **2018**, *82*, 488–501. [CrossRef]

9.  Degener, S.; Watson, J. O&M Best Practices Guidelines. Available online: http://alectris.com/guidelines/om-best-practices-guidelines/ (accessed on 24 November 2017).

10. Moreno-Garcia, I.; Palacios-Garcia, E.; Pallares-Lopez, V.; Santiago, I.; Gonzalez-Redondo, M.; Varo-Martinez, M.; Real-Calvo, R. Real-time monitoring system for a utility-scale photovoltaic power plant. *Sensors* **2016**, *16*, 770. [CrossRef] [PubMed]

11. Rezk, H.; Tyukhov, I.; Al-Dhaifallah, M.; Tikhonov, A. Performance of data acquisition system for monitoring PV system parameters. *Meas. J. Int. Meas. Confed.* **2017**, *104*, 204–211. [CrossRef]

12. Woyte, A.; Ritcher, M.; Moser, D.; Reich, N.; Green, M.; Mau, S.; Garrad Hassan, G.; Beyer, H. Analytical Monitoring of Grid-Connected Photovoltaic Systems. Good Practices for Monitoring and Performance Analysis. 2014. Available online: http://iea-pvps.org/index.php?id=276 (accessed on 25 November 2017).

13. Haeberlin, H.; Beutler, C. Normalized representation of energy and power for analysis of performance and on-line error detection in PV-systems. In Proceedings of the 13th EU PV Conference on Photovoltaic Solar Energy Conversion, Nice, France, 23–27 October 1995.

14. Blaesser, G.; Munro, D. *Guidelines for the Assessment of Photovoltaic Plants, Document A: Photovoltaic System Monitoring*; Commission of the European Communities, Joint Research Centre: Ispra, Italy, 1995.

15. Blaesser, G.; Munro, D. *Guidelines for the Assessment of Photovoltaic Plants, Document B: Analysis and Presentation of Monitoring Data*; Commission of the European Communities, Joint Research Centre: Ispra, Italy, 1996.

16. *International Standard IEC 61724: 1998 Photovoltaic System Performance Monitoring-Guidelines for Measurement, Data Exchange and Analysis. Edition 1.0*; International Electrotechnical Commission: Geneva, Switzerland, 1998.

17. *International Standard IEC 61724-1:2017 Photovoltaic System Performace-Part 1: Monitoring*; International Electrotechnical Commission: Geneva, Switzerland, 2017.

18. Stefan, M.; Lopez, J.G.; Andreasen, M.H.; Olsen, R.L. Visualization Techniques for Electrical Grid Smart Metering Data: A Survey. In Proceedings of the 2017 IEEE Third International Conference on Big Data Computing Service and Applications (BigDataService), San Francisco, CA, USA, 6–9 April 2017; pp. 165–171. [CrossRef]

19. Moreno-Muñoz, A.; Flores-Arias, J.M.; Gil-De-Castro, A.; De La Rosa, J.J.G. Hypermedia user-interface integration in distribution power systems SCADA. In Proceedings of the 7th IEEE International Conference on Industrial Informatics (INDIN 2009), Cardiff, Wales, UK, 23–26 June 2009; pp. 136–141.

20. Murugesan, L.K.; Hoda, R.; Salcic, Z. Design criteria for visualization of energy consumption: A systematic literature review. *Sustain. Cities Soc.* **2015**, *18*, 1–12. [CrossRef]

21. Harrison, J.; Uhomoibhi, J. Engineering study of tidal stream renewable energy generation and visualization: Issues of process modelling and implementation. In *Lecture Notes in Computer Science (Including Subseries Lecture Notes in Artificial Intelligence and Lecture Notes in Bioinformatics)*; Springer International Publishing AG: Cham, Switzerland, 2016.

22. Filali-Yachou, S.; González-González, C.S.; Lecuona-Rebollo, C. HMI/ SCADA standards in the design of data center interfaces: A network operations center case study. *DYNA* **2015**, *82*, 180–186. [CrossRef]

23. US Department of Energy and NASPI. *NASPI Synchrophasor Technical Report Phasor Tools Visualization Workshop Technical Summary*; US Department of Energy and NASPI: Washington, DC, USA, 2014.

24. National Energy Technology Laboratory for the US Department of Energy. *Office of Electricity Delivery and Energy Reliability Improved Interfaces and Decision Support*; National Energy Technology Laboratory for the US Department of Energy: Pittsburgh, PA, USA, 2007.

25. Garoudja, E.; Harrou, F.; Sun, Y.; Kara, K.; Chouder, A.; Silvestre, S. Statistical fault detection in photovoltaic systems. *Sol. Energy* **2017**, *150*, 485–499. [CrossRef]

26. Ventura, C.; Tina, G.M. Utility scale photovoltaic plant indices and models for on-line monitoring and fault detection purposes. *Electr. Power Syst. Res.* **2016**, *136*, 43–56. [CrossRef]

27. Ventura, C.; Tina, G.M. Development of models for on-line diagnostic and energy assessment analysis of PV power plants: The study case of 1 MW Sicilian PV plant. *Energy Procedia* **2015**, *83*, 248–257. [CrossRef]

28. Chouder, A.; Silvestre, S. Automatic supervision and fault detection of PV systems based on power losses analysis. *Energy Convers. Manag.* **2010**, *51*, 1929–1937. [CrossRef]

29. Chouder, A.; Silvestre, S.; Taghezouit, B.; Karatepe, E. Monitoring, modelling and simulation of PV systems using LabVIEW. *Sol. Energy* **2013**, *91*, 337–349. [CrossRef]

30. Chouder, A.; Silvestre, S.; Sadaoui, N.; Rahmani, L. Modeling and simulation of a grid connected PV system based on the evaluation of main PV module parameters. *Simul. Model. Pract. Theory* **2012**, *20*, 46–58. [CrossRef]

31. Chine, W.; Mellit, A.; Pavan, A.M.; Kalogirou, S.A. Fault detection method for grid-connected photovoltaic plants. *Renew. Energy* **2014**, *66*, 99–110. [CrossRef]

32. de Lima, L.C.; de Araujo Ferreira, L.; de Lima Morais, F.H.B. Performance analysis of a grid connected photovoltaic system in northeastern Brazil. *Energy Sustain. Dev.* **2017**, *37*, 79–85. [CrossRef]

33. Malvoni, M.; Leggieri, A.; Maggiotto, G.; Congedo, P.M.; De Giorgi, M.G. Long term performance, losses and efficiency analysis of a 960 kW P photovoltaic system in the Mediterranean climate. *Energy Convers. Manag.* **2017**, *145*, 169–181. [CrossRef]

34. Ma, T.; Yang, H.; Lu, L. Long term performance analysis of a standalone photovoltaic system under real conditions. *Appl. Energy* **2016**. [CrossRef]

35. Elhadj Sidi, C.E.B.; Ndiaye, M.L.; El Bah, M.; Mbodji, A.; Ndiaye, A.; Ndiaye, P.A. Performance analysis of the first large-scale (15 MWp) grid-connected photovoltaic plant in Mauritania. *Energy Convers. Manag.* **2016**, *119*, 411–421. [CrossRef]

36. Attari, K.; El Yaakoubi, A.; Asselman, A. Comparative Performance investigation between photovoltaic systems from two different cities. *Procedia Eng.* **2017**, *181*, 810–817. [CrossRef]

37. Attari, K.; Elyaakoubi, A.; Asselman, A. Performance analysis and investigation of a grid-connected photovoltaic installation in Morocco. *Energy Rep.* **2016**, *2*, 261–266. [CrossRef]

38. Bhakta, S.; Mukherjee, V. Performance indices evaluation and techno economic analysis of photovoltaic power plant for the application of isolated India's island. *Sustain. Energy Technol. Assess.* **2017**, *20*, 9–24. [CrossRef]

39. Kumar, M.; Kumar, A. Performance assessment and degradation analysis of solar photovoltaic technologies: A review. *Renew. Sustain. Energy Rev.* **2017**, *78*, 554–587. [CrossRef]

40. Shiva Kumar, B.; Sudhakar, K. Performance evaluation of 10 MW grid connected solar photovoltaic power plant in India. *Energy Rep.* **2015**, *1*, 184–192. [CrossRef]

41. Sundaram, S.; Babu, J.S.C. Performance evaluation and validation of 5MWp grid connected solar photovoltaic plant in South India. *Energy Convers. Manag.* **2015**, *100*, 429–439. [CrossRef]

42. Micheli, D.; Alessandrini, S.; Radu, R.; Casula, I. Analysis of the outdoor performance and efficiency of two grid connected photovoltaic systems in northern Italy. *Energy Convers. Manag.* **2014**, *80*, 436–445. [CrossRef]

43. Padmavathi, K.; Daniel, S.A. Performance analysis of a 3MWp grid connected solar photovoltaic power plant in India. *Energy Sustain. Dev.* **2013**, *17*, 615–625. [CrossRef]

44. Wittkopf, S.; Valliappan, S.; Liu, L.; Ang, K.S.; Cheng, S.C.J. Analytical performance monitoring of a 142.5 kWp grid-connected rooftop BIPV system in Singapore. *Renew. Energy* **2012**, *47*, 9–20. [CrossRef]

45. Díez-Mediavilla, M.; Alonso-Tristán, C.; Rodríguez-Amigo, M.C.; García-Calderón, T.; Dieste-Velasco, M.I. Performance analysis of PV plants: Optimization for improving profitability. *Energy Convers. Manag.* **2012**, *54*, 17–23. [CrossRef]

46. Başoğlu, M.E.; Kazdaloğlu, A.; Erfidan, T.; Bilgin, M.Z.; Çakir, B. Performance analyzes of different photovoltaic module technologies under İzmit, Kocaeli climatic conditions. *Renew. Sustain. Energy Rev.* **2015**, *52*, 357–365. [CrossRef]

47. Dabou, R.; Bouchafaa, F.; Arab, A.H.; Bouraiou, A.; Draou, M.D.; Neçaibia, A.; Mostefaoui, M. Monitoring and performance analysis of grid connected photovoltaic under different climatic conditions in south Algeria. *Energy Convers. Manag.* **2016**, *130*, 200–206. [CrossRef]

48. Shravanth Vasisht, M.; Srinivasan, J.; Ramasesha, S.K. Performance of solar photovoltaic installations: Effect of seasonal variations. *Sol. Energy* **2016**, *131*, 39–46. [CrossRef]

49. Edalati, S.; Ameri, M.; Iranmanesh, M. Comparative performance investigation of mono- and poly-crystalline silicon photovoltaic modules for use in grid-connected photovoltaic systems in dry climates. *Appl. Energy* **2015**, *160*, 255–265. [CrossRef]

50. Tripathi, B.; Yadav, P.; Rathod, S.; Kumar, M. Performance analysis and comparison of two silicon material based photovoltaic technologies under actual climatic conditions in Western India. *Energy Convers. Manag.* **2014**, *80*, 97–102. [CrossRef]

51. Milosavljević, D.D.; Pavlović, T.M.; Piršl, D.S. Performance analysis of A grid-connected solar PV plant in Nis, republic of Serbia. *Renew. Sustain. Energy Rev.* **2015**, *44*, 423–435. [CrossRef]

52. Congedo, P.M.; Malvoni, M.; Mele, M.; De Giorgi, M.G. Performance measurements of monocrystalline silicon PV modules in South-eastern Italy. *Energy Convers. Manag.* **2013**, *68*, 1–10. [CrossRef]

53. Drif, M.; Pérez, P.J.; Aguilera, J.; Almonacid, G.; Gomez, P.; de la Casa, J.; Aguilar, J.D. Univer Project. A grid connected photovoltaic system of 200 kWp at Jaen University. Overview and performance analysis. *Sol. Energy Mater. Sol. Cells* **2007**, *91*, 670–683. [CrossRef]

54. Trillo-Montero, D.; Santiago, I.; Luna-Rodriguez, J.J.; Real-Calvo, R. Development of a software application to evaluate the performance and energy losses of grid-connected photovoltaic systems. *Energy Convers. Manag.* **2014**, *81*. [CrossRef]

55. Fernández-Pacheco, D.G.; Molina-Martínez, J.M.; Ruiz-Canales, A.; Jiménez, M. A new mobile application for maintenance tasks in photovoltaic installations by using GPS data. *Energy Convers. Manag.* **2012**, *57*, 79–85. [CrossRef]

56. Silvestre, S.; Chouder, A.; Karatepe, E. Automatic fault detection in grid connected PV systems. *Sol. Energy* **2013**, *94*, 119–127. [CrossRef]

57. Gokmen, N.; Karatepe, E.; Silvestre, S.; Celik, B.; Ortega, P. An efficient fault diagnosis method for PV systems based on operating voltage-window. *Energy Convers. Manag.* **2013**, *73*, 350–360. [CrossRef]

58. Kymakis, E.; Kalykakis, S.; Papazoglou, T.M. Performance analysis of a grid connected photovoltaic park on the island of Crete. *Energy Convers. Manag.* **2009**, *50*, 433–438. [CrossRef]

59. Al-Sabounchi, A.M.; Yalyali, S.A.; Al-Thani, H.A. Design and performance evaluation of a photovoltaic grid-connected system in hot weather conditions. *Renew. Energy* **2013**, *53*, 71–78. [CrossRef]

60. Koehl, M.; Heck, M.; Wiesmeier, S.; Wirth, J. Modeling of the nominal operating cell temperature based on outdoor weathering. *Sol. Energy Mater. Sol. Cells* **2011**, *95*, 1638–1646. [CrossRef]

61. Skoplaki, E.; Boudouvis, A.G.; Palyvos, J.A. A simple correlation for the operating temperature of photovoltaic modules of arbitrary mounting. *Sol. Energy Mater. Sol. Cells* **2008**, *92*, 1393–1402. [CrossRef]

62. Skoplaki, E.; Palyvos, J.A. Operating temperature of photovoltaic modules: A survey of pertinent correlations. *Renew. Energy* **2009**, *34*, 23–29. [CrossRef]

63. Fuentes, M. *A Simplified Thermal Model for Flat-Plate Photovoltaic Arrays*; Report SAND-85-0330; United States Department of Commerce: Washington, DC, USA, 1987.

64. Chatzipanagi, A.; Frontini, F.; Dittmann, S. Investigation of the influence of module working temperatures on the performance of BiPV modules. In Proceedings of the 27th European Photovoltaic Solar Energy Conference and Exhibition, Frankfurt, Germany, 24–28 September 2012; pp. 4192–4197.

65. Reich, N.H.; Mueller, B.; Armbruster, A.; Van Sark, W.G.J.H.M.; Kiefer, K.; Reise, C. Performance ratio revisited: Is PR > 90% realistic? *Prog. Photovolt. Res. Appl.* **2012**, *20*, 717–726. [CrossRef]

66. Khalid, A.M.; Mitra, I.; Warmuth, W.; Schacht, V. Performance ratio—Crucial parameter for grid connected PV plants. *Renew. Sustain. Energy Rev.* **2016**, *65*, 1139–1158. [CrossRef]

67. Perpiñan, O. *Energía Solar Fotovoltaica*; Creative Commons: Boston, MA, USA, 2012.

68. Phyton. Available online: https://www.python.org/ (accessed on 1 October 2017).

69. HTML. Available online: https://www.w3schools.com/html/ (accessed on 1 October 2017).

70. CSS. Available online: https://www.w3schools.com/css/default.asp (accessed on 1 October 2017).

71. TypeScript. Available online: https://www.typescriptlang.org/ (accessed on 1 October 2017).

72. Solar del Valle, SL Company. Available online: http://www.solardelvalle.es (accessed on 17 November 2017).

73. SMA Company. Available online: https://www.sma.de/en.html (accessed on 17 November 2017).

74. Gobierno de España Real Decreto 1578/2008, de 26 de septiembre, de retribución de la actividad de producción de energía eléctrica mediante tecnología solar fotovoltaica para instalaciones posteriores a la fecha límite de mantenimiento de la retribución del Real Decreto 661/2008. *Boletín Ofical del Estado* **2008**, *234*, 39117–39125.

75. Chegaar, M.; Mialhe, P. Effect of atmospheric parameters on the silicon solar cells performance. *J. Electron Devices* **2008**, *6*, 173–176.

76. Cañete, C.; Carretero, J.; Sidrach-de-Cardona, M. Energy performance of different photovoltaic module technologies under outdoor conditions. *Energy* **2014**, *65*, 295–302. [CrossRef]

77. Touati, F.; Al-Hitmi, M.A.; Chowdhury, N.A.; Hamad, J.A.; San Pedro Gonzales, A.J.R. Investigation of solar PV performance under Doha weather using a customized measurement and monitoring system. *Renew. Energy* **2016**, *89*, 564–577. [CrossRef]

78. Lupangu, C.; Bansal, R.C. A review of technical issues on the development of solar photovoltaic systems. *Renew. Sustain. Energy Rev.* **2017**, *73*, 950–965. [CrossRef]

79. Rezk, H.; Fathy, A.; Abdelaziz, A.Y. A comparison of different global MPPT techniques based on meta-heuristic algorithms for photovoltaic system subjected to partial shading conditions. *Renew. Sustain. Energy Rev.* **2017**, *74*, 377–386. [CrossRef]

80. Ram, J.P.; Babu, T.S.; Rajasekar, N. A comprehensive review on solar PV maximum power point tracking techniques. *Renew. Sustain. Energy Rev.* **2017**, *67*, 826–847. [CrossRef]

81. Dileep, G.; Singh, S.N. Application of soft computing techniques for maximum power point tracking of SPV system. *Sol. Energy* **2017**, *141*, 182–202. [CrossRef]

*Article*

# Fuzzy Logic Based MPPT Controller for a PV System

Carlos Robles Algarín *, John Taborda Giraldo and Omar Rodríguez Álvarez

Facultad de Ingeniería, Universidad del Magdalena, Carrera 32 No. 22-08, 470004 Santa Marta, Colombia; jatabordag@gmail.com (J.T.G.); omarfro@gmail.com (O.R.Á.)
* Correspondence: carlosarturo.ing@gmail.com; Tel.: +57-5-421-7940

Received: 19 October 2017; Accepted: 22 November 2017; Published: 2 December 2017

**Abstract:** The output power of a photovoltaic (PV) module depends on the solar irradiance and the operating temperature; therefore, it is necessary to implement maximum power point tracking controllers (MPPT) to obtain the maximum power of a PV system regardless of variations in climatic conditions. The traditional solution for MPPT controllers is the perturbation and observation (P&O) algorithm, which presents oscillation problems around the operating point; the reason why improving the results obtained with this algorithm has become an important goal to reach for researchers. This paper presents the design and modeling of a fuzzy controller for tracking the maximum power point of a PV System. Matlab/Simulink (MathWorks, Natick, MA, USA) was used for the modeling of the components of a 65 W PV system: PV module, buck converter and fuzzy controller; highlighting as main novelty the use of a mathematical model for the PV module, which, unlike diode based models, only needs to calculate the curve fitting parameter. A P&O controller to compare the results obtained with the fuzzy control was designed. The simulation results demonstrated the superiority of the fuzzy controller in terms of settling time, power loss and oscillations at the operating point.

**Keywords:** fuzzy logic controller; maximum power point tracking (MPPT); dc-dc converter; photovoltaic system

## 1. Introduction

In recent years, the use of photovoltaic (PV) energy has experienced significant progress as an alternative to solve energy problems in places with high solar density, which is due to pollution caused by fossil fuels and the constant decrease of prices of the PV modules. Unfortunately, the energy conversion efficiency of the PV modules is low, which reduces the cost-benefit ratio of PV systems.

The maximum power that a PV module can supply is determined by the product of the current and the voltage at the maximum power point, which depends on the operating temperature and the solar irradiance. The short-circuit current of a PV module is directly proportional to the solar irradiance, decreasing considerably as the irradiation decreases, while the open circuit voltage varies moderately due to changes in irradiation. In contrast, the voltage decreases considerably when the temperature increases, while the short circuit current increases moderately.

In summary, increases in solar irradiation produce increases in the short-circuit current, while increases in temperature decrease the open circuit voltage, which affects the output power of the PV module. This variability of the output power means that in the absence of a coupling device between the PV module and the load, the system does not operate at the maximum power point (MPP).

According to the previous context, the use of maximum power point (MPPT) controllers is currently increasing [1]. These devices are responsible for regulating the charge of the batteries, controlling the point at which the PV modules produces the greatest amount of energy possible, regardless of variations in climatic conditions. The use of MPPT controllers in PV systems has the following advantages: 1. They yield more power, depending on weather and temperature; 2. They allow the connection of PV modules in series to increase the voltage of the system, which reduces the wiring gauge and adds flexibility; 3. They offer a cost savings in the transmission wire needed for the installation of the PV system.

In contrast to MPPT controllers, traditional controllers make a direct connection of the PV modules to the batteries, which requires that the modules operate in a voltage range that is below to the voltage in maximum power point. For example, in the case of a 12 V system, the battery voltage can vary between 11 V and 15 V, but the voltage at the maximum power point is a typical value between 16 V and 17 V. Due to this situation, with the traditional controllers the energy that the PV modules can deliver is not maximized.

Taking into account the above, different researches have been carried out using traditional algorithms for the modeling and implementation of MPPT controllers [2], of which the following are highlighted: perturb and observe (P&O) [3,4], modified P&O [5,6], fractional short circuit current [7], fractional open circuit voltage [8], sliding mode control [9,10] and incremental conductance [11]. The P&O algorithm has been used traditionally, but it has been shown that this method has problems for tracking the MPP when there are sudden changes in solar irradiance [12].

Also, algorithms based on artificial intelligence techniques such as fuzzy logic [13–19] and neural networks [20–22] have been used, as well as the implementation of optimization algorithms such as glowworm swarm [23], ant colony [24,25] and bee colony [26–28]. These algorithms are part of soft computing techniques and have the advantage of being easily implemented using embedded systems. Additionally, MPPT controllers are widely used in hybrid power systems, in which different control techniques based on neural networks, fuzzy logic and particle swarm optimization have been evaluated. In [29–31], the effectiveness of these control techniques was demonstrated in order to achieve a fast and stable response for real power control and power system applications. The implementation of new control and optimization techniques that are detailed in [32–35] for electrical power and energy systems can be studied in the modeling and implementation of MPPT controllers.

This paper presents the design and modeling of a fuzzy controller to track the maximum power point of a PV module, using the characteristics of fuzzy logic to represent a problem through linguistic expressions [36]. This paper presents as a novelty the use of the mathematical model proposed in [37,38] for modeling the PV module, which, unlike diode based models, only needs to calculate the curve fitting parameter. The results were compared with the P&O controller, which demonstrated that the proposed approach presents less energy losses and ensures MPP in all cases evaluated in simulation. It is worth mentioning that this work is part of a set of intelligent control techniques being evaluated in the research group Magma Ingeniería of the Universidad del Magdalena in order to implement a MPPT controller of low cost and high efficiency.

The main objective of this work is the design, modeling and simulation of a fuzzy logic controller and a dc-dc converter for an off-grid PV system. In a second stage, the fuzzy logic controller will be implemented using the low-cost Arduino platform [38], taking as a reference the input variables, output, fuzzification, inference system and defuzzification evaluated during the modeling stage. The dc-dc converter will also be implemented according to the design conditions evaluated in the simulations.

This work is structured as follows: Section 2 presents the design and modeling of PV system. Section 3 shows the simulation results for different operating conditions established in Matlab-Simulink. Finally, Section 4 summarizes the main conclusions.

## 2. Design and Modeling of PV System

Figure 1 shows the general diagram of the PV system, which is composed of the 65 W PV module, the buck converter, the battery and the MPPT algorithm (fuzzy or P&O).

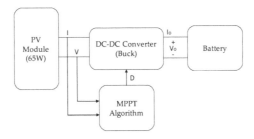

**Figure 1.** Block diagram of the photovoltaic (PV) system.

### 2.1. Modeling of the PV Module

In Equation (1) the mathematical model of the PV module is shown [37,38]. With this model, it is only necessary to calculate the curve fitting parameter that can be obtained directly from the Equation (1). The other parameters are obtained from the electrical data of the PV module.

$$I(V) = \frac{I_x}{1-e^{\left(\frac{-1}{b}\right)}}\left[1 - e^{\left(\frac{V}{bV_x}-\frac{1}{b}\right)}\right] \tag{1}$$

where $V_x$ and $I_x$ are the open circuit voltage and short circuit current with dynamic values for solar irradiance and temperature, which are defined by Equations (2) and (3); b is the characteristic constant, it does not have units and is the unique parameter that has to be calculated.

$$V_x = s\frac{E_i}{E_{iN}}TC_v(T - T_N) + sV_{max} - s(V_{max} - V_{min})e^{\left(\frac{E_i}{E_{iN}}\ln\left(\frac{V_{max}-V_{oc}}{V_{max}-V_{min}}\right)\right)} \tag{2}$$

$$I_x = p\frac{E_i}{E_{iN}}[I_{sc} + TC_i(T - T_N)] \tag{3}$$

where; s: number of PV modules connected in series; p: number of PV modules connected in parallel; $E_i$: effective irradiation of the PV module; $E_{iN}$: irradiation constant of 1000 W/m$^2$; T: temperature of the PV module; $T_N$: temperature constant of 25 °C; $T_{cv}$: temperature coefficient of voltage; $T_{ci}$: temperature coefficient of current; $V_{oc}$: open circuit voltage; $I_{sc}$: short-circuit current; $V_{max}$: voltage for irradiations under 200 W and operating temperature of 25 °C (this value is 103% of $V_{oc}$); $V_{min}$: voltage for irradiations over 1200 W and operating temperature of 25 °C (this value is 85% of $V_{oc}$).

The electrical parameters of the 65 W PV module (Yingli Solar, Baoding, China) are illustrated in Table 1. To find b, Equation (1) and the parameters of Table 1 were used. Knowing that the value of b is in the range of 0.01 to 0.18 [39], the approximation of Equation (4) can be done.

$$1 - e^{\left(\frac{-1}{b}\right)} \approx 1 \tag{4}$$

Therefore, for $V_x$ = 21.7 V; $I_x$ = 4 A; I = 3.71 A and V = 17.5 V; the value of b is 0.07375.

**Table 1.** Electrical parameters of the PV module type YL65P-17b.

| Parameter | Value |
|---|---|
| Short-circuit current ($I_{sc}$) | 4 A |
| Open circuit voltage ($V_{oc}$) | 21.7 V |
| Voltage at $P_{max}$ ($V_{mpp}$) | 17.5 V |
| Current at $P_{max}$ ($I_{mpp}$) | 3.71 A |
| Temperature coefficient of voltage ($T_{cv}$) | −0.0802 V/°C |
| Temperature coefficient of current ($T_{ci}$) | 0.0024 A/°C |
| Maximum voltage ($V_{max}$) | 22.35 V |
| Minimum voltage ($V_{min}$) | 18.44 V |

Figure 2a shows the modeling of the PV module with the Simulink function blocks (MathWorks, Natick, MA, USA). Figure 2b presents the PV module in a subsystem, which was evaluated for different values of solar irradiance and temperature.

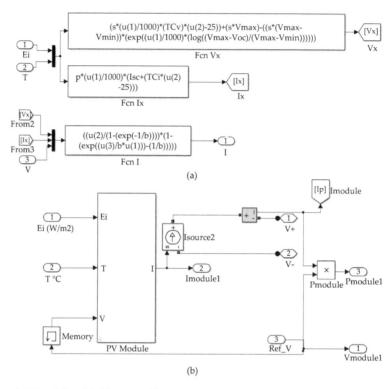

(a)

(b)

**Figure 2.** PV module in Matlab. (**a**) Model implemented with Simulink function blocks; (**b**) Subsystem implemented for the simulation.

Table 2 shows the values obtained with the mathematical model of the PV module, using variable solar irradiance and operating temperature of 25 °C. It can be seen that the values obtained for standard test conditions ($E_i$ = 1000 W/m$^2$, T = 25 °C) correspond to the electrical parameters of the PV module presented in Table 1. Additionally, it is worth noting that the decreases in the solar irradiance considerably affect the short-circuit current, while the open circuit voltage is affected in smaller proportion.

**Table 2.** Parameters of the PV module for variable solar irradiance.

| Parameter | 1000 W/m$^2$ | 800 W/m$^2$ | 600 W/m$^2$ | 400 W/m$^2$ | 200 W/m$^2$ |
|---|---|---|---|---|---|
| Short-circuit current $I_{sc}$ (A) | 4.0 | 3.2 | 2.4 | 1.6 | 0.8 |
| Open circuit voltage $V_{oc}$ (V) | 21.70 | 21.42 | 21.02 | 20.44 | 19.62 |
| Voltage at $P_{max}$ $V_{mpp}$ (V) | 17.66 | 17.55 | 17.37 | 16.78 | 16.08 |
| Current at $P_{max}$ $I_{mpp}$ (A) | 3.679 | 2.924 | 2.171 | 1.459 | 0.730 |
| Maximum Power Point (W) | 64.98 | 51.31 | 37.72 | 24.48 | 11.75 |

Table 3 shows the data obtained with the mathematical modeling of the PV module for solar irradiance of 1000 W/m$^2$ and variable temperature. In this case, it can be noted that increases in temperature considerably affect the open circuit voltage, while the short-circuit current is affected in a smaller proportion. Tables 2 and 3 will be used as references in the results and discussion section, in which a comparison with the fuzzy and P&O controllers will be made; with variations of the solar irradiance and the operating temperature of the PV module.

**Table 3.** Parameters of the PV module for variable temperature.

| Parameter | 0 °C | 25 °C | 50 °C | 75 °C |
|---|---|---|---|---|
| Short-circuit current $I_{sc}$ (A) | 3.94 | 4.00 | 4.06 | 4.12 |
| Open circuit voltage $V_{oc}$ (V) | 23.71 | 21.7 | 19.69 | 17.69 |
| Voltage at $P_{max}$ $V_{mpp}$ (V) | 19.39 | 17.66 | 16.47 | 14.47 |
| Current at $P_{max}$ $I_{mpp}$ (A) | 3.606 | 3.679 | 3.617 | 3.771 |
| Maximum Power Point (W) | 69.92 | 64.98 | 59.59 | 54.55 |

## 2.2. DC-DC Converter Model

A buck converter as control device was used. Figure 3 shows the circuit that was designed to ensure that the converter operates in the continuous conduction mode (CCM); in order to avoid that, the current in the inductor reaches zero during a time interval.

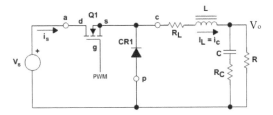

**Figure 3.** Buck converter circuit.

In the CCM, when the transistor is conducting, the diode is in open circuit ($T_{on}$). Using Equation (5), the ripple of the inductor is obtained as shown in Equation (6).

$$V_L = \frac{L\Delta I_L}{\Delta t} \quad (5)$$

$$\Delta I_L(+) = \frac{(V_s - V_{DS} - I_L R_L) - V_o}{L} T_{on} \quad (6)$$

The inductor current decreases during the off state as shown in Equation (7).

$$\Delta I_L(-) = \frac{V_o + (V_d + I_L R_L)}{L} T_{off} \quad (7)$$

Assuming that $V_d$, $R_L$ y $V_{DS}$ are very small values, Equations (8) and (9) are obtained.

$$\Delta I_L(+) = \frac{(V_s - V_o)}{L} T_{on} \quad (8)$$

$$\Delta I_L(-) = \frac{V_o}{L} T_{off} \quad (9)$$

Equating Equations (8) and (9); using $T_s = T_{off} + T_{on}$, Equation (10) for the duty cycle D is obtained.

$$D = \frac{T_{on}}{T_s} = \frac{V_o}{V_s} \quad (10)$$

## 2.2.1. Inductor Design

The inductor was designed to maintain the balance volts per second of the converter and to reduce ripple in the output current. Using an improper inductor produces an alternating current ripple in the direct current output, causing a change between continuous and discontinuous conduction modes. To operate in the continuous conduction mode, the critical output current must be greater than or equal to half the inductor current ripple. See Equation (11) and Figure 4.

$$i_o(\text{crit}) \geq \frac{\Delta I_L}{2} \tag{11}$$

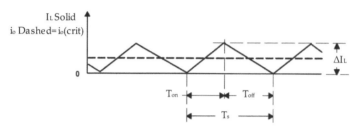

**Figure 4.** Critical output current.

Replacing Equation (8) in (12), using $T_{on} = DT_s$, the Equation (12) for the design of the inductor is obtained.

$$L_{min} \geq \frac{V_o(1 - \frac{V_o}{V_s})T_s}{2i_o(\text{crit})} \tag{12}$$

To calculate L, the maximum power and voltage according to the MPP of the PV module were used: $V_s = 17.71$ V, $P_{max} = 64.984$ W, $i_o = 5.41$ A, $f_s = 20$ KHz. Using a ripple value of 10% for a maximum output current, Equation (13) is obtained.

$$\Delta I_L = 0.1 \times i_o(\text{max}) = 0.541 \text{ A} \tag{13}$$

Using Equations (12) and (13), the minimum value of the inductor as shown in Equation (14) is obtained.

$$L \geq \frac{12 \times (1 - \frac{12}{17.71}) \times 50 \text{ μS}}{2 \times 0.2705} \geq 357.57 \text{ μH} \tag{14}$$

## 2.2.2. Capacitor Design

The current in the capacitor is defined as the variation of the charge with respect to time. See Equation (15).

$$i = \frac{\Delta Q}{\Delta t} = C\frac{\Delta V_c}{\Delta t} \tag{15}$$

Using Figure 5 and Equation (15), the expression for the variation of the load $\Delta Q$ is obtained. See Equation (16).

$$\Delta Q = \frac{\Delta I_L T_s}{8} \tag{16}$$

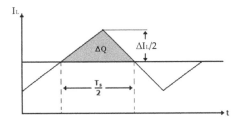

**Figure 5.** Time variation of the current in the inductor.

Therefore, Equation (17) for the design of the capacitor is obtained.

$$C \geq \frac{\Delta I_L T_s}{8 \Delta V_c} \tag{17}$$

Using a ripple value of 0.1%, Equation (18) is obtained.

$$\Delta V = (0.001)(V_o) = 0.012 \text{ V} \tag{18}$$

From Equations (13), (17) and (18), the minimum value of the capacitor is obtained. See Equation (19).

$$C \geq \frac{\Delta I_L T_s}{8 \Delta V_c} \geq 279.63 \text{ } \mu F \tag{19}$$

### 2.2.3. Modelling of Buck Converter

Figure 6 shows the buck converter that was modeled using the fundamental blocks of Simulink.

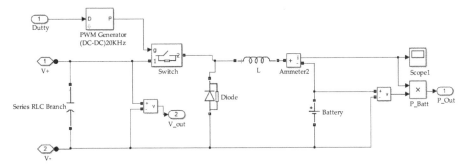

**Figure 6.** Buck converter modeled in Simulink.

Figure 7 shows the current in the battery with the buck converter in open loop, with solar irradiance of 200 W/m² and temperature of 25 °C; in which it is observed that the converter works in the CCM according to that established in the design conditions.

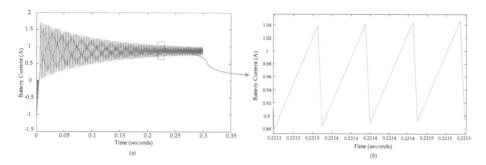

**Figure 7.** (**a**) Battery current for open loop; (**b**) Extended section for range (0.2213–0.2215 s).

*2.3. Fuzzy Controller Design*

Fuzzy control is a method that allows the construction of nonlinear controllers from heuristic information that comes from the knowledge of an expert. Figure 8 shows the block diagram of a fuzzy controller. The fuzzification block is responsible for processing the input signals and assign them a fuzzy value. The set of rules allows a linguistic description of the variables to be controlled and is based on the knowledge of the process. The inference mechanism is responsible for making an interpretation of the data taking into account the rules and their membership functions. With the defuzzification block, the fuzzy information coming from the inference mechanism is converted into non-fuzzy information that is useful for the process to be controlled.

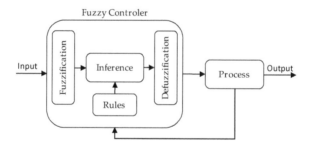

**Figure 8.** Block diagram for a fuzzy controller.

Taking into account the above, the design of fuzzy controller for this work is presented. A fuzzy controller with two inputs and one output was designed. The two input variables are Error (E) and Change of Error (CE), which are shown in Equations (20) and (21) for sample times k.

$$E(k) = \frac{P(k) - P(k-1)}{V(k) - V(k-1)} = \frac{\Delta P}{\Delta V} \tag{20}$$

$$CE(k) = E(k) - E(k-1) = \Delta E \tag{21}$$

The input E(k) is the slope of the P-V curve and defines the location of the MPP in the PV module. The CE(k) input defines whether the movement of the operating point is in the MPP direction or not. The output variable is the increment in duty cycle (ΔD), which can take positive or negative values depending on the location of the operating point. This output is sent to the dc-dc converter to drive the load. Using the value of ΔD delivered by the controller, an accumulator was made to obtain the value of the duty cycle. See Equation (22).

$$D(k) = D(k-1) + \Delta D(k) \tag{22}$$

2.3.1. Membership Functions

Triangular membership functions for the fuzzification process were used. For the inputs E, CE and for the output ΔD, 5 membership functions were defined in terms of the following linguistic variables: Very Low (MB), Low (B), Neutral (N), High (A) and Very High (MA). The range for the error is (−60 to 100), for the change of error is (−10 to 10) and for the increment in duty cycle is (−0.01 to 0.01). Figure 9 shows the membership functions for the inputs and outputs of the controller.

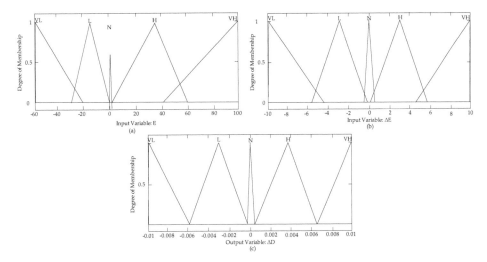

**Figure 9.** Membership functions. (**a**) Error; (**b**) Change of error; (**c**) Increment of duty cycle.

2.3.2. Fuzzy Rules

Table 4 shows the 25 fuzzy rules applied in the controller. The rows and columns represent the two inputs E and ΔE. The output ΔD is a variable located at the intersection of a row with a column.

**Table 4.** Fuzzy associative matrix.

| E/ΔE | Very Low | Low | Neutral | High | Very High |
|------|----------|-----|---------|------|-----------|
| Very Low | VH | VH | H | VL | VL |
| Low | H | H | H | VL | L |
| Neutral | H | H | N | L | L |
| High | H | H | L | L | VL |
| Very High | H | H | L | L | VL |

2.3.3. Fuzzy Controller Modelling

The controller was modeled with the Matlab Fuzzy Logic Toolbox (MathWorks, Natick, MA, USA). A Mamdani controller with the centroid defuzzification method was used. This procedure was carried out using the fuzzy inference system editor (FIS editor) (MathWorks, Natick, MA, USA). Figure 10 shows the controller modeled in Simulink, for which a subsystem was performed to calculate ΔV and ΔP in order to obtain the inputs E and ΔE.

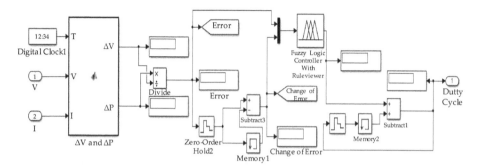

**Figure 10.** Fuzzy logic controller.

*2.4. P&O Controller Design*

The P&O algorithm consists of modifying the operating point of the PV module by increasing or decreasing the duty cycle of a dc-dc converter in order to measure the output power before and after the perturbance. If the power increases, the algorithm perturbs the system in the same direction; otherwise the system is perturbed in the opposite direction. Figure 11 shows the 4 possible options that are presented during the tracking of the MPP, with point 1 being the previous position and point 2 being the current position of each case (A, B, C and D).

- Case A: $\Delta P < 0$ y $\Delta V < 0$.
- Case B: $\Delta P < 0$ y $\Delta V > 0$.
- Case C: $\Delta P > 0$ y $\Delta V > 0$.
- Case D: $\Delta P > 0$ y $\Delta V < 0$.

In cases A and C, the duty cycle must decrease, causing the PV module voltage to increase; while in cases B and D the duty cycle must be increased so that the voltage of the PV module decreases. The flowchart implemented for the P&O controller is shown in Figure 12.

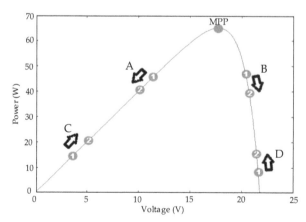

**Figure 11.** Movement of the maximum power point on the P-V curve of the PV module.

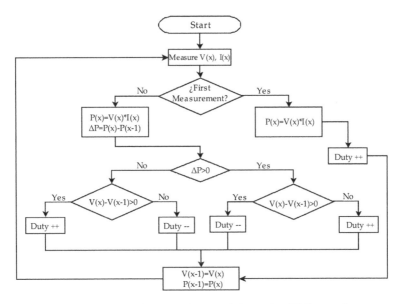

**Figure 12.** Flowchart of the perturbation and observation (P&O) controller.

## 2.5. PV System Modelling

Figure 13 shows the PV system implemented in Matlab/Simulink, which is composed of the PV module, the buck converter and the fuzzy/P&O controller. The signal builder block was used to generate the temperature and irradiance signals in order to evaluate the controller performance. Additionally, this system was used to evaluate the standard P&O controller and perform the comparison with the fuzzy controller.

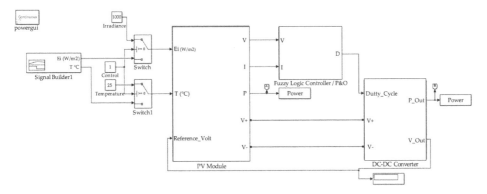

**Figure 13.** PV system modelling.

## 2.6. Limitations

The dc-dc converter and fuzzy control were designed based on the electrical parameters of the PV module under study; for this reason, the calculations made apply to PV modules with powers up to 65 W. One of the inputs of the fuzzy controller is the change of error, which requires a differentiation operation that increases the complexity in the calculations and can generate errors when measuring small powers that are sensitive to noise.

## 3. Results and Discussion

To test the performance of the PV system, different scenarios were simulated in which the traditional P&O control is evaluated in comparison with the fuzzy controller. Four scenarios that simulate sudden changes in solar irradiance and operating temperature of the PV module are presented. In all cases, the following elements were used: 65 W PV module, 12 V battery, inductor of 416 µH and capacitor of 500 µF; with a sampling frequency of 20 KHz for the dc-dc converter.

- Case 1: standard test conditions

In this case, the two controllers were evaluated for solar irradiance of 1000 W/m² and temperature of 25 °C. Figure 14a shows the results obtained for the power delivered to the battery with a simulation time of 0.03 s. It can be seen that the two controllers extract the maximum power of 65 W with a good stabilization time of 0.005 s, which is consistent with the results obtained in [15,16]. In Figure 14b, it is observed that the duty cycle of the P&O control presents small oscillations between 0.6926 and 0.7, in contrast to the fuzzy control that is stabilized at a value of D = 0.694.

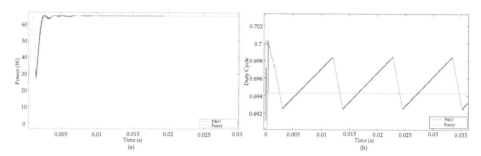

**Figure 14.** PV system with $E_i$ = 1000 W/m² and T = 25 °C. (**a**) Output power of the PV module; (**b**) Duty cycle.

- Case 2: changes in solar irradiance, temperature of 25 °C

In this case, the performance of the controllers was evaluated with an operating temperature of 25 °C and sudden changes in solar irradiance. Initially, an irradiance signal was used with increments of 200 W/m², starting at 200 W/m² and ending at 1000 W/m². Changes in irradiance were made every 0.2 s with a total simulation time of 1 s (see Figure 15a). Subsequently, a test signal with decreases in solar irradiance between 1000 W/m² and 200 W/m² was used (see Figure 15b).

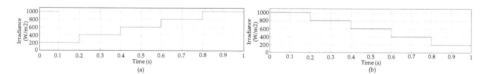

**Figure 15.** (**a**) Increases in solar irradiance; (**b**) Decreases in solar irradiance.

Figure 16a shows the output power for increments in the irradiance signal. In general terms it can be noted that the two controllers present a good performance in the different instants of time. The power obtained is between 11.7 W and 64.9 W, which corresponds to the values presented in Table 2. However, it should be noted that the P&O controller presented small oscillations for $E_i$ = 200 W/m², which is evidenced in the duty cycle of Figure 16b, in times between 0 and 0.2 s.

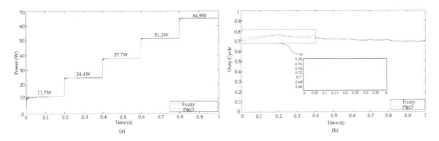

**Figure 16.** (**a**) Output power of the PV module for increases in solar irradiance; (**b**) Duty cycle.

In Figure 17a, the output power for decreases in the irradiance signal is shown. As with the increase signal, the two controllers exhibit good performance with output power between 64.9 W and 11.7 W. Figure 17b shows the duty cycle, in which it is noted that the P&O controller presents the oscillations that characterize this method, but that do not significantly affect the performance of the system.

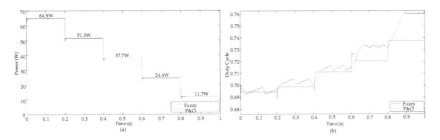

**Figure 17.** (**a**) Output power of the PV module for decreases in solar irradiance; (**b**) Duty cycle.

With the tests carried out in case 2, it is evident that the two controllers exhibit a similar behavior for sudden changes in solar irradiance. In addition, it was found that the P&O control has small oscillations that do not significantly affect the power delivered to the battery.

- Case 3: changes in temperature, solar irradiance of 1000 W/m$^2$

At this point, the performance of the system was evaluated for sudden changes in temperature with a constant solar irradiance of 1000 W/m$^2$. Initially, the signal shown in Figure 18a was used, with temperature increases every 0.2 s between 0 °C and 100 °C, for a test time of 1 s. Subsequently, the signal shown in Figure 18b was used, with decreases in temperature between 100 °C and 0 °C.

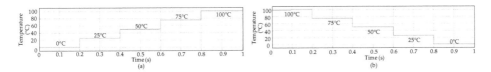

**Figure 18.** (**a**) Increases in temperature; (**b**) Decreases in temperature.

In Figure 19a, the power delivered to the battery is shown, where it is evident the oscillations and power losses that are obtained with the P&O control. Sudden changes in temperature significantly affect the P&O control, which is confirmed by the duty cycle signal shown in Figure 19b. In contrast, the fuzzy control delivers stable power with duty cycle values that adapt to changes in the operating temperature of the PV module. With the P&O control, there are average power losses of 3.15 W, 2.13 W, 2.84 W, 4.12 W and 6.38 W for each of the simulation intervals. The losses were calculated taking as reference the power obtained with the fuzzy controller for the five operating temperatures between 0 °C and 100 °C, which correspond to the values presented in Table 3.

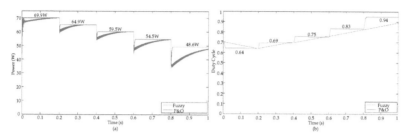

**Figure 19.** (**a**) Output power of the PV module para increases in temperature; (**b**) Duty cycle.

Figure 20 shows the power obtained for decreases in operating temperature. As in the scenario proposed in Figure 19, the P&O control presents oscillations. The worst scenario occurs when the temperature drops from 100 °C to 75 °C, where the system does not reach stabilization and there are oscillations between 20 W and 52 W. With the P&O control, there are average power losses of 46.18 W, 17.32 W, 0 W, 1.11 W and 1.2 W.

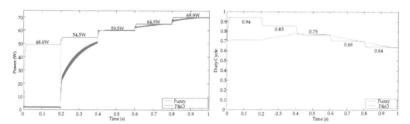

**Figure 20.** (**a**) Output power of the PV module for decreases in temperature; (**b**) Duty cycle.

- Case 4: variations in solar irradiance and temperature

Finally, the performance of the system was evaluated for sudden changes in temperature and solar irradiance in different time values between 0 and 1 s, as seen in Figure 21.

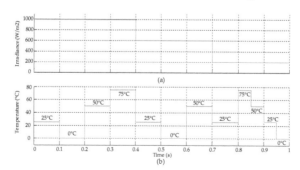

**Figure 21.** Irradiance and temperature signals to evaluate the performance of fuzzy and P&O controllers. (**a**) Increases in solar irradiance; (**b**) Variable temperature.

For the described test conditions, the power obtained from the PV module using fuzzy and P&O controllers is shown in Figure 22. The results prove that the fuzzy controller tracks the MPP without oscillations and power losses. In contrast, the P&O controller exhibits power losses and oscillations for changes in solar irradiance and temperature. The worst case scenario, for the P&O control, is between 0.3 and 0.4 s when the temperature changes from 50 to 75 °C with a solar irradiance of 1000 W/m$^2$, in which the power oscillates between 11.5 and 37.5 W. The highest average power losses with the P&O control occurred between the times 0.2 s to 0.3 s and 0.3 s to 0.4 s with values of 8.52 W and 30.48 W, respectively.

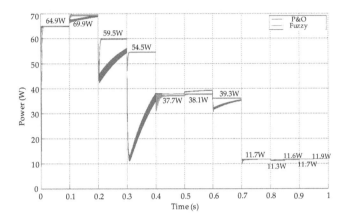

**Figure 22.** Output power of the PV module with the fuzzy and P&O controllers.

## 4. Conclusions

In this paper, a fuzzy controller to track the maximum power point of a PV module was presented, for which their performance was compared with a P&O controller. All components of the PV system were modeled in Matlab/Simulink (PV module, buck converter, fuzzy and P&O controllers). In this way, different test scenarios with signals of temperature and solar irradiance variables were used in order to evaluate the performance of the PV system. It was demonstrated that the fuzzy controller has an excellent performance when there are sudden changes in the operating temperature of the PV module, in contrast with P&O control that is considerably affected, presenting power losses up to 46.18 W. On the other hand, it was also evidenced that in the presence of variations in solar irradiance the two controllers presented a good performance and extracted the maximum power according to the electrical characteristics of the PV module; although the P&O control presented the well-known oscillations, mainly in the sudden changes of irradiance. In this way, the main contribution of this manuscript is the guarantee of supplying the maximum possible power to a battery in an off-grid PV system, using a fuzzy controller. As future work, in the second stage of the project, the fuzzy controller will be implemented in the low-cost Arduino platform; emphasizing that fuzzy control offers the advantage of being easily programmed in microcontrollers.

**Acknowledgments:** This work was supported by the Vicerrectoría de Investigación of the Universidad del Magdalena.

**Author Contributions:** Carlos Robles Algarín conceived and modeled the PV module and the fuzzy controller. John Taborda Giraldo designed the P&O controller and dc-dc converter. Omar Rodríguez Álvarez contributed in the design of the dc-dc converter and data analysis. All authors contributed in writing the manuscript.

**Conflicts of Interest:** The authors declare no conflict of interest.

## Abbreviations

| | |
|---|---|
| $V_x$ | Open circuit voltage for variable values of solar irradiance and operating temperature. |
| $I_x$ | Short-circuit current for variable values of solar irradiance and operating temperature. |
| MPP | Maximum power point of the PV module. |
| $P_{max}$ | Maximum power of the PV module. |
| $V_{mpp}$ | Voltage at $P_{max}$. |
| $I_{mpp}$ | Current at $P_{max}$. |
| s | Number of PV modules connected in series. |
| p | Number of PV modules connected in parallel. |
| $E_i$ | Effective irradiation of the PV module. |
| $E_{iN}$ | Irradiation constant of 1000 W/m$^2$. |
| T | Temperature of the PV module. |
| $T_N$ | Temperature constant of 25 °C. |
| $T_{cv}$ | Temperature coefficient of voltage. |
| $T_{ci}$ | Temperature coefficient of current. |
| $V_{oc}$ | Open circuit voltage. |
| $I_{sc}$ | Short-circuit current. |
| $V_{max}$ | Voltage for irradiations under 200 W and operating temperature of 25 °C. |
| $V_{min}$ | Voltage for irradiations over 1200 W and operating temperature of 25 °C. |
| $V_L$ | Voltage in the inductor. |
| $R_L$ | Internal resistance of the inductor. |
| $R_c$ | Internal resistance of the capacitor. |
| $T_{on}$ | The on time in the dc-dc converter. |
| $T_{off}$ | The off time in the dc-dc converter. |
| $T_s$ | Sampling time. |
| D | Duty cycle. |
| $V_s$ | Input voltage in dc-dc converter. |
| $V_{dc}$ | Transistor voltage in the on mode. |
| $V_d$ | Diode forward voltage. |
| $V_o$ | Output voltage of the dc-dc converter. |
| $\Delta I_L$ | Ripple current in the inductor. |
| $\Delta I_L(+)$ | Ripple current in $T_{on}$. |
| $\Delta I_L(-)$ | Ripple current in $T_{off}$. |
| $I_o$ | Critical output current. |
| $\Delta Q$ | Charge variation in the capacitor. |
| $\Delta V$ | Voltage variation in the capacitor. |

## References

1. Karami, N.; Moubayed, N.; Outbib, R. General review and classification of different MPPT Techniques. *Renew. Sustain. Energy Rev.* **2017**, *68*, 1–18. [CrossRef]
2. Mohapatra, A.; Nayak, B.; Das, P.; Mohanty, K.B. A review on MPPT techniques of PV system under partial shading condition. *Renew. Sustain. Energy Rev.* **2017**, *80*, 854–867. [CrossRef]
3. Bianconi, E.; Calvente, J.; Giral, R.; Mamarelis, E.; Petrone, G.; Ramos, C.A.; Spagnuolo, G.; Vitelli, M. Perturb and Observe MPPT algorithm with a current controller based on the sliding mode. *Int. J. Electr. Power* **2013**, *44*, 346–356. [CrossRef]
4. Chen, M.; Ma, S.; Wu, J.; Huang, L. Analysis of MPPT Failure and Development of an Augmented Nonlinear Controller for MPPT of Photovoltaic Systems under Partial Shading Conditions. *Appl. Sci.* **2017**, *7*, 95. [CrossRef]
5. Kwan, T.H.; Wu, X. High performance P&O based lock-on mechanism MPPT algorithm with smooth tracking. *Sol. Energy* **2017**, *155*, 816–828. [CrossRef]
6. Alik, R.; Jusoh, A. Modified Perturb and Observe (P&O) with checking algorithm under various solar irradiation. *Sol. Energy* **2017**, *148*, 128–139. [CrossRef]

7. Bounechba, H.; Bouzid, A.; Snani, A.; Lashab, A. Real time simulation of MPPT algorithms for PV energy system. *Int. J. Electr. Power* **2016**, *83*, 67–78. [CrossRef]

8. Huang, Y.P.; Hsu, S.Y. A performance evaluation model of a high concentration photovoltaic module with a fractional open circuit voltage-based maximum power point tracking algorithm. *Comput. Electr. Eng.* **2016**, *51*, 331–342. [CrossRef]

9. Cortajarena, J.A.; Barambones, O.; Alkorta, P.; De Marcos, J. Sliding mode control of grid-tied single-phase inverter in a photovoltaic MPPT application. *Sol. Energy* **2017**, *155*, 793–804. [CrossRef]

10. Tobón, A.; Peláez-Restrepo, J.; Villegas-Ceballos, J.P.; Serna-Garcés, S.I.; Herrera, J.; Ibeas, A. Maximum Power Point Tracking of Photovoltaic Panels by Using Improved Pattern Search Methods. *Energies* **2017**, *10*, 1316. [CrossRef]

11. Loukriz, A.; Haddadi, M.; Messalti, S. Simulation and experimental design of a new advanced variable step size Incremental Conductance MPPT algorithm for PV systems. *ISA Trans.* **2016**, *62*, 30–38. [CrossRef] [PubMed]

12. Mellit, A.; Rezzouk, H.; Messai, A.; Medjahed, B. FPGA-based real time implementation of MPPT-controller for photovoltaic systems. *Renew. Energy* **2011**, *36*, 1652–1661. [CrossRef]

13. Ramalu, T.; Mohd Radzi, M.A.; Mohd Zainuri, M.A.A.; Abdul Wahab, N.I.; Abdul Rahman, R.Z. A Photovoltaic-Based SEPIC Converter with Dual-Fuzzy Maximum Power Point Tracking for Optimal Buck and Boost Operations. *Energies* **2016**, *9*, 604. [CrossRef]

14. Hassan, S.Z.; Li, H.; Kamal, T.; Arifoğlu, U.; Mumtaz, S.; Khan, L. Neuro-Fuzzy Wavelet Based Adaptive MPPT Algorithm for Photovoltaic Systems. *Energies* **2017**, *10*, 394. [CrossRef]

15. Nabipour, M.; Razaz, M.; Seifossadat, S.; Mortazavi, S. A new MPPT scheme based on a novel fuzzy approach. *Renew. Sustain. Energy Rev.* **2017**, *74*, 1147–1169. [CrossRef]

16. Bendib, B.; Krim, F.; Belmili, H.; Almi, M.F.; Boulouma, S. Advanced Fuzzy MPPT Controller for a Stand-alone PV System. *Energy Procedia* **2014**, *50*, 383–392. [CrossRef]

17. Belaidi, R.; Haddouche, A.; Fathi, M.; Larafi, M.M.; Kaci, G.M. Performance of grid-connected PV system based on SAPF for power quality improvement. In Proceedings of the International Renewable and Sustainable Energy Conference (IRSEC), Marrakech, Morocco, 14–17 November 2016; pp. 1–4.

18. Chekired, F.; Larbes, C.; Rekioua, D.; Haddad, F. Implementation of a MPPT fuzzy controller for photovoltaic systems on FPGA circuit. *Energy Procedia* **2011**, *6*, 541–549. [CrossRef]

19. Na, W.; Chen, P.; Kim, J. An Improvement of a Fuzzy Logic-Controlled Maximum Power Point Tracking Algorithm for Photovoltic Applications. *Appl. Sci.* **2017**, *7*, 326. [CrossRef]

20. Messaltia, S.; Harrag, A.; Loukriz, A. A new variable step size neural networks MPPT controller: Review, simulation and hardware implementation. *Renew. Sustain. Energy Rev.* **2017**, *68*, 221–233. [CrossRef]

21. Dounis, A.I.; Kofinas, P.; Papadakis, G.; Alafodimos, C. A direct adaptive neural control for maximum power point tracking of photovoltaic system. *Sol. Energy* **2015**, *115*, 145–165. [CrossRef]

22. Muthuramalingam, M.; Manoharan, P.S. Comparative analysis of distributed MPPT controllers for partially shaded stand alone photovoltaic systems. *Energy Convers. Manag.* **2014**, *86*, 286–299. [CrossRef]

23. Jin, Y.; Hou, W.; Li, G.; Chen, X. A Glowworm Swarm Optimization-Based Maximum Power Point Tracking for Photovoltaic/Thermal Systems under Non-Uniform Solar Irradiation and Temperature Distribution. *Energies* **2017**, *10*, 541. [CrossRef]

24. Titri, S.; Larbes, C.; Toumi, K.Y.; Benatchba, K. A new MPPT controller based on the Ant colony optimization algorithm for Photovoltaic systems under partial shading conditions. *Appl. Soft Comput.* **2017**, *58*, 465–479. [CrossRef]

25. Jiang, L.L.; Maskell, D.L.; Patra, J.C. A novel ant colony optimization-based maximum power point tracking for photovoltaic systems under partially shaded conditions. *Energy Build.* **2013**, *58*, 227–236. [CrossRef]

26. Benyoucef, A.S.; Chouder, A.; Kara, K.; Silvestre, S.; Sahed, O.A. Artificial bee colony based algorithm for maximum power point tracking (MPPT) for PV systems operating under partial shaded conditions. *Appl. Soft Comput.* **2015**, *32*, 38–48. [CrossRef]

27. Fathy, A. Reliable and efficient approach for mitigating the shading effect on photovoltaic module based on Modified Artificial Bee Colony algorithm. *Renew. Energy* **2015**, *81*, 78–88. [CrossRef]

28. Atawi, I.E.; Kassem, A.M. Optimal Control Based on Maximum Power Point Tracking (MPPT) of an Autonomous Hybrid Photovoltaic/Storage System in Micro Grid Applications. *Energies* **2017**, *10*, 643. [CrossRef]

29. Ou, T.-C.; Hong, C.-M. Dynamic operation and control of microgrid hybrid power systems. *Energy* **2014**, *66*, 314–323. [CrossRef]

30. Hong, C.-M.; Ou, T.-C.; Lu, K.-H. Development of intelligent MPPT (maximum power point tracking) control for a grid-connected hybrid power generation system. *Energy* **2013**, *50*, 270–279. [CrossRef]

31. Shiau, J.-K.; Wei, Y.-C.; Lee, M.-Y. Fuzzy Controller for a Voltage-Regulated Solar-Powered MPPT System for Hybrid Power System Applications. *Energies* **2015**, *8*, 3292–3312. [CrossRef]

32. Ou, T.-C.; Su, W.-F.; Liu, X.-Z.; Huang, S.-J.; Tai, T.-Y. A Modified Bird-Mating Optimization with Hill Climbing for Connection Decisions of Transformers. *Energies* **2016**, *9*, 671. [CrossRef]

33. Ou, T.-C. A novel unsymmetrical faults analysis for microgrid distribution systems. *Electr. Power Energy Syst.* **2012**, *43*, 1017–1024. [CrossRef]

34. Ou, T.-C. Ground fault current analysis with a direct building algorithm for microgrid distribution. *Electr. Power Energy Syst.* **2013**, *53*, 867–875. [CrossRef]

35. Ou, T.-C.; Lu, K.-H.; Huang, C.-J. Improvement of Transient Stability in a Hybrid Power Multi-System Using a Designed NIDC (Novel Intelligent Damping Controller). *Energies* **2017**, *10*, 488. [CrossRef]

36. Robles Algarín, C.; Callejas Cabarcas, J.; Polo Llanos, A. Low-Cost Fuzzy Logic Control for Greenhouse Environments with Web Monitoring. *Electronics* **2017**, *6*, 71. [CrossRef]

37. Ortiz, E. Modeling and Analysis of Solar Distributed Generation. Ph.D. Thesis, Michigan State University, Michigan, MI, USA, 2006.

38. Gil, O. Modelado Y Simulación de Dispositivos Fotovoltaicos. Master's Thesis, Universidad de Puerto Rico, San Juan, Puerto Rico, 2008.

39. Robles, C.; Ospino, A.; Casas, J. Dual-Axis Solar Tracker for Using in Photovoltaic Systems. *Int. J. Renew. Energy Res. IJRER* **2017**, *10*, 137–145.

*Article*

# Modeling and Stability Analysis of a Single-Phase Two-Stage Grid-Connected Photovoltaic System

Liying Huang, Dongyuan Qiu *, Fan Xie, Yanfeng Chen and Bo Zhang

School of Electric Power Engineering, South China University of Technology, Guangzhou 510641, China;
ephly@mail.scut.edu.cn (L.H.); epfxie@scut.edu.cn (F.X.); eeyfchen@scut.edu.cn (Y.C.);
epbzhang@scut.edu.cn (B.Z.)
* Correspondence: epdyqiu@scut.edu.cn; Tel.: +86-20-8711-2508

Received: 30 November 2017; Accepted: 14 December 2017; Published: 19 December 2017

**Abstract:** The stability issue of a single-phase two-stage grid-connected photovoltaic system is complicated due to the nonlinear $v$-$i$ characteristic of the photovoltaic array as well as the interaction between power converters. Besides, even though linear system theory is widely used in stability analysis of balanced three-phase systems, the application of the same theory to single-phase systems meets serious challenges, since single-phase systems cannot be transformed into linear time-invariant systems simply using Park transformation as balanced three-phase systems. In this paper, (1) the integrated mathematical model of a single-phase two-stage grid-connected photovoltaic system is established, in which both DC-DC converter and DC-AC converter are included also the characteristic of the PV array is considered; (2) an observer-pattern modeling method is used to eliminate the time-varying variables; and (3) the stability of the system is studied using eigenvalue sensitivity and eigenvalue loci plots. Finally, simulation results are given to validate the proposed model and stability analysis.

**Keywords:** photovoltaic system; modeling; stability analysis; grid-connected

## 1. Introduction

Under pressure from the energy crisis, photovoltaic (PV) energy has been more and more attractive for generating electricity. At the end of 2016, the total PV installation capacity around the world amounted to 305 GW [1]. The majority of PV installations are grid-connected PV systems, since they can deliver power to the grid directly and are more cost-effective than stand-alone systems [2]. Whereas large commercial PV systems are connected to the three-phase grid, single-phase topology is advantageous in small-scale PV systems such as residential systems due to its simplicity [3,4].

A typical grid-connected PV system is a two-stage system, where the first stage is normally a DC-DC converter for extracting power from the PV array and the second stage is a DC-AC converter for delivering power to the grid. However, the stability of such a system is a major concern. There are two main factors that make the stability analysis of a two-stage grid-connected PV system more difficult than other power electronic systems. First is the characteristic of the PV array. The $v$-$i$ characteristic of a PV array is nonlinear and changes with the light intensity or temperature, thus the dynamics of a PV system can be vastly different from a traditional power electronic system fed from a constant voltage source. In some studies on stability analysis of PV systems, the PV array is replaced by a constant voltage source [5,6] or a constant current source [7]. These methods neglect the nonlinear characteristics of the PV array and may cause deviation between the theoretical analysis and the behavior of the real system [8]. Some studies take into account this characteristic of the PV array by using the $v$-$i$ curve calculated from numerical techniques with the aid of a computer [8,9]. However, specific parameters of the PV array such as shunt resistance and series resistance are necessary for implementing the calculation. These parameters usually cannot be obtained from the datasheet. Second, a DC-DC converter and a DC-AC converter are connected in cascade. Even in a single power converter there exists complex behaviors such as bifurcation and chaos [10–13]. In this case, the behavior of the overall system may be more complicated than only one converter, since the interconnected converters will influence each other [14–17]. Hence, it is necessary to establish an integrated mathematical model for the entire single-phase two-stage grid-connected photovoltaic system that is able to describe the characteristic of the PV array as well as the interactions between two converters.

For a balanced three-phase system, application of Park transformation facilitates modeling of the system. The system can be first transformed into a Multiple-Input-Multiple-Output (MIMO) system in d-q reference frame and then be linearized around a fixed stead-state operating point [18]. Finally, the balanced three-phase system can be described using a linear time invariant (LTI) model. Thus, vast LTI theory tools can be applied to completing the controller design and stability analysis [19,20]. However, it is hard to put a single-phase system within the framework of an LTI model. The main difficulty for this is that linearization process must be performed around a fixed steady-state operating point rather than a steady-state time-periodic trajectory [18]. To deal with this problem, an observer-pattern modeling method [21,22] is proposed that eliminates the effect of time-variance.

In this paper, the stability analysis of the whole single-phase two-stage grid-connected PV system is presented. Both DC-DC converter and DC-AC converter will be included in the model. Also, the characteristic of the PV array will be considered. To avoid the lack of specific parameters of the PV array, the proposed model uses the basic parameters that are provided in all datasheets of PV arrays. The application of observer-pattern modeling method successfully transforms the system into time-invariant. With the proposed model, the stability of the system can be studied by calculating the eigenvalues of the Jacobian of the system.

## 2. System Description and Nonlinear Averaged Equations

Figure 1 presents the diagram of a single-phase two-stage grid-connected photovoltaic system. In this figure, $C_{in}$ is the capacitance of the input filter, $L_b$ and $C_{dc}$ are the inductance and the capacitance of the boost converter, respectively, and $L_f$ is the inductance of the output filter. The PV array generates electricity from solar radiation. A boost converter with an input filter connects the PV array to the DC bus in order to raise the output voltage of PV array to the voltage level of DC bus while implementing maximum power point tracking (MPPT). The boost converter is designed to operate in continuous conduction mode (CCM).

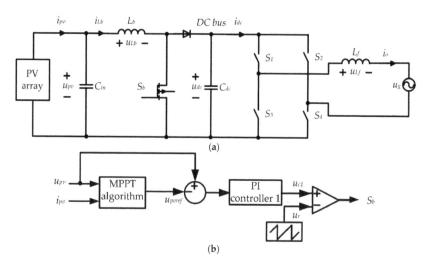

(a)

(b)

**Figure 1.** *Cont.*

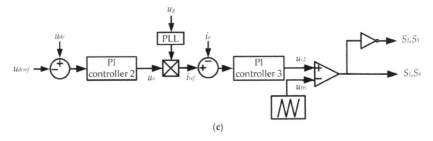

(c)

**Figure 1.** Diagram of a single-phase two-stage grid-connected photovoltaic system: (a) Power stage circuit; (b) MPPT controller; (c) double loop controller.

In this study, the Perturb & Observe (P&O) method is adopted for MPPT, since it is one of the most popular MPPT algorithms [23]. A full bridge inverter with an L filter supplies the power to the AC grid [24]. The full bridge inverter is controlled by a double loop controller, which is comprised of a voltage control loop and a current control loop. The voltage control loop regulates the DC bus voltage $u_{dc}$ and generates the reference current $i_{ref}$ for the current control loop. Then, the current control loop regulates the output current of full bridge inverter $i_o$. In order to facilitate the analysis, the boost converter and the full bridge inverter are assumed to have same switching frequency $f_s$.

In order to develop an integrated mathematical model, the system equations are derived for three parts: PV array part as given in Section 2.1, power stage part as described in Section 2.2, and controller part as detailed in Section 2.3.

### 2.1. PV Array

The output current $i_{pv}$ of the PV array is related to the output voltage $u_{pv}$ and also affected by other parameters of PV panel itself [25]. However, the manufacturer's datasheet do not provide some of these parameters, such as equivalent series resistance and equivalent parallel resistance. Basic parameters provided in all datasheets of PV array are open circuit voltage $U_{OC}$, short circuit current $I_{SC}$, the current at maximum power point (MPP) $I_M$, and the voltage at maximum power point $U_M$. These parameters are all with the standard test condition (STC). To overcome the lack of detailed information of the PV array, a simplified model proposed in [26] is used in this study, which describes the terminal characteristic of the PV array with STC in the following equation:

$$i_{pv} = I_{SC}\left[1 - A_1\left(e^{\frac{u_{pv}}{A_2 U_{OC}}} - 1\right)\right] \tag{1}$$

where $A_1 = \left(1 - \frac{I_M}{I_{SC}}\right)e^{-\frac{U_M}{A_2 U_{OC}}}$, $A_2 = \left(\frac{U_M}{U_{OC}} - 1\right)\left[ln\left(1 - \frac{I_M}{I_{SC}}\right)\right]^{-1}$.

### 2.2. Power Stage Circuit

To describe the power stage circuit, four sate variables are used: the output voltage of PV array $u_{pv}$, the current of $L_b$ which denoted as $i_{L_b}$, the DC bus voltage $u_{dc}$ and the output current of full bridge inverter $i_o$. The averaged equations of the input filter and boost converter can be derived as follows [19]:

$$\begin{cases} \frac{du_{pv}}{dt} = \frac{1}{C_{in}}\left(i_{pv} - i_{L_b}\right) \\ \frac{di_{L_b}}{dt} = \frac{1}{L_b}\left[u_{pv} - (1 - d_1)u_{dc}\right] \\ \frac{du_{dc}}{dt} = \frac{1}{C_{dc}}\left[(1 - d_1)i_{L_b} - i_{dc}\right] \end{cases} \tag{2}$$

where $d_1$ is the duty cycle of the boost converter and $i_{dc}$ is the input current of full bridge inverter.

According to the operation principle of the full bridge inverter, following equations can be derived [24]:

$$\begin{cases} i_{dc} = (2d_2 - 1)i_o \\ \frac{di_o}{dt} = \frac{1}{L_f}[(2d_2 - 1)u_{dc} - u_g] \end{cases} \tag{3}$$

where $d_2$ is the duty cycle of $S_1$ and $S_4$ in the full bridge inverter and $u_g$ is the grid voltage.

Combine (2) and (3), then the averaged equations of the power stage circuit can be derived as presented in (4).

$$\begin{cases} \frac{du_{pv}}{dt} = \frac{1}{C_{in}}(i_{pv} - i_{L_b}) \\ \frac{di_{L_b}}{dt} = \frac{1}{L_b}[u_{pv} - (1 - d_1)u_{dc}] \\ \frac{du_{dc}}{dt} = \frac{1}{C_{dc}}[(1 - d_1)i_{L_b} - (2d_2 - 1)i_o] \\ \frac{di_o}{dt} = \frac{1}{L_f}[(2d_2 - 1)u_{dc} - u_g] \end{cases} \tag{4}$$

*2.3. Controller*

The state equations corresponding to the Proportional-Integral (PI) controller can be expressed as follow:

$$\frac{dm(t)}{dt} = K_p \frac{de(t)}{dt} + \frac{K_p}{T_i}e(t) \tag{5}$$

where $m(t)$ and $e(t)$ are the output and the input signals of the PI controller, respectively, $K_p$ is the gain of the PI controller, and $T_i$ is the time constant of the PI controller.

Since the perturbation step size of P&O method is small, the reference voltage $u_{pvref}$ given by the algorithm is assumed to be constant. Therefore, the state equation of PI controller 1 can be derived as

$$\frac{du_{c1}}{dt} = K_{p1} \frac{du_{pv}}{dt} + \frac{K_{p1}}{T_{i1}}\left(u_{pv} - u_{pvref}\right) \tag{6}$$

where $u_{c1}$ is the output signal of PI controller 1 and $K_{p1}$ and $T_{i1}$ are the gain and the time constant of the PI controller. $u_{c1}$ is compared to a ramp $u_r$ in the Pulse Width Modulation (PWM) comparator to produce driving signal. To complete the model of MPPT controller, the duty cycle $d_1$ is expressed as following equation:

$$d_1 = \frac{u_{c1}}{U_{M1}} \tag{7}$$

where $U_{M1}$ is the peak-to-peak amplitude of ramp $u_r$.

The state equation of PI controller 2 can be derived as

$$\frac{du_e}{dt} = K_{p2} \frac{du_{dc}}{dt} + \frac{K_{p2}}{T_{i2}}\left(u_{dc} - u_{dcref}\right) \tag{8}$$

where $u_{dcref}$ is the reference voltage of DC bus, $u_e$ is the output signal of PI controller 2, $K_{p2}$ and $T_{i2}$ is the gain and the time constant of the PI controller. Then the reference current $i_{ref}$ can be expressed as:

$$i_{ref} = u_e sin(\omega t) \tag{9}$$

where $\omega$ is the angular frequency of power grid.

Thus the state equation of PI controller 3 can be derived as

$$\frac{du_{c2}}{dt} = K_{p3}\left(u_e \omega cos(\omega t) + sin(\omega t)\frac{du_e}{dt} - \frac{di_o}{dt}\right) + \frac{K_{p3}}{T_{i3}}(u_e sin(\omega t) - i_o) \tag{10}$$

where $u_{c2}$ is the output signal of PI controller 3 and $K_{p3}$ and $T_{i3}$ are the gain and the time constant of the PI controller. $u_{c2}$ is compared to a triangular wave $u_{tri}$ in the PWM comparator to generate driving signal. The duty cycle $d_2$ is expressed as following equation:

$$d_2 = \frac{1}{2}\left(1 + \frac{u_{c2}}{U_{M2}}\right) \tag{11}$$

where $U_{M2}$ is the peak-to-peak amplitude of triangular wave $u_{tri}$.

Combine (1), (4) and (6)–(11), the nonlinear averaged equations of the system can be written as

$$
\begin{cases}
\frac{du_{pv}}{dt} = \frac{1}{C_{dc}}\left(I_{SC}\left[1 - A_1\left(e^{\frac{u_{pv}}{A_2 U_{OC}}} - 1\right)\right] - i_{L_b}\right) \\
\frac{di_{L_b}}{dt} = \frac{1}{L_b}\left[u_{pv} - \left(1 - \frac{u_{c1}}{U_{M1}}\right)u_{dc}\right] \\
\frac{du_{dc}}{dt} = \frac{1}{C_{dc}}\left[\left(1 - \frac{u_{c1}}{U_{M1}}\right)i_{L_b} - \frac{u_{c2}i_o}{U_{M2}}\right] \\
\frac{di_o}{dt} = \frac{1}{L_f}\left[\frac{u_{c2}}{U_{M2}}u_{dc} - U_{gm}\sin(\omega t)\right] \\
\frac{du_{c1}}{dt} = K_{p1}\frac{du_{pv}}{dt} + \frac{K_{p1}}{T_{i1}}\left(u_{pv} - u_{pvref}\right) \\
\frac{du_e}{dt} = K_{p2}\frac{du_{dc}}{dt} + \frac{K_{p2}}{T_{i2}}\left(u_{dc} - u_{dcref}\right) \\
\frac{du_{c2}}{dt} = K_{p3}\left(u_e\omega\cos(\omega t) + \sin(\omega t)\frac{du_e}{dt} - \frac{di_o}{dt}\right) + \frac{K_{p3}}{T_{i3}}\left(u_e\sin(\omega t) - i_o\right)
\end{cases}
\tag{12}
$$

where $U_{gm}$ is the amplitude of $u_g$.

## 3. Observer-Pattern Model

The equations shown in (12) contain $sin(\omega t)$ and $cos(\omega t)$, which means that the single-phase two-stage grid-connected photovoltaic system is a time-variant nonlinear system. This time-variance is the main difficulty in stability analysis of the system. To eliminating the effect of time-variance, the system is transformed into a time-invariant one using an observer-pattern modeling method [21,22]. Notice that $sin(\omega t)$ and $cos(\omega t)$ only exist in the expressions of $di_o/dt$ and $du_{c2}/dt$. Thus, only $i_o$ and $u_{c2}$ need to be processed.

First, the time-variance originated from fundamental frequency of the grid is removed by Park transformation. Implementation of Park transformation needs at least two orthogonal variables, so the concept of Imaginary Orthogonal Circuit is introduced [27]. Denote the corresponding Imaginary Orthogonal Circuit variables to $i_o$ and $u_{c2}$ as $i_{oI}$ and $u_{c2I}$. Since $i_o$ and $u_{c2}$ are sinusoidal, $i_{oI}$ and $u_{c2I}$ maintain 90° phase shift with $i_o$ and $u_{c2}$. The Park transformation can be expressed as

$$
\begin{cases}
i_{odq} = \begin{bmatrix} i_{od} \\ i_{oq} \end{bmatrix} = T\begin{bmatrix} i_o \\ i_{oI} \end{bmatrix} \\
u_{c2dq} = \begin{bmatrix} u_{c2d} \\ u_{c2q} \end{bmatrix} = T\begin{bmatrix} u_{c2} \\ u_{c2I} \end{bmatrix}
\end{cases}
\tag{13}
$$

where $T$ is the transformation matrix given by (14).

$$
T = \begin{bmatrix} \cos(\omega t) & \sin(\omega t) \\ -\sin(\omega t) & \cos(\omega t) \end{bmatrix}
\tag{14}
$$

Apply inverse transformation of (13) to the state equations of $i_o$ and $u_{c2}$, resulting in the following equations:

$$
\begin{cases}
\frac{d}{dt}\left(T^{-1}i_{odq}\right) = \frac{u_{dc}}{L_f U_{M2}}T^{-1}u_{c2dq} - \frac{U_{gm}}{L_f}\begin{bmatrix} \sin(\omega t) \\ -\cos(\omega t) \end{bmatrix} \\
\frac{d}{dt}\left(T^{-1}u_{c2dq}\right) = K_{p3}u_e\omega\begin{bmatrix} \cos(\omega t) \\ \sin(\omega t) \end{bmatrix} - \frac{K_{p3}}{L_f}\left(\frac{u_{dc}}{L_f U_{M2}}T^{-1}u_{c2dq} - \frac{U_{gm}}{L_f}\begin{bmatrix} \sin(\omega t) \\ -\cos(\omega t) \end{bmatrix}\right) \\
\quad + K_{p3}\frac{du_e}{dt}\begin{bmatrix} \sin(\omega t) \\ -\cos(\omega t) \end{bmatrix} + \frac{K_{p3}}{T_{i3}}u_e\begin{bmatrix} \sin(\omega t) \\ -\cos(\omega t) \end{bmatrix} - \frac{K_{p3}}{T_{i3}}T^{-1}i_{odq}
\end{cases}
\tag{15}
$$

Multiplying (15) by $T$ gives

$$
\begin{cases}
T\frac{d}{dt}\left(T^{-1}i_{odq}\right) = \frac{u_{dc}}{L_f \bar{U}_{M2}}u_{c2dq} - \frac{U_{gm}}{L_f}T\begin{bmatrix} sin(\omega t) \\ -cos(\omega t) \end{bmatrix} \\
T\frac{d}{dt}\left(T^{-1}u_{c2dq}\right) = \frac{K_{p3}}{T_{i3}}u_e T\begin{bmatrix} sin(\omega t) \\ -cos(\omega t) \end{bmatrix} - \frac{K_{p3}}{L_f}\left(\frac{u_{dc}}{L_f \bar{U}_{M2}}u_{c2dq} - \frac{U_{gm}}{L_f}T\begin{bmatrix} sin(\omega t) \\ -cos(\omega t) \end{bmatrix}\right) \\
\quad - \frac{K_{p3}}{T_{i3}}i_{odq} + K_{p3}T\left(u_e\omega\begin{bmatrix} cos(\omega t) \\ sin(\omega t) \end{bmatrix} + \frac{du_e}{dt}\begin{bmatrix} sin(\omega t) \\ -cos(\omega t) \end{bmatrix}\right)
\end{cases}
\tag{16}
$$

Since $T\frac{d}{dt}\left(T^{-1}i_{odq}\right) = T\frac{dT^{-1}}{dt}i_{odq} + \frac{di_{odq}}{dt}$ and replacing $i_{odq}$ with $u_{c2dq}$ satisfies this equation as well, (16) can be expressed as

$$
\begin{cases}
\frac{d}{dt}\begin{bmatrix} i_{od} \\ i_{oq} \end{bmatrix} = \frac{u_{dc}}{L_f \bar{U}_{M2}}\begin{bmatrix} u_{c2d} \\ u_{c2q} \end{bmatrix} - \frac{U_{gm}}{L_f}\begin{bmatrix} 0 \\ -1 \end{bmatrix} + \begin{bmatrix} 0 & \omega \\ -\omega & 0 \end{bmatrix}\begin{bmatrix} i_{od} \\ i_{oq} \end{bmatrix} \\
\frac{d}{dt}\begin{bmatrix} u_{c2d} \\ u_{c2q} \end{bmatrix} = \begin{bmatrix} 0 & \omega \\ -\omega & 0 \end{bmatrix}\begin{bmatrix} u_{c2d} \\ u_{c2q} \end{bmatrix} + \frac{U_{gm}K_{p3}}{L_f}\begin{bmatrix} 0 \\ -1 \end{bmatrix} + \frac{K_{p3}}{T_{i3}}u_e\begin{bmatrix} 0 \\ -1 \end{bmatrix} \\
\quad - \frac{K_{p3}}{T_{i3}}\begin{bmatrix} i_{od} \\ i_{oq} \end{bmatrix} + K_{p3}\left(u_e\omega\begin{bmatrix} 1 \\ 0 \end{bmatrix} + \frac{du_e}{dt}\begin{bmatrix} 0 \\ -1 \end{bmatrix} - \frac{u_{dc}}{L_f \bar{U}_{M2}}\begin{bmatrix} u_{c2d} \\ u_{c2q} \end{bmatrix}\right)
\end{cases}
\tag{17}
$$

Notice that

$$
\begin{cases}
u_{c2} = u_{c2d}cos(\omega t) - u_{c2q}sin(\omega t) \\
i_o = i_{od}cos(\omega t) - i_{oq}sin(\omega t)
\end{cases}
\tag{18}
$$

Thus, the product term $u_{c2}i_o$ in the averaged Equation (12) can be substituted by

$$
u_{c2}i_o = \frac{1+cos(2\omega t)}{2}u_{c2d}i_{od} + \frac{1-cos(2\omega t)}{2}u_{c2q}i_{oq} \\
\quad - \frac{sin(2\omega t)}{2}(u_{c2d}i_{oq} + u_{c2q}i_{od})
\tag{19}
$$

Since the Equation (19) contains $cos(2\omega t)$ and $sin(2\omega t)$, the equation is still time-variant and needs to be further processed. Assuming that $g_1 = cos(2\omega t)$, $g_2 = sin(2\omega t)$, the following equations can be constructed:

$$
\begin{cases}
\frac{dg_1}{dt} = -2\omega g_2 \\
\frac{dg_2}{dt} = 2\omega g_1
\end{cases}
\tag{20}
$$

Combining Equations (12), (17), (19) and (20), the observer-pattern model of the system can be written as

$$
\begin{cases}
\frac{du_{pv}}{dt} = \frac{1}{C_{in}}\left(I_{SC}\left[1 - A_1\left(e^{\frac{u_{pv}}{A_2 U_{OC}}} - 1\right)\right] - i_{L_b}\right) \\
\frac{di_{L_b}}{dt} = \frac{1}{L_b}\left[u_{pv} - \left(1 - \frac{u_{c1}}{\bar{U}_{M1}}\right)u_{dc}\right] \\
\frac{du_{dc}}{dt} = \frac{1}{C_{dc}}\left[\left(1 - \frac{u_{c1}}{\bar{U}_{M1}}\right)i_{L_b} - \frac{1+g_1}{2}u_{c2d}i_{od} - \frac{1-g_1}{2}u_{c2q}i_{oq} + \frac{g_2}{2}(u_{c2d}i_{oq} + u_{c2q}i_{od})\right] \\
\frac{di_{od}}{dt} = \frac{u_{dc}u_{c2d}}{L_f \bar{U}_{M2}} + \omega i_{oq} \\
\frac{di_{oq}}{dt} = \frac{u_{dc}u_{c2q}}{L_f \bar{U}_{M2}} + \frac{U_{gm}}{L_f} - \omega i_{od} \\
\frac{du_{c1}}{dt} = K_{p1}\frac{du_{pv}}{dt} + \frac{K_{p1}}{T_{i1}}\left(u_{pv} - u_{pvref}\right) \\
\frac{du_e}{dt} = K_{p2}\frac{du_{dc}}{dt} + \frac{K_{p2}}{T_{i2}}\left(u_{dc} - u_{dcref}\right) \\
\frac{du_{c2d}}{dt} = K_{p3}u_e\omega - \frac{K_{p3}}{L_f \bar{U}_{M2}}u_{dc}u_{c2d} - \frac{K_{p3}}{T_{i3}}i_{od} + \omega u_{c2q} \\
\frac{du_{c2q}}{dt} = -K_{p3}\frac{du_e}{dt} - \frac{K_{p3}}{L_f \bar{U}_{M2}}u_{dc}u_{c2q} - \frac{U_{gm}K_{p3}}{L_f} - \frac{K_{p3}}{T_{i3}}u_e - \frac{K_{p3}}{T_{i3}}i_{oq} - \omega u_{c2d} \\
\frac{dg_1}{dt} = -2\omega g_2 \\
\frac{dg_2}{dt} = 2\omega g_1
\end{cases}
\tag{21}
$$

In order to verify the proposed model, simulation results obtained from PSIM are compared to the solutions of the model obtained from MATLAB. The simulation parameters, including PV array parameters and converter parameters, are given in Tables 1 and 2. Figures 2 and 3 show the simulation waveforms of DC bus voltage $u_{dc}$ and output current $i_o$, respectively. In both pictures, it can be seen obviously that observer-pattern model and simulation give almost the same results in steady state.

**Table 1.** Photovoltaic (PV) array parameters.

| Parameter | Symbol | Quantity |
|---|---|---|
| Voltage at MPP | $U_M$ | 119.6 V |
| Current at MPP | $I_M$ | 8.36 A |
| Open circuit voltage | $U_{OC}$ | 149.2 V |
| Short circuit current | $I_{SC}$ | 8.81 A |

**Table 2.** Converter parameters.

| Parameter | Symbol | Quantity |
|---|---|---|
| Capacitance of input filter | $C_{in}$ | 1000 μF |
| Inductance of boost converter | $L_b$ | 10 mH |
| Capacitance of boost converter | $C_{dc}$ | 1500 μF |
| Inductance of output filter | $Lf$ | 25 mH |
| Amplitude of grid voltage | $U_{gm}$ | $220\sqrt{2}$ V |
| Switching frequency | $f_s$ | 10 kHz |
| Gain of PI controller 1 | $K_{p1}$ | 0.05 |
| Time constant of PI controller 1 | $T_{i1}$ | 0.1 |
| Gain of PI controller 2 | $K_{p2}$ | 0.02 |
| Time constant of PI controller 2 | $T_{i2}$ | 0.01 |
| Gain of PI controller 3 | $K_{p3}$ | 1 |
| Time constant of PI controller 3 | $T_{i3}$ | 0.2 |

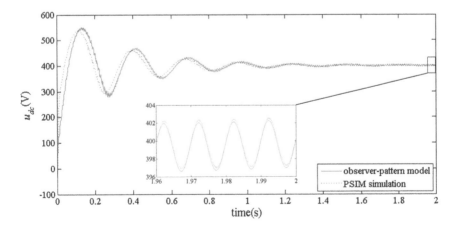

**Figure 2.** DC bus voltage waveform.

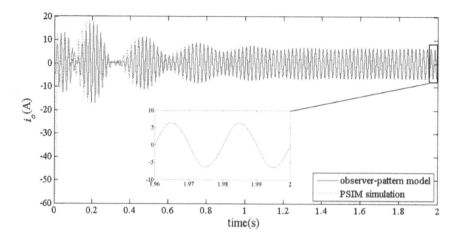

**Figure 3.** Output current waveform.

## 4. Stability Analysis

For simplicity, the observer-pattern model of the system can be written as

$$\dot{X} = F(X) \qquad (F : R^{11} \to R^{11}) \tag{22}$$

where $X = [u_{pv}, i_{L_b}, u_{dc}, i_{od}, i_{oq}, u_{c1}, u_e, u_{c2d}, u_{c2q}, g_1, g_2]^T$. By setting all the differential items in (21) to zero, the equilibrium point $X^e = [u_{pv}{}^e, i_{L_b}{}^e, u_{dc}{}^e, i_{od}{}^e, i_{oq}{}^e, u_{c1}{}^e, u_e{}^e, u_{c2d}{}^e, u_{c2q}{}^e, g_1{}^e, g_2{}^e]^T$ is obtained. Then. the Jacobian $A$ of the observer-pattern model at the equilibrium point can be derived as

$$A = \left. \frac{\partial F}{\partial X} \right|_{X = X^e} \tag{23}$$

Detailed description of Jacobian $A$ is presented in Appendix A.

To analyze the stability of the system, the eigenvalues of Jacobian are used, which can be calculated by

$$det[\lambda I - A] = 0 \tag{24}$$

where $I$ is unit matrix.

The gain and the time constant of three PI controllers are key parameters that affect the performance of the system. As these parameters change, the eigenvalues of the system also change. For the purpose of estimating the direction and size of the eigenvalue movement due to variations in system parameters, a sensitivity analysis is often used [28,29]. The eigenvalue sensitivity of $\lambda_i$ with respect to an uncertain parameter $\mu$ can be calculated as follow:

$$\frac{\partial \lambda_i}{\partial \mu} = \frac{\omega_i{}^T (\partial A / \partial \mu) v_i}{\omega_i{}^T v_i} \tag{25}$$

where $\omega_i$ and $v_i$ are the left and right eigenvectors corresponding to the eigenvalue $\lambda_i$, respectively. If the real part of $\partial \lambda_i / \partial \mu$ is positive, an increase in the parameter $\mu$ causes the eigenvalue $\lambda_i$ to move towards right in horizontal direction. The size of the horizontal movement is decided by the magnitude of the real part of $\partial \lambda_i / \partial \mu$. Similarly, the imaginary part of $\partial \lambda_i / \partial \mu$ is associated with the movement in vertical direction.

Table 3 lists the eigenvalue sensitivities with respect to PI controller parameters calculated by (25). For $\lambda_{1,2}$, the most sensitive parameter is $K_{p3}$ as the real part of the sensitivity of $\lambda_{1,2}$ with respect to

$K_{p3}$ is the largest. The decrease in $K_{p3}$ makes $\lambda_{1,2}$ move toward right in the s-plane. For $\lambda_{3,4}$, the most sensitive parameters are $T_{i1}$ and $K_{p1}$. A negative perturbation in $T_{i1}$ makes $\lambda_{3,4}$ move towards right in the s-plane. In contrast, the decrease in $K_{p3}$ makes $\lambda_{3,4}$ move to left. For $\lambda_5$, the most critical parameter is $T_{i1}$. The increase in $T_{i1}$ leads to $\lambda_5$ moving towards right-half plane. For $\lambda_{6,7}$, the most sensitive parameter is $K_{p2}$. When $K_{p2}$ decreases, $\lambda_{6,7}$ moves towards right in the s-plane. For $\lambda_{8,9}$, the most sensitive parameters is $T_{i3}$. The increase in $T_{i3}$ leads to $\lambda_{8,9}$ moving towards right in the s-plane. $\lambda_{10,11}$ are insensitive to all these parameters listed in Table 3.

**Table 3.** Eigenvalue sensitivities.

| | $\lambda_{1,2}$ | $\lambda_{3,4}$ | $\lambda_5$ | $\lambda_{6,7}$ | $\lambda_{8,9}$ | $\lambda_{10,11}$ |
|---|---|---|---|---|---|---|
| $K_{p1}$ | $-8.93 \times 10^{-4} \pm j7.47 \times 10^{-5}$ | $5.57 \pm j1.38 \times 10^4$ | $-9.31$ | $-0.91 \pm j0.145$ | $-2.45 \times 10^{-5} \pm j2.42 \times 10^{-4}$ | 0 |
| $T_{i1}$ | $-2.78 \times 10^{-7} \pm j2.88 \times 10^{-8}$ | $-47.5 \pm j0.21$ | $94.9$ | $0.0286 \pm j0.193$ | $3.85 \times 10^{-6} \pm j3.28 \times 10^{-7}$ | 0 |
| $K_{p2}$ | $-937 \pm j35.6$ | $-0.605 \pm j0.188$ | $-0.664$ | $-134 \pm j553$ | $0.00347 \pm j0.00126$ | 0 |
| $T_{i2}$ | $-11.8 \pm j0.68$ | $-0.0211 \pm j0.0846$ | $1.47$ | $11 \pm j1144$ | $1.22 \times 10^{-4} \pm j0.00224$ | 0 |
| $K_{p3}$ | $-1.6 \times 10^4 \pm j0.977$ | $7.45 \times 10^{-4} \pm j0.00145$ | $-1.35 \times 10^{-4}$ | $0.236 \pm j0.0336$ | $0.00154 \pm j0.00409$ | 0 |
| $T_{i3}$ | $-25 \pm j7.68 \times 10^{-4}$ | $2.62 \times 10^{-5} \pm j1.36 \times 10^{-5}$ | $2.21 \times 10^{-5}$ | $0.00141 \pm j0.0192$ | $25 \pm j0.0208$ | 0 |

Figure 4 illustrates the loci of the eigenvalues with respect to various PI controller parameters. It is obviously that the eigenvalue loci in Figure 4 match the sensitivity analysis results above. Especially note that in Figure 4b, $\lambda_{3,4}$ move across the imaginary axis from the left-half plane to the right-half plane when $T_{i1}$ decreases to 0.01, which means the system becomes unstable.

The eigenvalues when $T_{i1}$ equals to 0.01 and 0.03 are listed in Table 4 as a comparison. It can be seen clearly that $\lambda_{10,11}$, which originated from Equation (20), always equals to $\pm j628$. Therefore, $\lambda_{10,11}$ is only associated with the angular frequency of power grid and has no influence on the stability of the system. Actually, $\lambda_{1,2}$ and $\lambda_{8,9}$ are different when $T_{i1}$ varies. However, the slight differences are disregarded in this paper. Neglecting $\lambda_{10,11}$, all eigenvalues are in the left half of the s-plane when $T_{i1} = 0.03$, and the system operates in a stable state. However, when $T_{i1}$ decreases to 0.01, $\lambda_{3,4}$ becomes $26.8 \pm j1453$, which indicates the system is unstable and a low-frequency oscillation occurs. The frequency of the oscillation can be calculated according to the magnitude of the imaginary part of $\lambda_{3,4}$ as follows:

$$f_{osc} = \frac{1453}{2\pi} = 231 \text{ Hz} \tag{26}$$

**Table 4.** Eigenvalues when $T_{i1} = 0.01$ and $T_{i1} = 0.03$.

| $T_{i1}$ | $\lambda_{1,2}$ | $\lambda_{3,4}$ | $\lambda_5$ | $\lambda_{6,7}$ | $\lambda_{8,9}$ | $\lambda_{10,11}$ |
|---|---|---|---|---|---|---|
| 0.01 | $-16016 \pm j314$ | $26.8 \pm j1453$ | $-94.7$ | $-2.947 \pm j22.55$ | $-5 \pm j314$ | $\pm j628$ |
| 0.03 | $-16016 \pm j314$ | $-4.743 \pm j1451$ | $-31.6$ | $-2.927 \pm j22.56$ | $-5 \pm j314$ | $\pm j628$ |

Figure 5 presents the simulation results of $u_{dc}$ obtained by PSIM when $T_{i1}$ equals to 0.03 and 0.01. In Figure 5a, the waveform of $u_{dc}$ is sinusoidal and fluctuates around the nominal value. It can be seen clearly that $u_{dc}$ contains a DC component and a ripple at 100 Hz. The ripple at 100 Hz is due to pulsating output power of the single-phase inverter [30]. In Figure 5c, the waveform of $u_{dc}$ is non-sinusoidal. It can be seen from Figure 5d that the waveform contains a ripple at 100 Hz and a component at 230.5 Hz, which matches the theoretical analysis given by the observer-pattern model.

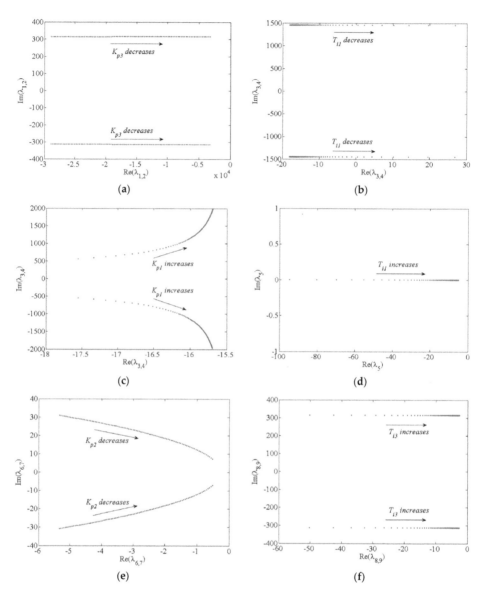

**Figure 4.** Loci of the eigenvalues with respect to various PI controller parameters: (**a**) $\lambda_{1,2}$ when $K_{p3}$ is varied within the range (0.2, 1.8); (**b**) $\lambda_{3,4}$ when $T_{i1}$ is varied within the range (0.01, 0.19); (**c**) $\lambda_{3,4}$ when $K_{p1}$ is varied within the range (0.005, 0.095); (**d**) $\lambda_5$ when $T_{i1}$ is varied within the range (0.01, 0.19); (**e**) $\lambda_{6,7}$ when $K_{p2}$ is varied within the range (0.002, 0.038); (**f**) $\lambda_{8,9}$ when $T_{i3}$ is varied within the range (0.02, 0.38).

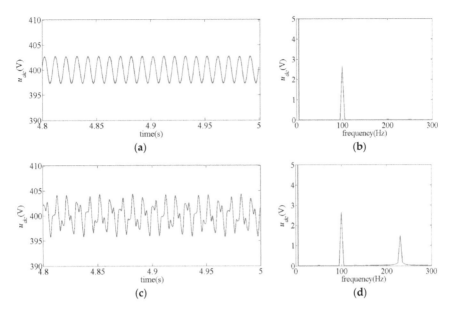

**Figure 5.** Simulation results of $u_{dc}$: (**a**) Time domain waveforms when $T_{i1} = 0.03$; (**b**) Fast Fourier Transformation (FFT) analysis when $T_{i1} = 0.03$; (**c**) Time domain waveforms when $T_{i1} = 0.01$; (**d**) FFT analysis when $T_{i1} = 0.01$.

## 5. Conclusions

Modeling and stability analysis of a single-phase two-stage grid-connected photovoltaic system have been presented in this paper. (1) An integrated mathematical model including both DC-DC converter and DC-AC converter is developed to capture the dynamics of the system, also the nonlinear characteristic of the PV array is considered in the model; (2) An observer-pattern modeling method is applied to transform the system into time-invariant; (3) Critical controller parameters that influence the stability of the system are identified using eigenvalue sensitivity and eigenvalue loci plots. It is found that $T_{i1}$ is closely related to $\lambda_{3,4}$. The decrease in $T_{i1}$ makes $\lambda_{3,4}$ move to left in the s-plane and is disadvantageous to the stability of the system. The theoretical results have been validated by PSIM simulations.

**Acknowledgments:** This project was supported by the National Natural Science Foundation of China (Grant Nos. 51507068 and 51277079) and the Team Program of Natural Science Foundation of Guangdong Province, China (Grant No. 2017B030312001).

**Author Contributions:** Liying Huang established the model, implemented the simulation and wrote this article; Dongyuan Qiu guided and revised the paper; Dongyuan Qiu, Fan Xie, Yanfeng Chen, and Bo Zhang guided the research.

**Conflicts of Interest:** The authors declare no conflict of interest.

## Appendix

The Jacobian matrix $A$ is given as:

$$A = \begin{pmatrix} A_{1,1} & A_{1,2} & 0 & 0 & 0 & 0 & 0 & 0 & 0 & 0 & 0 \\ A_{2,1} & 0 & A_{2,3} & 0 & 0 & A_{2,6} & 0 & 0 & 0 & 0 & 0 \\ 0 & A_{3,2} & 0 & A_{3,4} & A_{3,5} & A_{3,6} & 0 & A_{3,8} & A_{3,9} & A_{3,10} & A_{3,11} \\ 0 & 0 & A_{4,3} & 0 & A_{4,5} & 0 & 0 & A_{4,8} & 0 & 0 & 0 \\ 0 & 0 & A_{5,3} & A_{5,4} & 0 & 0 & 0 & 0 & A_{5,9} & 0 & 0 \\ A_{6,1} & A_{6,2} & 0 & 0 & 0 & 0 & 0 & 0 & 0 & 0 & 0 \\ 0 & A_{7,2} & A_{7,3} & A_{7,4} & A_{7,5} & A_{7,6} & 0 & A_{7,8} & A_{7,9} & A_{7,10} & A_{7,11} \\ 0 & 0 & A_{8,3} & A_{8,4} & 0 & 0 & A_{8,7} & A_{8,8} & A_{8,9} & 0 & 0 \\ 0 & A_{9,2} & A_{9,3} & A_{9,4} & A_{9,5} & A_{9,6} & A_{9,7} & A_{9,8} & A_{9,9} & A_{9,10} & A_{9,11} \\ 0 & 0 & 0 & 0 & 0 & 0 & 0 & 0 & 0 & 0 & A_{10,11} \\ 0 & 0 & 0 & 0 & 0 & 0 & 0 & 0 & 0 & A_{10,11} & 0 \end{pmatrix}$$

where $A_{1,1} = -\frac{I_{SC}A_1}{C_{in}A_2U_{OC}}e^{u_{pv}{}^e/(A_2U_{OC})}$, $A_{1,2} = -\frac{1}{C_{in}}$, $A_{2,1} = \frac{1}{L_b}$, $A_{2,3} = \left(\frac{u_{c1}{}^e}{U_{M1}} - 1\right)/L_b$, $A_{2,6} = \frac{u_{dc}{}^e}{L_b}$,

$A_{3,2} = \left(1 - \frac{u_{c1}{}^e}{U_{M1}}\right)/C_{dc}$, $A_{3,4} = \frac{1}{2C_{dc}}\left(g_2{}^e u_{c2q}{}^e - (1+g_1{}^e)u_{c2d}{}^e\right)$, $A_{3,5} = \frac{1}{2C_{dc}}\left(g_2{}^e u_{c2d}{}^e + (g_1{}^e - 1)u_{c2q}{}^e\right)$,

$A_{3,6} = -\frac{i_{L_b}{}^e}{C_{dc}}$, $A_{3,8} = \frac{1}{2C_{dc}}\left(g_2{}^e i_{oq}{}^e - (1+g_1{}^e)i_{od}{}^e\right)$, $A_{3,9} = -\frac{1}{2C_{dc}}\left(g_2{}^e i_{od}{}^e + (g_1{}^e - 1)i_{oq}{}^e\right)$,

$A_{3,10} = \frac{1}{2C_{dc}}\left(-i_{od}{}^e u_{c2d}{}^e + i_{oq}{}^e u_{c2q}{}^e\right)$, $A_{3,11} = \frac{1}{2C_{dc}}\left(i_{od}{}^e u_{c2q}{}^e + i_{oq}{}^e u_{c2d}{}^e\right)$, $A_{4,3} = \frac{u_{c2d}{}^e}{L_f U_{M2}}$, $A_{4,5} = \omega$,

$A_{4,8} = \frac{u_{dc}{}^e}{L_f U_{M2}}$, $A_{5,3} = \frac{u_{c2q}{}^e}{L_f U_{M2}}$, $A_{5,4} = -\omega$, $A_{5,9} = \frac{u_{dc}{}^e}{L_f U_{M2}}$, $A_{6,1} = -\frac{K_{p1}I_{SC}A_1}{C_{in}A_2U_{OC}}e^{u_{pv}{}^e/(A_2U_{OC})} + \frac{K_{p1}}{T_{i1}}$,

$A_{6,2} = -\frac{K_{p1}}{C_{in}}$, $A_{7,2} = K_{p2}\left(1 - \frac{u_{c1}{}^e}{U_{M1}}\right)/C_{dc}$, $A_{7,3} = \frac{K_{p2}}{T_{i2}}$, $A_{7,4} = \frac{K_{p2}}{2C_{dc}}\left(g_2{}^e u_{c2q}{}^e - (1+g_1{}^e)u_{c2d}{}^e\right)$,

$A_{7,5} = \frac{K_{p2}}{2C_{dc}}\left(g_2{}^e u_{c2d}{}^e + (g_1{}^e - 1)u_{c2q}{}^e\right)$, $A_{7,6} = -\frac{K_{p2}i_{L_b}{}^e}{C_{dc}}$, $A_{7,8} = \frac{K_{p2}}{2C_{dc}}\left(g_2{}^e i_{oq}{}^e - (1+g_1{}^e)i_{od}{}^e\right)$,

$A_{7,9} = -\frac{K_{p2}}{2C_{dc}}\left(g_2{}^e i_{od}{}^e + (g_1{}^e - 1)i_{oq}{}^e\right)$, $A_{7,10} = \frac{K_{p2}}{2C_{dc}}\left(-i_{od}{}^e u_{c2d}{}^e + i_{oq}{}^e u_{c2q}{}^e\right)$,

$A_{7,11} = \frac{K_{p2}}{2C_{dc}}\left(i_{od}{}^e u_{c2q}{}^e + i_{oq}{}^e u_{c2d}{}^e\right)$, $A_{8,3} = -\frac{K_{p3}u_{c2d}{}^e}{L_f U_{M2}}$, $A_{8,4} = -\frac{K_{p3}}{T_{i3}}$, $A_{8,7} = K_{p2}\omega$,

$A_{8,8} = -\frac{K_{p3}u_{dc}{}^e}{L_f U_{M2}}$, $A_{8,9} = \omega$, $A_{9,2} = -K_{p2}K_{p3}\left(1 - \frac{u_{c1}{}^e}{U_{M1}}\right)/C_{dc}$, $A_{9,3} = -\frac{K_{p2}K_{p3}}{T_{i3}} - \frac{K_{p3}u_{c2d}{}^e}{L_f U_{M2}}$,

$A_{9,4} = \frac{K_{p2}K_{p3}}{2C_{dc}}\left(g_2{}^e u_{c2q}{}^e - (1+g_1{}^e)u_{c2d}{}^e\right)$, $A_{9,5} = -\frac{K_{p3}}{T_{i3}} - \frac{K_{p2}K_{p3}}{2C_{dc}}\left(g_2{}^e u_{c2d}{}^e + (g_1{}^e - 1)u_{c2q}{}^e\right)$,

$A_{9,6} = \frac{K_{p2}K_{p3}i_{L_b}{}^e}{C_{dc}}$, $A_{9,7} = -\frac{K_{p3}}{T_{i3}}$, $A_{9,8} = -\omega - \frac{K_{p2}K_{p3}}{2C_{dc}}\left(g_2{}^e i_{oq}{}^e - (1+g_1{}^e)i_{od}{}^e\right)$,

$A_{9,9} = -\frac{K_{p3}u_{dc}{}^e}{L_f U_{M2}} - \frac{K_{p2}K_{p3}}{2C_{dc}}\left(g_2{}^e i_{od}{}^e + (g_1{}^e - 1)i_{oq}{}^e\right)$, $A_{9,10} = \frac{K_{p2}K_{p3}}{2C_{dc}}\left(i_{od}{}^e u_{c2d}{}^e - i_{oq}{}^e u_{c2q}{}^e\right)$,

$A_{9,11} = -\frac{K_{p2}K_{p3}}{2C_{dc}}\left(i_{od}{}^e u_{c2q}{}^e + i_{oq}{}^e u_{c2d}{}^e\right)$, $A_{10,11} = -2\omega$, $A_{11,10} = 2\omega$.

## References

1. Cucchiella, F.; D'Adamo, I.; Gastaldi, M. Economic analysis of a photovoltaic system: A resource for residential households. *Energies* **2017**, *10*, 814. [CrossRef]
2. Kouro, S.; Leon, J.I.; Vinnikov, D.; Franquelo, L.G. Grid-Connected Photovoltaic Systems: An Overview of Recent Research and Emerging PV Converter Technology. *IEEE Ind. Electron. Mag.* **2015**, *9*, 47–61. [CrossRef]
3. Schimpf, F.; Norum, L. Effective use of film capacitors in single-phase PV-inverters by active power decoupling. In Proceedings of the 36th Annual Conference on IEEE Industrial Electronics Society, Glendale, AZ, USA, 7–10 November 2010; pp. 2784–2789.
4. Nižetić, S.; Papadopoulos, A.M.; Tina, G.M.; Rosa-Clot, M. Hybrid energy scenarios for residential applications based on the heat pump split air-conditioning units for operation in the Mediterranean climate conditions. *Energy Build.* **2017**, *140*, 110–120. [CrossRef]
5. Xiong, X.; Chi, K.T.; Ruan, X. Bifurcation Analysis of Standalone Photovoltaic-Battery Hybrid Power System. *IEEE Trans. Circuits Syst. I Regul. Pap.* **2013**, *60*, 1354–1365. [CrossRef]

6.   Deivasundari, P.; Uma, G.; Poovizhi, R. Analysis and experimental verification of Hopf bifurcation in a solar photovoltaic powered hysteresis current-controlled cascaded-boost converter. *IET Power Electron.* **2013**, *6*, 763–773. [CrossRef]

7.   Zhioua, M.; Aroudi, A.E.; Belghith, S.; Bosquemoncusí, J.M.; Giral, R.; Hosani, K.A.; Alnumay, M. Modeling, Dynamics, Bifurcation Behavior and Stability Analysis of a DC–DC Boost Converter in Photovoltaic Systems. *Int. J. Bifurc. Chaos* **2016**, *26*, 1650166. [CrossRef]

8.   Al-Hindawi, M.M.; Abusorrah, A.; Al-Turki, Y.; Giaouris, D.; Mandal, K.; Banerjee, S. Nonlinear Dynamics and Bifurcation Analysis of a Boost Converter for Battery Charging in Photovoltaic Applications. *Int. J. Bifurc. Chaos* **2014**, *24*, 373–491. [CrossRef]

9.   Abusorrah, A.; Al-Hindawi, M.M.; Al-Turki, Y.; Mandal, K.; Giaouris, D.; Banerjee, S.; Voutetakis, S.; Papadopoulou, S. Stability of a boost converter fed from photovoltaic source. *Sol. Energy* **2013**, *98*, 458–471. [CrossRef]

10.  Li, X.; Tang, C.; Dai, X.; Hu, A.; Nguang, S. Bifurcation Phenomena Studies of a Voltage Controlled Buck-Inverter Cascade System. *Energies* **2017**, *10*, 708. [CrossRef]

11.  Tse, C.K.; Bernardo, M.D. Complex behavior in switching power converters. *Proc. IEEE* **2002**, *90*, 768–781. [CrossRef]

12.  Banerjee, S.; Chakrabarty, K. Nonlinear modeling and bifurcations in the boost converter. *IEEE Trans. Power Electron.* **1998**, *13*, 252–260. [CrossRef]

13.  Deane, J.H.B.; Hamill, D.C. Instability, subharmonics, and chaos in power electronic systems. *IEEE Trans. Power Electron.* **1989**, *5*, 260–268. [CrossRef]

14.  Aroudi, A.E.; Giaouris, D.; Mandal, K.; Banerjee, S. Complex non-linear phenomena and stability analysis of interconnected power converters used in distributed power systems. *IET Power Electron.* **2016**, *9*, 855–863. [CrossRef]

15.  Saublet, L.M.; Gavagsaz-Ghoachani, R.; Martin, J.P.; Nahid-Mobarakeh, B.; Pierfederici, S. Asymptotic Stability Analysis of the Limit Cycle of a Cascaded DC–DC Converter Using Sampled Discrete-Time Modeling. *IEEE Trans. Ind. Electron.* **2016**, *63*, 2477–2487. [CrossRef]

16.  Zadeh, M.K.; Gavagsaz-Ghoachani, R.; Pierfederici, S.; Nahid-Mobarakeh, B.; Molinas, M. Stability Analysis and Dynamic Performance Evaluation of a Power Electronics-Based DC Distribution System with Active Stabilizer. *IEEE J. Emerg. Sel. Top. Power Electron.* **2016**, *4*, 93–102. [CrossRef]

17.  Xie, F.; Zhang, B.; Qiu, D.; Jiang, Y. Non-linear dynamic behaviours of DC cascaded converters system with multi-load converters. *IET Power Electron.* **2016**, *9*, 1093–1102. [CrossRef]

18.  Salis, V.; Costabeber, A.; Cox, S.M.; Zanchetta, P. Stability Assessment of Power-Converter-Based AC Systems by LTP Theory: Eigenvalue Analysis and Harmonic Impedance Estimation. *IEEE J. Emerg. Sel. Top. Power Electron.* **2017**, *5*, 1513–1525. [CrossRef]

19.  Cai, H.; Xiang, J.; Wei, W. Modelling, analysis and control design of a two-stage photovoltaic generation system. *IET Renew. Power Gener.* **2016**, *10*, 1195–1203. [CrossRef]

20.  Zadeh, M.K.; Gavagsaz-Ghoachani, R.; Nahid-Mobarakeh, B.; Pierfederici, S.; Molinas, M. Stability analysis of hybrid AC/DC power systems for more electric aircraft. In Proceedings of the Applied Power Electronics Conference and Exposition, Long Beach, CA, USA, 20–24 May 2016; pp. 446–452.

21.  Zhang, H.; Wan, X.; Li, W.; Ding, H.; Yi, C. Observer-Pattern Modeling and Slow-Scale Bifurcation Analysis of Two-Stage Boost Inverters. *Int. J. Bifurc. Chaos* **2017**, *27*, 1750096. [CrossRef]

22.  Zhang, H.; Li, W.; Ding, H.; Yi, C.; Wan, X. Observer-Pattern Modeling and Nonlinear Modal Analysis of Two-stage Boost Inverter. *IEEE Trans. Power Electron.* **2017**. [CrossRef]

23.  Ram, J.P.; Babu, T.S.; Rajasekar, N. A comprehensive review on solar PV maximum power point tracking techniques. *Renew. Sustain. Energy Rev.* **2017**, *67*, 826–847. [CrossRef]

24.  Fadil, H.E.; Giri, F.; Guerrero, J.M. Grid-connected of photovoltaic module using nonlinear control. In Proceedings of the IEEE International Symposium on Power Electronics for Distributed Generation Systems, Aalborg, Denmark, 25–28 June 2012; pp. 119–124.

25.  Franzitta, V.; Orioli, A.; Gangi, A.D. Assessment of the Usability and Accuracy of the Simplified One-Diode Models for Photovoltaic Modules. *Energies* **2016**, *9*, 1019. [CrossRef]

26.  Khouzam, K.; Ly, C.; Chen, K.K.; Ng, P.Y. Simulation and real-time modelling of space photovoltaic systems. In Proceedings of the 1994 IEEE 1st World Conference on Photovoltaic Energy Conversion, Waikoloa, HI, USA, 5–9 December 1994; Volume 2, pp. 2038–2041.

27. Zhang, R.; Cardinal, M.; Szczesny, P.; Dame, M. A grid simulator with control of single-phase power converters in D-Q rotating frame. In Proceedings of the 2002 IEEE 33rd Annual IEEE Power Electronics Specialists Conference, Cairns, Australia, 23–27 June 2002; Volume 3, pp. 1431–1436.
28. Gao, F.; Zheng, X.; Bozhko, S.; Hill, C.I.; Asher, G. Modal Analysis of a PMSG-Based DC Electrical Power System in the More Electric Aircraft Using Eigenvalues Sensitivity. *IEEE Trans. Transp. Electrification* **2015**, *1*, 65–76.
29. Yang, L.; Xu, Z.; Østergaard, J.; Dong, Z.Y.; Wong, K.P.; Ma, X. Oscillatory Stability and Eigenvalue Sensitivity Analysis of A DFIG Wind Turbine System. *IEEE Trans. Energy Convers.* **2011**, *26*, 328–339. [CrossRef]
30. Shi, Y.; Liu, B.; Duan, S. Low-Frequency Input Current Ripple Reduction Based on Load Current Feedforward in a Two-Stage Single-Phase Inverter. *IEEE Trans. Power Electron.* **2016**, *31*, 7972–7985. [CrossRef]

*Article*

# A New PV Array Fault Diagnosis Method Using Fuzzy C-Mean Clustering and Fuzzy Membership Algorithm

Qiang Zhao [1], Shuai Shao [1], Lingxing Lu [2], Xin Liu [1] and Honglu Zhu [2,3,*]

[1] School of Control and Computer Engineering, North China Electric Power University, Beijing 102206, China; zhaoqiang@ncepu.edu.cn (Q.Z.); shaoshuai93@163.com (S.S.); liuxing93619@163.com (X.L.)

[2] School of Renewable Energy, North China Electric Power University, Beijing 102206, China; LingXingLu@ncepu.edu.cn

[3] State Key Laboratory of Alternate Electrical Power System with Renewable Energy Sources, North China Electric Power University, Changping District, Beijing 102206, China

* Correspondence: hongluzhu@126.com; Tel.: +86-186-0055-1350

Received: 1 November 2017; Accepted: 16 January 2018; Published: 19 January 2018

**Abstract:** Photovoltaic (PV) power station faults in the natural environment mainly occur in the PV array, and the accurate fault diagnosis is of particular significance for the safe and efficient PV power plant operation. The PV array's electrical behavior characteristics under fault conditions is analyzed in this paper, and a novel PV array fault diagnosis method is proposed based on fuzzy C-mean (FCM) and fuzzy membership algorithms. Firstly, clustering analysis of PV array fault samples is conducted using the FCM algorithm, indicating that there is a fixed relationship between the distribution characteristics of cluster centers and the different fault, then the fault samples are classified effectively. The membership degrees of all fault data and cluster centers are then determined by the fuzzy membership algorithm for the final fault diagnosis. Simulation analysis indicated that the diagnostic accuracy of the proposed method was 96%. Field experiments further verified the correctness and effectiveness of the proposed method. In this paper, various types of fault distribution features are effectively identified by the FCM algorithm, whether the PV array operation parameters belong to the fault category is determined by fuzzy membership algorithm, and the advantage of the proposed method is it can classify the fault data from normal operating data without foreknowledge.

**Keywords:** PV array; FCM algorithm; cluster analysis; fault diagnosis; membership algorithm

## 1. Introduction

The photovoltaic (PV) power plant works under a tough natural environment, and PV array faults are complicated and various, seriously affecting safe-stable operation and economic benefits of the power station in a very complex and dynamic manner. The DC monitor resolution available to PV power plants has reached the PV array level. The resolution of certain smart PV power stations has even reached the module level. There is critical significance in identifying and early warning for DC faults using the PV module/output data array of the PV power station in regards to intelligent predictive maintenance of the PV power plant and improving the overall operation level of the station [1].

The classification and diagnosis of PV array faults has become a popular research subject in recent years. Model-based algorithms and intelligence-based algorithms have drawn increasing attentions recently. Model-based and multivariable statistical monitoring methods are the common methods for fault identification, for example, Stellbogen [2] compared actual and expected values through detection equipment for fault analysis; however, they did not establish any method for setting thresholds between them. The model using PCA and other multivariate statistical monitoring methods for fault diagnosis is difficult to establish [3]. Model-based fault diagnosis algorithms depend on the analysis and calculation for the equivalent circuit models of PV array, and the modeling results will affect the results of diagnosis results, but the difficulty of modelling of PV array restricts such methods. The problem of multivariable statistical monitoring methods is that the selection and division of fault samples rely on human prior knowledge. In addition, quite a few researchers presented applicable intelligent recognition algorithms for PV array fault analysis and diagnosis. Karatepe et al. [4], Noguchi et al. [5], Chowdhury et al. [6], Miyatake et al. [7] and Wang et al. [8] used a particle swarm intelligent algorithm and Fibonacci search technique for PV array fault diagnosis. Zhong [9] and Cheng et al. [10] conducted PV array fault detection by analyzing the function relationship of the parameters in the PV array system according to data fusion, but this method mainly focuses on heat pot phenomenon and neglects other relevant factors. Chao et al. [11] identified PV array fault locations by processing the acquired power data and environmental parameters via an intelligent learning technique. Xu et al. [12] proposed a PV array fault location method based on a Gaussian process that is effective in its own regard, but not suitable for fault detection in large-scale PV generation systems. Zhao et al. [13] used a decision tree algorithm for PV array fault diagnosis, but this method is too dependent on the measured fault sample data to be fully practical. Li et al. [14] proposed a neural network-based fault diagnosis technique that is similarly disadvantaged by the difficulty in obtaining fault samples and training the fault model. Wang et al. [15] established a back propagation (BP) neural network method, they analyzed and classified some faults of PV array based on L-M algorithm. Some researchers expanded upon the using the support vector machine (SVM) [16,17]. The SVM algorithm can realize the fault diagnosis using small amounts of sample data set, but the accuracy of SVM algorithm still relies on classified of fault samples. Compared with model algorithms, the intelligent-based algorithms can identify the faults types of PV array without additional equipment and complicated calculation model, but the accuracy of algorithms need better setting of the algorithms parameters and good classified sample data, which making intelligent-based algorithms is of poor certainty and stability. The researchers have did lots work in the classification of fault samples, despite these valuable contributions to the literature, there is yet no technique for the scientific division of fault thresholds. Previous researchers have also had success in combining neural network, SVM, cluster analysis, and other model recognition techniques with fuzzy mathematics, rough set, data analysis, and other mathematical methods in identifying fault thresholds and diagnosing various types of fault [16–19]. Krishnapuram and Keller [20] combined the fuzzy clustering algorithm with the three-ratio method for reliable, accurate transformer fault diagnosis. The FCM algorithm converges rapidly with relatively few training samples. Du W et al. [21] successfully applied the FCM theory for analog circuit fault diagnosis. The FCM algorithm is easily computed and quickly operated as-applied to fault diagnosis; it does not require a large labeled sample. However, FCM algorithm sample typicality is not reflected in the constraint of membership matrix $U \sum_{j=1}^{c} \mu_{ik} = 1$ and the algorithm does not work well when there is large discrepancy among multiple specimen classes. In the fuzzy case index, the cluster centers of various fault states are obtained based on the FCM algorithm, however, how to make rules according to the clustering results, is a major concern. Li et al. [22] used similarity computation based on the membership function to secure fuzzy numbers and fuzzy linguistics for calculating global similarity weights; this process allows for quick and efficient case retrieval on different types of demand. In short: the membership function algorithm can effectively distinguish different categories quantitatively. Although fuzzy theory and its related technologies have been widely used in fault diagnosis, there has been little research on their application in PV arrays. During PV array fault diagnosis, there is complex randomness and uncertainty between the causes and external characteristics of different types of fault as well as significant differences among various samples. Therefore, the combination of FCM algorithm and fuzzy membership algorithm provides research ideas for division and identification of the PV fault samples. The combination of FCM and fuzzy membership algorithm has the following advantages: the FCM algorithm requires little computation and can be quickly operated to classify fault samples. The membership algorithm can be used to distinguish the influence degree of fault classifications based on a wide variety of data;

it classifies individual sample points to realize fault diagnosis. This paper presents a novel approach to PV array fault diagnosis based on FCM and fuzzy membership algorithms. The randomness and uncertainty of PV array fault characteristics are solved by the introduction of fuzzy theory, simulation and experimental analyses demonstrate that the proposed method scientifically classifies fault sample data for efficient, accurate PV array fault diagnosis.

This paper is organized as follows: Section 2 discusses the fault characteristics of PV arrays. Section 3 provides an introduction to basic theories, principles, and application methods relevant to the FCM algorithm and fuzzy membership algorithm for fault diagnosis. Section 4 reports simulation tests and Section 5 reports experimental tests conducted to validate the proposed diagnosis method. Section 6 provides concluding remarks.

## 2. Characteristics Analysis of Typical Faults in PV Arrays

### 2.1. Generation Mechanism of Typical Faults in PV Arrays

The actual operation of the PV power station is affected by multiple external factors such as solar radiation intensity, temperature, humidity, dust, hail, and snow constituting a harsh (and highly fault-prone) environment [23]. The PV array is an integral part of the PV system; its cost can account for about 40% of the power system as a whole. In this paper, four common faults or abnormal condition of the laboratory PV plant is studied respectively, which is configured as three parallel PV strings of 13 PV modules in series (regard this as the research object), as shown in Figure 1. Three common faults of the PV array include the hot spot phenomenon (partial blockage), open circuits, or short circuiting of the PV module caused by junction box error. Long-term shadow shadings and module mismatch also accelerate the rate of degradation and introduce corresponding aging faults.

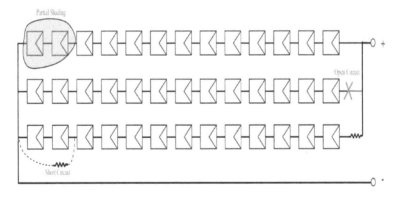

**Figure 1.** Simulated faults in laboratory PV array.

For simplicity, six failure modes of the PV array are referred to in this paper: a one module short circuit, two module short circuit, local shadow shading in one string group, local shadow shading in two string groups, and one module open circuit. These modes are marked F1, F2, F3, F4, F5, and F6, respectively. As shown in Table 1, based on typical fault mode of PV array set, the physics-based mathematical model of the PV cell is established according to "Accurate model simulation research on PV cells, modules and arrays" for different fault type [24]. The modeling results are shown in Figure 2, where I-V and P-V curves describe the distribution characteristics of the PV array's electrical parameters under different fault conditions. The fault features are summarized in Table 1.

**Table 1.** PV array fault features.

| Fault Types | Descriptions | Electrical Characteristics |
|---|---|---|
| F1 | Normal | — |
| F2 | One module shorted | Current normal, voltage decreases |
| F3 | Two modules shorted | Current normal, voltage decreases |
| F4 | One module shaded | Current normal, voltage decreases |
| F5 | Two modules shaded | Current normal, voltage decreases |
| F6 | One module opened | No current, no voltage |

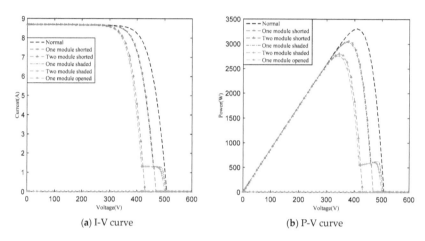

(a) I-V curve        (b) P-V curve

**Figure 2.** Output characteristic curves of PV array under different fault conditions, (**a**) I-V curve under different fault conditions; (**b**) P-V curve under different fault conditions.

### 2.2. Fault Characteristic Parameter Selection

Changes in the PV array are similar under different fault modes and the same test conditions (light intensity and temperature), as shown in the I-V and P-V curves in Figure 2. This suggests that it is not feasible to diagnose fault in the PV array only by analyzing the I-V and P-V curves. Additional fault parameters must be selected to describe the working conditions of the PV power system.

Changes in the electrical parameters under different fault conditions were determined as shown in Figure 3 using the actual external environment input excitation simulation model.

As shown in Figure 3, under different fault conditions, the output characteristics of one or more PV arrays change dramatically. To this effect, the output characteristics of the PV array may serve as fault characteristic parameters under different fault states and environments: $U_{oc}$, $I_{sc}$, $U_m$, $I_m$ and $P_m$, expressed as the form of the fault eigenvectors ($U_{oc}$, $I_{sc}$, $U_m$, $I_m$, $P_m$) (the description of the parameters is given in Table 2).

**Table 2.** Selected fault characteristic parameters.

| Fault Parameters | Name | Descriptions |
|---|---|---|
| $U_{oc}$ | Open-circuit voltage of the array | Short-circuit fault caused the decline of $U_m$, $U_{oc}$. |
| $I_{sc}$ | Short-circuit current of the array | |
| $U_m$ | Maximum power-point voltage of the array | Open-circuit fault caused the decline of $I_{sc}$, $I_m$. |
| $I_m$ | Maximum power-point current of the array | |
| $P_m$ | Maximum power of the array | Shadow fault caused the decline of $U_m$, $I_m$ |

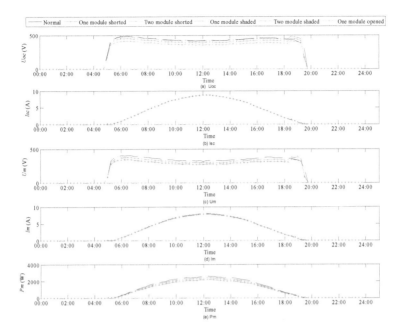

**Figure 3.** PV array electrical parameters change under different fault conditions, (**a**) Change in the electrical parameter $U_{oc}$ under different fault conditions; (**b**) Change in the electrical parameter $I_{sc}$ under different fault conditions; (**c**) Change in the electrical parameter $U_m$ under different fault conditions; (**d**) Change in the electrical parameter $I_m$ under different fault conditions; (**e**) Change in the electrical parameter $P_m$ under different fault conditions.

## 3. Basic Theories Supporting the Algorithm

### 3.1. Fuzzy C-Mean Clustering Algorithm

Fuzzy clustering is commonly applied within knowledge discovery, pattern recognition, and many other research fields. The FCM algorithm is one of the most widely used and successful algorithms for fuzzy clustering, which improves Hard C-mean clustering (HCM) algorithm, and represents the foundation upon which other fuzzy clustering analysis methods have been developed in theory and application.

FCM is a classification method as well as a clustering algorithm. The membership degree of individual sample points is obtained iteratively by optimizing the objective function. The class of sample points is determined to achieve the automatic classification of sample data. As discussed above, this method is commonly used in the fault diagnosis field [20].

Set up n data sample as $X = |x_i, i = 1, 2, \ldots, n| = \{x_1, x_2, \ldots, x_n\}$, divide n data vectors $X_i$ into a c fuzzy group, then calculate the c cluster center $v = \{v_1, v_2, \ldots, v_n\}$. This produces the minimum value of objective functions. Next, determine the level of each data point belonging to each group according to the membership degree, which is any value in the [0, 1] interval. The sum of membership values of each sample point to each cluster center is 1. The following two principles must be satisfied:

$$\mu_{ik} \in [0, 1] \tag{1}$$

$$\sum_{j=1}^{c} \mu_{ik} = 1, i = 1, 2, \cdots, n \tag{2}$$

The general form of FCM algorithm's objective function $J_b$ can be expressed as follows:

$$J_b(U, v) = \sum_{i=1}^{n} \sum_{k=1}^{c} (\mu_{ik})^b (d_{ik})^2 \tag{3}$$

where $n$ is the number of samples, $c$ ($2 \leq c \leq n$) is the number of cluster centers; $\mu_{ik}$ is the membership degree between sample $x_i$ and class $A_k$; $d_{ik}$ is a Euclidean measurement distance between the $i$ sample $x_i$ and $k$ central point, $d_{ik} = d(x_i - v_k) = \sqrt{\sum_{j=1}^{m} (x_{ij} - v_{kj})^2}$; $m$ is the characteristic number of samples; $b$ is a weighted parameter $\infty$ in the range $1 \leq b \leq \infty$.

The membership degree $\mu_{ik}$ between the sample $x_i$ and class $A_k$ is calculated as follows:

$$\mu_{ik} = \frac{1}{\sum_{j=1}^{c} \left(\frac{d_{ik}}{d_{lk}}\right)^{\frac{2}{b-1}}} \tag{4}$$

Set up $I_k = \{i | 2 \leq c < n; d_{i,k} = 0\}$, for all $i$ classes, $i \in I_k$, $\mu_{ik} = 0$. The cluster centers are calculated as follows:

$$v_i = \frac{\sum_{k=1}^{n} (\mu_{ik})^b x_{kj}}{\sum_{k=1}^{n} (\mu_{ik})^b} \tag{5}$$

We modify the cluster centers and membership repeatedly according to Equations (3) and (4). When the algorithm converges, the cluster center and membership degree of each sample to each class can be obtained successfully and the fuzzy clustering division is complete. The analysis shows that FCM algorithm is a simple iteration process, the general steps of determining cluster center and membership matrix based on the FCM algorithm [20,25] are as shown in Figure 4.

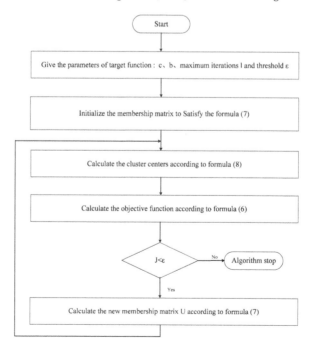

**Figure 4.** FCM flow chart.

The FCM algorithm converges rapidly with relatively few training samples; it can thus facilitate fault diagnosis very efficiently. MATLAB software (R2015b) also provides a rich functions for the FCM algorithm and is easily operable for fault diagnosis personnel [20].

### 3.2. Membership Function Algorithm Based on Fuzzy Normal Distribution

Fuzzy sets are completely described by their corresponding membership functions. In classical sets, the membership degree between sets and elements can only be 0; in fuzzy sets, the membership degree between sets and elements can be any value in the [0, 1] interval. It can thus be used to describe the extent to which an element belongs to the concept in the domain $U$. The membership function is the most fundamental concept of fuzzy mathematics as it quantizes the necessary fuzzy sets [26–28].

To define the fuzzy set, make sure that fuzzy subset A in the domain $U$ encompasses the characteristics of membership function $\mu_A$ and construct the following map:

$$\mu_A : U \rightarrow [0,1] \tag{6}$$

where $\mu_A$ is the membership function of the fuzzy subset; $\mu_A(x)$ is the membership degree of $U$ to $A$, which represents the degree of the element u belonging to its fuzzy subset $A$ in the domain $U$ with continuous variables on a closed interval [0, 1]. The closer $\mu_A(x)$ is to 1, the greater the extent to which u belongs to A. The closer $\mu_A(x)$ is to 0, the lesser the degree of u belonging to A.

For the fault diagnosis of PV array, the characteristic parameters change in a certain range, PV array is under healthy conditions with these parameters in a certain scope, and PV array is under faulty conditions while these parameters out the scope, so the typical normal distribution function is selected to calculate membership degree between diagnosis samples and the cluster center in the PV system to diagnose the PV array directly and clearly according to the membership degree. Figure 5 shows the curve of Normal Distribution Membership Function. The membership function of normal distribution-Gaussian function is used to calculate the membership degree of each parameter:

$$\mu(x) = e^{-\frac{(x-\mu)^2}{2\sigma^2}} \times 100\% \tag{7}$$

where $\mu(x)$ is the membership degree of the parameter x; $\mu$ is expected value of the distribution; $\sigma$ is the width of the Gaussian function.

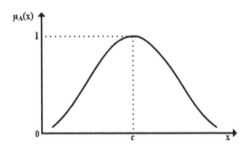

**Figure 5.** The curve of Normal Distribution Membership Function.

According to the Gaussian function characteristics, 99.73% of the area under the function curve is within three standard deviations ($3\sigma$) of the expected value $\mu$. In this paper, $6\sigma$ is used as the function domain. The value of $\sigma$ was obtained as follows:

$$\sigma = \frac{\mu_{max} - \mu_{min}}{6} \tag{8}$$

where $\mu_{max}$, $\mu_{min}$ are the maximum and minimum values of the parameters.

### 3.3. Fault Diagnosis Based on FCM Algorithm and Fuzzy Membership Algorithm

The relationship between the fault category and the fault eigenvectors is established to improve robustness of the fault diagnosis methods based on the FCM algorithm, considering the randomness and uncertainty of the fault eigenvectors. The fuzzy membership algorithm (membership function based on fuzzy normal distribution) is a distance algorithm. It is used to quantize fuzzy sets to diagnose fault samples. By measuring the membership degree between fault samples and each fault mode, the fault diagnosis is finished by fuzzy membership algorithms.

The purpose of this study, as stated above, was to establish a novel PV array fault diagnosis technique based on the FCM clustering and fuzzy membership algorithms. The proposed method effectively exploits the advantages of FCM (excellent classification ability) as well as the membership function algorithm (excellent distance computing ability), improving the proposed method's accuracy. Firstly, the FCM clustering algorithm is used to conduct clustering analysis of PV array fault samples and give the cluster centers of various fault states. Then, the fuzzy membership function is designed and used to carry out the fault diagnosis of PV array. Figure 6 shows the fault diagnosis framework of a PV array.

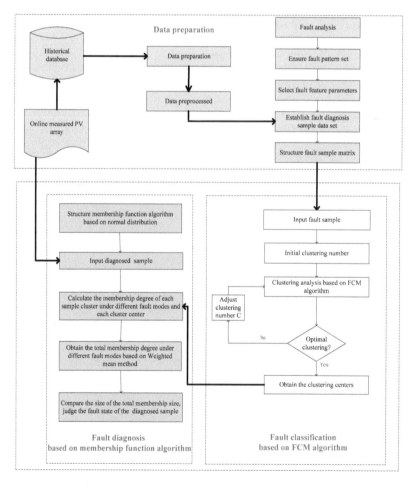

**Figure 6.** Framework of proposed fault diagnosis technique.

Step 1: Several fault feature parameters are selected through fault analysis, and fault samples are collected under various fault modes based on simulated (or measured) data, structure fault sample matrix. The fault sample sets are obtained. Meanwhile, the clustering number is also obtained by the types of the faults through fault analysis, which is set as the input parameter of the FCM algorithm.

Step 2: The FCM clustering algorithm is used to classify the selected fault samples. The optimal cluster centers of various fault states are obtained by adjusting clustering number C and the fault data sets are clustered under different fault types based on the FCM algorithm, which means that the fault classification is complete. The data classification process based on FCM algorithm is that the FCM classifies the existing fault data into several classes, comparable with established the number of fault types, known as cluster centers. In the process, the changes of the fault characteristics caused the changes of clustering center. When the new data was input, we can adjust some parameters of the FCM algorithm to obtain the new clustering center based on the process shown in Figure 6.

Step 3: The membership function algorithm based on fuzzy normal distribution is used to diagnose the faults using operation data, then calculate the degree of membership according to the cluster centers obtained by step 2. By transforming the fuzzy membership function into the distance function, it quantize the faults, calculate the membership degree of fault samples between each fault mode and each cluster center to complete the diagnosis, then the total membership degree of each failure mode is calculated via weighted mean method.

Step 4: The total membership degrees under various fault modes are sorted, then the largest membership degree is selected as the fault state of the diagnosed sample. The fault diagnosis is complete.

## 4. Simulation Study

### 4.1. Formation of Fault Sample Data Sets

To simulate fault characteristics in different light intensities and temperatures in a typical PV system, an $3 \times 13$ PV array simulation model was built in MATLAB/Simulink (R2015b) as shown in Figure 7.

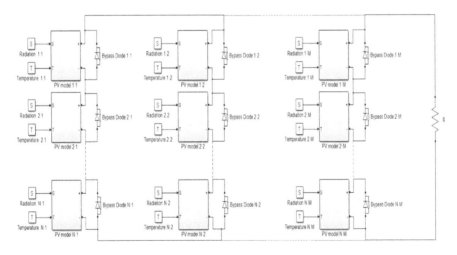

**Figure 7.** Simulated PV array model.

The irradiance of the model was set to range from 900 W/m$^2$ and 1000 W/m$^2$ and the temperature from 25 °C and 45 °C to simulate six different fault modes of the PV module; each fault mode's value of $U_{oc}$, $I_{sc}$, $U_m$, $I_m$, $P_m$ was obtained accordingly. Fault data samples under various fault modes were collected through multiple cycles of simulation. In each fault mode, 15 sample data points were randomly collected under different irradiation intensities to make a total of 90 data samples across six fault modes constituting fault sample matrix $X$:

$$X = \{x_1, x_2, \cdots, x_{90}\} \tag{9}$$

*4.2. FCM Algorithm Cluster Analysis*

The selection of clustering number $C$ for the FCM algorithm is very important. Generally, $C$ is significantly smaller than the total number of cluster samples, and clustering number $C > 1$. Through the analysis of Section 2.2, PV array is taken as basic fault diagnosis units, and fault is classified into six classes: a one module short circuit, two module short circuit, local shadow shading in one string group, local shadow shading in two string groups, and one module open circuit. The number of fault types equals to clustering number $C$ of FCM algorithm. So the parameters of the FCM algorithm (Equations (3) and (4)) include clustering number $C = 6$, weighted exponent $m = 2$, maximum iteration number $L = 1000$, and stopping iteration threshold $\varepsilon = 10^{-5}$. The cluster centers of six fault modes were obtained using the FCM function as shown in Table 3. Each cluster center is the typical value of each fault mode and can be plugged into the fault dictionary of the PV array diagnostic system.

Table 3. Clustering results of typical faults.

| Fault Types | $U_{oc}$/V | $I_{sc}$/A | $U_m$/V | $I_m$/A | $P_m$/W |
|---|---|---|---|---|---|
| F1 Normal | 426.4071 | 7.8172 | 333.4751 | 7.0820 | 2357.7497 |
| F2 One module shorted | 393.5892 | 7.8174 | 305.9612 | 7.1291 | 2176.4355 |
| F3 Two modules shorted | 360.7937 | 7.8170 | 281.5260 | 7.1009 | 1995.3092 |
| F4 One module shaded | 422.0471 | 7.8168 | 303.5406 | 7.1275 | 2159.3655 |
| F5 Two modules shaded | 417.7141 | 7.8159 | 276.3280 | 7.1116 | 1959.9594 |
| F6 One module opened | 0 | 0 | 0 | 0 | 0 |

The cluster center rules under different fault modes were quantified as shown in Table 4 according to the initial diagnosis and cluster centers of six fault modes in the PV array system (Table 3).

Table 4. Cluster center rules under different fault modes.

| Rules | Input (Electrical Characteristic Parameters) | Fault Mode | | | | | |
|---|---|---|---|---|---|---|---|
| | | F1 | F2 | F3 | F4 | F5 | F6 |
| R1 | $\Delta U_{oc} \approx 32$ V, $\Delta U_m \approx 28$ V, $\Delta P_m \approx 185$ W, | 0 | 1 | 0 | 0 | 0 | 0 |
| R2 | $\Delta U_{oc} \approx 65$ V, $\Delta U_m \approx 52$ V, $\Delta P_m \approx 380$ W, | 0 | 0 | 1 | 0 | 0 | 0 |
| R3 | $\Delta U_{oc} \approx 4$ V, $\Delta U_m \approx 30$ V, $\Delta P_m \approx 200$ W, | 0 | 0 | 0 | 1 | 0 | 0 |
| R4 | $\Delta U_{oc} \approx 9$ V, $\Delta U_m \approx 58$ V, $\Delta P_m \approx 420$ W, | 0 | 0 | 0 | 0 | 1 | 0 |
| R5 | All the characteristic parameters are zero | 0 | 0 | 0 | 0 | 0 | 1 |

Five rules correspond to the results in Table 4 in comparison against the normal operation state.

(1) When the open-circuit voltage drops about 32 V, the maximum-power-point voltage drops about 28 V and the maximum power drops 185 W. This is diagnosed as F2, i.e., one module short-circuit fault.

(2) When the open-circuit voltage drops about 65 V, the maximum-power-point voltage drops about 52 V and the maximum power drops 380 W. This is diagnosed as F3, i.e., two modules short-circuit fault.

(3) When the open-circuit voltage drops about 4 V, the maximum-power-point voltage drops about 30 V and the maximum power drops 200 W; this is an F4, or one module shaded fault.

(4) When the open-circuit voltage drops about 9 V, the maximum-power-point voltage drops about 58 V and the maximum power drops 420 W; this is an F5, or two modules shaded fault.

(5) When all the characteristic parameters are zero the fault is an F6, or one module opened fault.

### 4.3. Fault Diagnosis Using Fuzzy Membership Algorithm

The cluster centers of the PV array diagnosis system (Table 3) can be combined with the fuzzy membership function algorithm to calculate the membership degree between fault diagnosed samples and their cluster centers for complete fault diagnosis of the PV array. As mentioned above, the larger the membership degree, the more likely the diagnosis sample is to belong to the given fault state.

To apply the membership function algorithm based on fuzzy normal distribution to the PV array fault diagnosis, select a sample randomly and obtain the diagnosed parameters: $U_{oc} = 361.0858$ V, $I_{sc} = 7.7779$ A, $U_m = 281.8601$ V, $I_m = 7.0669$ A, $P_m = 1987.7488$ W. According to the deviation theory $\sigma^2$ introduced in Section 2.2, calculate the standard deviation of five fault characteristic parameters, then set up the membership function of the open-circuit voltage ($U_{oc}$), the short-circuit current ($I_{sc}$), the maximum-power-point voltage ($U_m$), the maximum-power-point current($I_m$), and maximum power ($P_m$):

$$\begin{cases} \mu(U_\infty) = e^{-\frac{(U_\infty - \mu_{U_\infty})^2}{693.7701}} \times 100\% \\ \mu(I_{sc}) = e^{-\frac{(I_{sc} - \mu_{I_{sc}})^2}{0.3716}} \times 100\% \\ \mu(U_m) = e^{-\frac{(U_m - \mu_{U_m})^2}{633.7355}} \times 100\% \\ \mu(I_m) = e^{-\frac{(I_m - \mu_{I_m})^2}{0.3639}} \times 100\% \\ \mu(P_m) = e^{-\frac{(P_m - \mu_{P_m})^2}{112202.4769}} \times 100\% \end{cases} \quad (10)$$

Plug the parameters of the measured sample into Formula (10) to obtain the membership degree of fault samples between each fault mode and each cluster center as shown in Table 5. The extent to which PV power impacts each characteristic parameter is the same; the weighted total membership degree in the last column of Table 5 is the average value of $U_{oc}, I_{sc}, U_m, I_m, P_m$.

**Table 5.** Membership degree between fault sample and each cluster center.

| Fault Types | $U_{oc}$ Membership | $I_{sc}$ Membership | $U_m$ Membership | $I_m$ Membership | $P_m$ Membership | Total Membership |
|---|---|---|---|---|---|---|
| F1 | $4.55 \times 10^{-6}$ | 0.991688 | 0.000223 | 0.998751 | 0.087140 | 0.415561 |
| F2 | 0.047568 | 0.991603 | 0.159910 | 0.978920 | 0.530140 | 0.541628 |
| F3 | 0.999754 | 0.991770 | 0.999648 | 0.993650 | 0.998982 | 0.996761 |
| F4 | $2.22 \times 10^{-5}$ | 0.991873 | 0.226865 | 0.980015 | 0.591565 | 0.558068 |
| F5 | $9.66 \times 10^{-5}$ | 0.992234 | 0.907935 | 0.989078 | 0.986329 | 0.775134 |
| F6 | $5.8 \times 10^{-164}$ | $3.9 \times 10^{-142}$ | $1.3 \times 10^{-109}$ | $6.3 \times 10^{-120}$ | $2.59 \times 10^{-31}$ | $5.18 \times 10^{-32}$ |

Once the total membership degrees are sorted (Table 5), select the largest membership degree as the fault state of the diagnosed sample. As shown in the last column of Table 5, $\mu_{F3} > \mu_{F5} > \mu_{F4} > \mu_{F2} > \mu_{F1} > \mu_{F6}$. Within the total membership degree sorting results, F4 two modules shorted fault comprises the largest proportion—that is, the diagnosed samples in the F4 fault state are consistent with the preset fault type, indicating that the proposed fault diagnosis method is effective and accurate.

Next, the selected range of the sample irradiance was expanded to 700 W/m$^2$ and 1000 W/m$^2$ and 150 fault samples of the PV array were selected based on the proposed method. Only six fault samples showed diagnostic errors out of the 150 sample. During actual diagnosis, some fault types are easily misjudged which may cause some errors, due to the similarity of faults types. The error analysis of fault samples is shown in Table 6. The diagnostic accuracy was 96%, effective, indicating that the proposed method is also highly feasible.

**Table 6.** Error analysis of fault samples.

| | | 1 | 2 | 3 | 4 | 5 | 6 | |
|---|---|---|---|---|---|---|---|---|
| Obtained Class | 1 | 25 | 0 | 0 | 0 | 0 | 0 | 100.00% |
| | | 16.67% | 0.00% | 0.00% | 0.00% | 0.00% | 0.00% | 0.00% |
| | 2 | 0 | 24 | 0 | 2 | 0 | 0 | 92.31% |
| | | 0.00% | 16.00% | 0.00% | 1.33% | 0.00% | 0.00% | 7.69% |
| | 3 | 0 | 0 | 24 | 0 | 2 | 0 | 92.31% |
| | | 0.00% | 0.00% | 16.00% | 0.00% | 1.33% | 0.00% | 7.69% |
| | 4 | 0 | 1 | 0 | 23 | 0 | 0 | 95.83% |
| | | 0.00% | 0.67% | 0.00% | 15.33% | 0.00% | 0.00% | 4.17% |
| | 5 | 0 | 0 | 1 | 0 | 23 | 0 | 95.83% |
| | | 0.00% | 0.00% | 0.67% | 0.00% | 15.33% | 0.00% | 4.17% |
| | 6 | 0 | 0 | 0 | 0 | 0 | 25 | 100.00% |
| | | 0.00% | 0.00% | 0.00% | 0.00% | 0.00% | 16.67% | 0.00% |
| | | 100.00% | 96.00% | 96.00% | 92.00% | 92.00% | 100.00% | 96.00% |
| | | 0.00% | 4.00% | 4.00% | 8.00% | 8.00% | 0.00% | 4.00% |
| | | 1 | 2 | 3 | 4 | 5 | 6 | |

Actual Class

Table 6 gives the misdiagnosis rate of the fault samples. Among them, the correct number of diagnosed samples is marked with green, the wrong number of diagnosed samples is marked with orange, the misdiagnosis rate of each sample is marked with gray, and the misdiagnosis rate of all samples is marked with blue.

To verify the adaptability of the proposed method when a new fault was coming, a new fault named F7 which six PV modules are shaded in one string is addressed. With a perfect scalability of FCM algorithm, and the clustering number C is changed to 7, new cluster center is obtained, and the faults types are identified by comparing the membership degree based on the fuzzy membership function algorithm. Seven faults are re-simulated based on simulated PV array model established in this paper, 175 fault samples of the PV array were selected based on the proposed method. And the diagnostic accuracy was 96.6%. The result shows that the proposed method has a good scalability and adaptability.

*4.4. Comparison of Algorithm Performance*

4.4.1. Comparison of Classification Algorithms

K-Means algorithm is popular as one of hard C-means (HCM) clustering algorithms. When the data set and clustering number are given, K-Mean classifies the data into different clustering domain iteratively according to specific distance function, and its membership degree can only be 0 or 1. FCM algorithm is the improvement of HCM algorithm and extends HCM algorithm to a fuzzy case, its membership degree can be any value in the [0, 1] interval. FCM algorithm is more suitable for extraction of the fault feature and classification of the fault data in the course of PV array fault diagnosis.

In order to verify the performance of FCM algorithm, K-Means algorithm and FCM algorithm are used for the classification of 6 types fault described in Table 1. Fifteen samples are given for each fault, and the total number of fault samples is 90.

The clustering result of different algorithm is shown in Figure 8, which describes six fault states of PV array fault modes. Figure 8a shows that there are mixings between different clustering result, and the clustering result of K-Means is not ideal. While Figure 8a shows the FCM clustering algorithm can divide the data into six groups, and six types of fault data are aggregated in cluster center. For PV array's fault data classification, compared with the K-Means algorithm, the FCM algorithm can cluster and classify the fault data accurately and effectively.

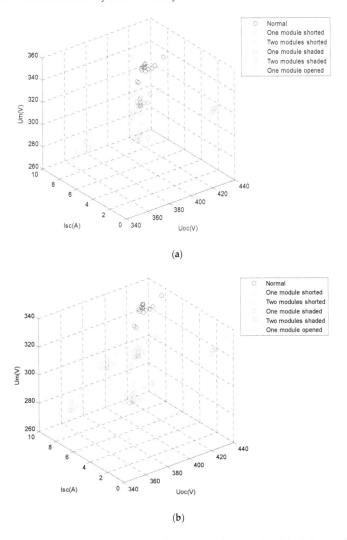

(a)

(b)

**Figure 8.** Comparison of clustering results, (a) Clustering analysis results of six fault states based on K-Means algorithm; (b) Clustering analysis results of six fault states based on FCM algorithm.

Table 7 shows the comparison of different algorithms by wrong classified number, running time and accuracy. The running time of FCM algorithm is larger than the K-Means algorithm in, but the clustering accuracy of FCM algorithm is much higher than the K-Means algorithm, which indicates that the FCM algorithm has a better clustering performance.

**Table 7.** The experimental comparison results of K-Means algorithm and FCM algorithm.

| Clustering Methods | Wrong Clustering Number | Running Time/s | Accuracy/% |
|---|---|---|---|
| K-Means | 15 | 0.128001 | 83.3 |
| FCM | 8 | 0.46744 | 91.1 |

4.4.2. Comparison of Diagnostic Algorithms

As a typical intelligent-based fault diagnosis method, the BP neural network is widely applied in the field of fault diagnosis, but its fault detection is mainly determined by its parameter setting and the training data. In order to verify the performance of the proposed diagnostic method, the diagnostic performance is compared between the BP neural network methods. The 90 fault samples describing six typical fault types are selected for the training of BP neural network and the proposed method. 24 typical fault data samples are selected for the testing of different method. The results are shown in Table 8. According to Table 8, the proposed method in this paper has one diagnosis error, and the BP neural network diagnosis method has three diagnostic errors. The contrast analysis shows that the proposed algorithm is more accurate the BP-based method. The reason for the low accuracy of BP neural network diagnosis method is that the general neural network needs a large amount of fault samples for the training process, but the sparsity of the fault samples in actual operation data leads to the limitation of the neural network diagnosis method.

**Table 8.** The comparison results of different fault diagnosis methods for testing samples.

| Group Number | F1 Membership | F2 Membership | F3 Membership | F4 Membership | F5 Membership | F6 Membership | Proposed Method | BP Neural Network | Actual Results |
|---|---|---|---|---|---|---|---|---|---|
| 1 | 0.9979 | 0.9108 | 0.7259 | 0.9311 | 0.7903 | $2.7 \times 10^{-11}$ | F1 | F1 | F1 |
| 2 | 0.9867 | 0.8975 | 0.7172 | 0.9209 | 0.7833 | $4.8 \times 10^{-11}$ | F1 | F1 | F1 |
| 3 | 0.9906 | 0.9381 | 0.7577 | 0.9523 | 0.8161 | $2.3 \times 10^{-11}$ | F1 | F1 | F1 |
| 4 | 0.9931 | 0.8879 | 0.6947 | 0.9084 | 0.7638 | $6.4 \times 10^{-12}$ | F1 | F1 | F1 |
| 5 | 0.8938 | 0.9978 | 0.9167 | 0.9717 | 0.9067 | $4.1 \times 10^{-10}$ | F2 | F2 | F2 |
| 6 | 0.8913 | 0.9987 | 0.9241 | 0.9751 | 0.9169 | $6.7 \times 10^{-10}$ | F2 | F2 | F2 |
| 7 | 0.8746 | 0.9882 | 0.9267 | 0.9652 | 0.9173 | $1.4 \times 10^{-9}$ | F2 | F2 | F2 |
| 8 | 0.8486 | 0.9570 | 0.9011 | 0.9365 | 0.8936 | $2.9 \times 10^{-9}$ | F2 | **F1** | F2 |
| 9 | 0.7138 | 0.9162 | 0.9993 | 0.8643 | 0.9131 | $9.9 \times 10^{-9}$ | F3 | F3 | F3 |
| 10 | 0.7078 | 0.9102 | 0.9989 | 0.8593 | 0.9149 | $1.4 \times 10^{-8}$ | F3 | F3 | F3 |
| 11 | 0.6863 | 0.8870 | 0.9864 | 0.8371 | 0.9014 | $2.8 \times 10^{-8}$ | F3 | F3 | F3 |
| 12 | 0.6562 | 0.8505 | 0.9512 | 0.8016 | 0.8665 | $5.4 \times 10^{-8}$ | F3 | F3 | F3 |
| 13 | 0.9240 | 0.9750 | 0.8635 | 0.9989 | 0.9309 | $5.4 \times 10^{-10}$ | F4 | F4 | F4 |
| 14 | 0.9187 | 0.9724 | 0.8685 | 0.9990 | 0.9377 | $8.8 \times 10^{-10}$ | F4 | F4 | F4 |
| 15 | 0.9004 | 0.9870 | 0.8686 | 0.9600 | 0.9366 | $1.8 \times 10^{-9}$ | **F2** | **F2** | F4 |
| 16 | 0.8699 | 0.9236 | 0.8390 | 0.9533 | 0.9082 | $3.9 \times 10^{-9}$ | F4 | F4 | F4 |
| 17 | 0.7683 | 0.8940 | 0.9081 | 0.9190 | 0.9959 | $2.0 \times 10^{-8}$ | F5 | F5 | F5 |
| 18 | 0.7905 | 0.9072 | 0.9102 | 0.9346 | 0.9915 | $2.7 \times 10^{-8}$ | F5 | **F4** | F5 |
| 19 | 0.7503 | 0.8721 | 0.9007 | 0.8992 | 0.9896 | $5.2 \times 10^{-8}$ | F5 | F5 | F5 |
| 20 | 0.7274 | 0.8427 | 0.8728 | 0.8720 | 0.9617 | $9.1 \times 10^{-8}$ | F5 | F5 | F5 |
| 21 | 0.3269 | 0.3995 | 0.3437 | 0.3689 | 0.3318 | 0.6135 | F6 | F6 | F6 |
| 22 | 0.3387 | 0.3995 | 0.3320 | 0.3794 | 0.3409 | 0.6117 | F6 | F6 | F6 |
| 23 | 0.3657 | 0.3696 | 0.2711 | 0.3991 | 0.3529 | 0.6105 | F6 | F6 | F6 |
| 24 | 0.3572 | 0.3697 | 0.2788 | 0.3997 | 0.3616 | 0.6181 | F6 | F6 | F6 |

*4.5. Dynamic Attribute of the Algorithm*

To illustrate the adaptability of the proposed method when transient faults come, a transient fault is set up in PV array 1 based on the simulation model showed in Figure 6. The introduced transient fault is a shadow fault occurs in PV array 1 within a period of time and other time is normal in a day. The simulation conditions are described as Figure 9.

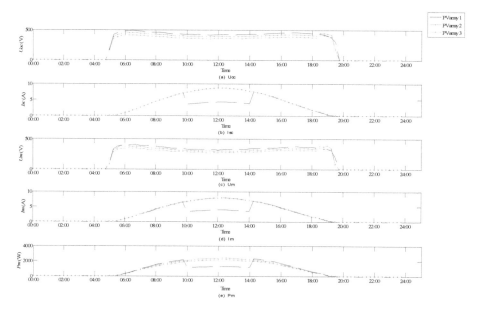

**Figure 9.** The simulation of transient fault.

Table 9 shows the dynamic adaptability and diagnosis results of the proposed method for transient fault. According to Figure 9, there are five status modes for different PV arrays: PV array 1 before the failure, PV array 1 in the failure, PV array 1 after the failure, PV array 2 and PV array 3. These modes are marked M1, M2, M3, M4 and M5, and the diagnosis results are shown in Table 9. Table 9 shows that the proposed method can identify the unknown transient faults effectively, and the faults can be classified according to the feature distribution of faults.

**Table 9.** The dynamic adaptability analysis of the algorithm.

| | The Results | | | | |
|---|---|---|---|---|---|
| M1 membership | 0.9979 | 0.6863 | 0.9906 | 0.9867 | 0.9931 |
| M2 membership | 0.9108 | 0.9864 | 0.9381 | 0.8975 | 0.8879 |
| M3 membership | 0.7259 | 0.887 | 0.7577 | 0.7172 | 0.6947 |
| M4 membership | 0.9311 | 0.8371 | 0.9523 | 0.9209 | 0.9084 |
| M5 membership | 0.7903 | 0.9014 | 0.8161 | 0.7833 | 0.7638 |
| Diagnostic results | Normal | Shaded | Normal | Normal | Normal |

## 5. Experiment Analysis

To verify the correctness and effectiveness of the proposed method, experiments were carried out under short circuit, open circuit, and partial occlusion conditions. The fault conditions of each data sample cover a wide range of work irradiances and temperatures. First, some labeled data samples under different fault conditions were collected on an experimental platform. Then, tests and analyses were carried out based on the proposed diagnosis method.

*5.1. Experimental Description*

In order to verify the correctness and effectiveness of the proposed diagnosis method in this paper under different environmental conditions, an empirical test platform for PV power generation is constructed. Figure 10 illustrates the system structure of the empirical test platform. The platform installed capacity is 9.555 kWp. Thirty nine PV modules are used and the electrical parameters are shown in Table 10. In order to better analyze the influence of external environment on the PV power generation and performance, the experimental platform includes a high-precision irradiator for measuring solar irradiance, a small weather station for measuring the external environmental parameters such as global solar irradiance, temperature, wind speed, a temperature sensor for measuring the operating temperature, a data collector for measuring the current and voltage, an I-V scanner, etc., then those data are stored in the computer through a Supervisory Control And Data Acquisition (SCADA) system, which can collect multiple operating parameters of PV plant such as the PV power generation, the current of AC and DC sides and the voltage of AC and DC sides. Table 10 illustrates the specific parameters of the experimental platform. Different tests under short circuit, open circuit, and partial shading conditions were run on an empirical platform.

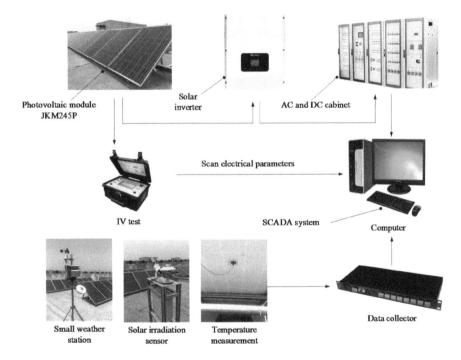

**Figure 10.** Outdoor experimental PV module platform.

**Table 10.** The description of the experimental equipment.

| Name | Model | Description of Parameters |
|---|---|---|
| 3 kW power station (3 × 13 serial-parallel module) | JKM245P | Maximum power: 254 Wp; Optimal operating voltage (Vmp): 30.1 V; Optimal operating current (Imp): 8.14 A; Module efficiency: 14.97%; Operating temperature range: −40~+85 °C; Cell operating temperature: 45 ± 2 °C. |
| I-V scanner | MP-11 | power measurement range: 10 W~18 kW; Voltage measurement range: 10~1000 V; Current measurement range: 100 mA~30 A. |
| Current and voltmeter | PZ72 | Voltage measurement range: 0~1000 V; Current measurement range: 0~10 A. |
| Backplane temperature sensor | WRM-101 | Temperature range: −50~200 °C; Measuring accuracy: ≤0.2 °C. |
| Solar irradiance meter | MS-80 | Irradiation measurement range:0~2000 W/m$^2$; Measuring accuracy: ≤±3%; Operating temperature: −40 °C~80 °C. |
| Weather station | WS200 | Temperature, wind, humidity and total, direct, scattered radiation observation. |

*5.2. Experimental Data Acquisition*

Five PV array faults (normal, one module shorted, two modules shorted, one module shaded, and two modules shaded) were tested (Table 11) based on the PV array experimental platform. 200 data samples were collected. The description of the selected data was shown in Table 12.

**Table 11.** Experiments description.

| Fault Types | Fault Descriptions | Fault Pictures |
|---|---|---|
| F1 Normal | Normal condition. | |
| F2 One module shorted F3 Two modules shorted | The short-circuit fault is tested by short-circuiting some PV modules. | |

**Table 11.** *Cont.*

| Fault Types | Fault Descriptions | Fault Pictures |
|---|---|---|
| F4 One module shaded<br>F5 Two modules shaded | The partial shading condition is tested by covering some PV modules with shield panels. |  |
| F6 One module opened | The open-circuit fault is tested by open-circuiting some PV modules. | |

**Table 12.** Experimental collection distribution.

|  | Normal | One Module Shorted | Two Modules Shorted | One Module Shaded | Two Modules Shaded |
|---|---|---|---|---|---|
| Data number | 20 | 40 | 40 | 50 | 50 |
| Proportion (%) | 10 | 20 | 20 | 25 | 25 |

## 5.3. Experimental Result Analysis

Five typical fault data samples were selected from the five fault modes mentioned above as shown in Table 13. The diagnosis results obtained via the proposed method are shown in Table 14.

**Table 13.** PV array fault diagnosis samples.

| Group Number | $U_{oc}$/V | $I_{sc}$/A | $U_m$/V | $I_m$/A | $P_m$/W | F1 | F2 | F3 | F4 | F5 |
|---|---|---|---|---|---|---|---|---|---|---|
| 1 | 426.7566 | 7.7779 | 333.8868 | 7.0476 | 2348.7680 | 1 | 0 | 0 | 0 | 0 |
| 2 | 393.9118 | 7.7779 | 306.4041 | 7.0933 | 2168.0670 | 0 | 1 | 0 | 0 | 0 |
| 3 | 361.0858 | 7.7779 | 281.8601 | 7.0669 | 1987.7490 | 0 | 0 | 1 | 0 | 0 |
| 4 | 422.3981 | 7.7776 | 303.9081 | 7.0934 | 2151.1730 | 0 | 0 | 0 | 1 | 0 |
| 5 | 418.0583 | 7.7772 | 276.8681 | 7.0732 | 1952.6170 | 0 | 0 | 0 | 0 | 1 |

**Table 14.** Proposed method diagnosis results.

| Group Number | F1 | F2 | F3 | F4 | F5 |
|---|---|---|---|---|---|
| 1 | 0.996579 | 0.534397 | 0.416860 | 0.695547 | 0.565517 |
| 2 | 0.532841 | 0.996478 | 0.552609 | 0.812129 | 0.541072 |
| 3 | 0.415561 | 0.541628 | 0.996761 | 0.558068 | 0.775134 |
| 4 | 0.695257 | 0.810263 | 0.569131 | 0.996675 | 0.708160 |
| 5 | 0.572519 | 0.526165 | 0.777834 | 0.700664 | 0.996326 |

Again, a larger membership degree indicates a greater likelihood that the diagnosis sample is in the given fault state. As shown in Table 14, the five diagnosed samples fell into F1, F2, F3, F4, and F5 states, once more indicating that the proposed method yields correct results.

## 6. Conclusions

This paper proposed a novel PV array fault diagnosis method based on the FCM and fuzzy membership algorithms. The proposed method effectively detects short circuit, open circuit, partial occlusion, and other faults. Simulation analysis indicated that the diagnostic accuracy of the proposed method was 96%. Field experiments further verified the correctness and effectiveness of the proposed method, and the method can complete the PV array diagnosis. The innovations of this paper can be summarized as follows:

(1) The FCM algorithm described the distribution characteristics of fault data effectively based on small amounts of fault samples data, and avoided the difficult for obtaining the fault samples.

(2) By using the membership function of vague math as the fault diagnosis function, it quantized the membership degree between fault samples and each fault mode, and described the degree of similarity between fault samples and each fault mode clearly and objectively.

(3) The proposed method effectively exploits the advantages of FCM (excellent classification ability) as well as the membership function algorithm (excellent distance computing ability). And the proposed method didn't need additional equipment support, concerned people can detect the fault module quickly by measuring voltage, current, power and other parameters.

(4) The distribution characteristics of the FCM cluster centers reflected the fault characteristics, and the distribution characteristics can be used for updating membership function.

(5) The clustering centers obtained by the FCM algorithm can be used as the typical value of each fault state, and then fault characteristic database can be established. Based on the fault characteristic database, combined with other intelligent methods, it will be much easier to develop new ideas for the PV array fault diagnosis.

**Acknowledgments:** The authors would like to acknowledge the Fundamental Research Funds for the Central Universities "2016MS52, 2016MS31" and the Research Funds from State Grid Corporation of China "SGHE0000KXJS1700074".

**Author Contributions:** All authors contributed to this work by collaboration. Qiang Zhao is the first author in this manuscript. All authors revised and approved for the publication.

**Conflicts of Interest:** The authors declare no conflict of interest.

## References

1. Zhang, X.; Gao, D.; Liu, H.; Ye, J.; Wang, S. A novel method for multi-sensor fault detection and positioning of photovoltaic array. *Renew. Energy Resour.* **2016**, *34*, 166–172.
2. Stellbogen, D. Use of PV circuit simulation for fault detection in PV array fields. In Proceedings of the 23th Photovoltaic Specialists Conference, Louisville, KY, USA, 6 August 2002; pp. 1302–1307.
3. Chuang, W.; Chen, C.; Yen, J.; Hsu, Y. Using MPCA of spectra model for fault detection in a hot strip mill. *J. Mater. Process. Technol.* **2009**, *209*, 4162–4168. [CrossRef]
4. Karatepe, E.; Boztepe, M.; Colak, M. Development of a suitable model for characterizing PV arrays with shaded solar cells. *Sol. Energy* **2007**, *81*, 977–992. [CrossRef]
5. Noguchi, T.; Togashi, S. Short-current pulse-based Maximum-power-point tracking method for multiple PV-and-converter module system. *IEEE Trans. Ind. Electron.* **2002**, *49*, 217–223. [CrossRef]
6. Chowdhury, S.R.; Saha, H. Maximum power point tracking of partially shaded solar PV arrays. *Sol. Energy Mater. Sol. Cells* **2010**, *94*, 1441–1447. [CrossRef]

7.    Miyatake, M.; Inada, T.; Hiratsuka, I.; Zhao, H.; Otsuka, H.; Nakano, M. Control characteristics of a fibonacci-search-based maximum power point tracker when a PV array is partially shaded. In Proceedings of the 4th International Power Electronics and Motion Control Conference, Xi'an, China, 14–16 August 2004; pp. 816–821.

8.    Wang, X.; Wang, S.; Du, H.; Wang, P. Fault Diagnosis of Chemical Industry Process Based on FRS and SVM. *Control Decis.* **2015**, 353–356. [CrossRef]

9.    Zhong, D. *Research on Fault Detection of PV Array Based on Composite Data Fusion*; Tianjin University: Tianjin, China, 2010.

10.   Cheng, Z.; Zhong, D.; Li, B.; Liu, Y. Research on fault detection of PV array based on data fusion and fuzzy mathematics. In Proceedings of the 2011 Asia-Pacific Power and Energy Engineering Conference, Wuhan, China, 25–28 March 2011; pp. 1–4.

11.   Chao, K.H.; Ho, S.H.; Wang, M.H. Modeling and fault diagnosis of a PV system. *Electr. Power Syst. Res.* **2008**, *78*, 97–105. [CrossRef]

12.   Xu, R.; Chen, H.; Hu, Y.; Sun, X. Fault Location of Photovoltaic Array Based on Gaussian Process. *Trans. China Electrotech. Soc.* **2013**, *28*, 249–256.

13.   Zhao, Y.; Yang, L.; Lehman, B.; de Palma, J.-F.; Mosesian, J.; Lyons, R. Decision tree-based fault detection and classification in solar PV arrays. In Proceedings of the 2012 27th Annual IEEE Applied Power Electronics Conference and Exposition, Orlando, FL, USA, 5–9 February 2012; pp. 93–99.

14.   Li, Z.; Wang, Y.; Zhou, D.; Wu, C. An Intelligent Method for Fault Diagnosis in Photovoltaic Array. In *System Simulation and Scientific Computing*; Springer: Berlin/Heidelberg, Germany, 2012; pp. 10–16.

15.   Wang, Y.; Wu, C.; Fu, L.; Zhou, D.; Li, Z. Fault diagnosis of PV array based on BP neural network. *Power Syst. Prot. Control* **2013**, *41*, 108–114.

16.   Zhang, X.; Zhou, J.; Huang, Z.; Li, C.; He, H. Vibrant Fault Diagnosis for Hydro-turbine Generating Unit Based on Rough sets and Multi-class Support Vector Machine. *Proc. CSEE* **2010**, *30*, 88–93.

17.   Han, S.; Zhu, J.; Mao, J.; Zhan, W. Fault Diagnosis of Transformer Based on Particle Swarm optimization-Based Support Vector Machine. *Electr. Meas. Instrum.* **2014**, *11*, 509–513.

18.   Dong, S.; Xu, X.; Chen, R. Application of fuzzy C-means method and classification model of optimized K-nearest neighbor for fault diagnosis of bearing. *J. Braz. Soc. Mech. Sci. Eng.* **2015**, *38*, 1–9. [CrossRef]

19.   Hu, Y.; Deng, Y.; He, X. A Summary on PV Array Fault Diagnosis Method. *Power Electron.* **2013**, *3*, 21–23.

20.   Krishnapuram, R.; Keller, J.M. A possibilistic approach to clustering. *IEEE Trans. Fuzzy Syst.* **2002**, *1*, 98–110. [CrossRef]

21.   Du, W.; Lie, X.; LV, F. Research on Transformer Fault Diagnosis Based on Fuzzy Clustering Algorithm. *Transformer* **2009**, *46*, 65–69.

22.   Li, J.; Qi, J.; Hu, J.; Peng, Y. Similarity Measurement Method Based on Membership Function and Its Application. *Appl. Res. Comput.* **2010**, *27*, 891–893.

23.   Hariharan, R.; Chakkarapani, M.; Ilango, G.S.; Nagamani, C. A Method to Detect Photovoltaic Array Faults and Partial Shading in PV Systems. *IEEE J. Photovol.* **2016**, *6*, 1278–1285. [CrossRef]

24.   Zhou, L.; Wang, G.; Jie, G. Simulation Research on Photovoltaic Cells, Modules, and Arrays Model. *Ship Electr. Technol.* **2011**, *31*, 25–29.

25.   Wang, Z.T.; Zhao, N.B.; Wang, W.Y.; Tang, R.; Li, S.Y. A Fault Diagnosis Approach for Gas Turbine Exhaust Gas Temperature Based on Fuzzy C-Means Clustering and Support Vector Machine. *Math. Probl. Eng.* **2015**, 1–11. [CrossRef]

26.   Wu, J.; Cai, Z.; Hu, C.; Cao, J. Status Evaluation of Protection Relays Based on The Membership Function in Fuzzy Normal Distribution. *Power Syst. Prot. Control* **2012**, *40*, 48–52.

27.   Liu, X.; Liu, X.; Wang, M. The Determination of Membership Function and Use. *Comput. Knowl. Technol.* **2010**, *6*, 8831–8832.

28.   Huang, L.; Hou, J.; Luo, L. Soft Fault Diagnosis of Analog Circuits with Tolerance Based on Fuzzy C-means Clustering Algorithm and Membership Algorithm. *Signal Process.* **2011**, *27*, 624–628.

Article

# Improvement of Shade Resilience in Photovoltaic Modules Using Buck Converters in a Smart Module Architecture

S. Zahra Mirbagheri Golroodbari *, Arjen. C. de Waal and Wilfried G. J. H. M. van Sark

Copernicus Institute, Utrecht University, Heidelberglaan 2, 3584 CS Utrecht, The Netherlands;
A.C.deWaal@uu.nl (A.C.d.W.); w.g.j.h.m.vansark@uu.nl (W.G.J.H.M.v.S.)
* Correspondence: s.z.mirbagherigolroodbari@uu.nl; Tel.: +31-302-537-681

Received: 13 December 2017; Accepted: 17 January 2018; Published: 19 January 2018

**Abstract:** Partial shading has a nonlinear effect on the performance of photovoltaic (PV) modules. Different methods of optimizing energy harvesting under partial shading conditions have been suggested to mitigate this issue. In this paper, a smart PV module architecture is proposed for improvement of shade resilience in a PV module consisting of 60 silicon solar cells, which compensates the current drops caused by partial shading. The architecture consists of groups of series-connected solar cells in parallel to a DC-DC buck converter. The number of cell groups is optimized with respect to cell and converter specifications using a least-squares support vector machine method. A generic model is developed to simulate the behavior of the smart architecture under different shading patterns, using high time resolution irradiance data. In this research the shading patterns are a combination of random and pole shadows. To investigate the shade resilience, results for the smart architecture are compared with an ideal module, and also ordinary series and parallel connected architectures. Although the annual yield for the smart architecture is 79.5% of the yield of an ideal module, we show that the smart architecture outperforms a standard series connected module by 47%, and a parallel architecture by 13.4%.

**Keywords:** photovoltaics; modules; shade resilience; buck converter; module architecture

## 1. Introduction

It is now commonly acknowledged that fossil fuel-based generation presents serious challenges to the environment, in terms of global warming, climate change, and society at large. It is also commonly acknowledged that renewable energy sources (RES) are viable, clean, and efficient alternatives. Amongst the RES, photovoltaic (PV) systems, which are maintenance and pollution free [1–3], have been increasingly used as the main source of power generation in both standalone and grid-connected residential and large-scale systems [3]. Every year the solar industry is breaking new records and the global PV market grew significantly to at least 74.4 GW in 2016 [4]. Moreover, in 2016 solar installations contributed 39% of all new electric generating capacity, for the first time more than all other technologies [5].

Energy harvesting from a PV module under uniform irradiation is simply possible by connecting it to an inverter that implements a maximum power point tracking (MPPT) algorithm along with a DC-AC converter. The most frequently used MPPT methods are gradient descent based methods such as perturb and observe (P&O) and incremental conductance (Inc. Cond); these conventional methods are also denoted as hill climbing algorithms [6]. MPPT algorithms are used to control the converter as an interface between the PV module and the grid and/or load. However, irradiation is not always uniform, and partial shading (PS) conditions lead to module mismatches, which is one of the main issues in urban PV installations due to adjacent obstacles in buildings. This will become even more relevant in building-integrated PV, in particular façade-based systems. Partial shading has a strongly non-linear effect on PV outputs [7]. For a series connected PV module system even with a highly efficient MPPT algorithm partial shading may lead to almost 70% of power loss [8]. Based on the module architecture and system topology, a power-voltage (P-V) curve as a PV module characteristic may change from a concave curve with one global maximum (GM) to a curve with multiple local maxima, of which one will be the global maximum. The main challenge for this partial shading condition is to find the GM for maximum performance. The aforementioned conventional MPPT algorithms may not perform well under PS conditions. It is possible for conventional algorithms to be trapped on a local maximum [9,10]. Many MPPT algorithms have been proposed for the PS condition [11–15], however they are complicated and/or require long tracking times.

An alternative way to mitigate the PS effects in a PV system is to change the configuration of the PV system depending on the variation of the shading pattern. PV system configurations have been suggested to be changed via different interconnections of individual PV modules, which are typically from one of the following configurations: (i) series-parallel (SP); (ii) total-cross-tied (TCT); and (iii) bridge-linked [9,16]. In [17] an electrical array reconfiguration is proposed, which uses a switching matrix for changing the position of the PV modules and find the best configuration between those. This method is at the module level and implemented to maximize the available DC power by grouping modules with similar shading patterns.

Dynamic reconfiguration methods that are implemented on the module level and depend on an optimization algorithm may be very complicated and may perform at a sluggish pace. As curves with multiple maxima occur because of the effect of bypass diodes (BPD), one way to mitigate PS effects is to divide the module into groups of a small number of cells, while we note that the best option is to have one diode for each cell. Although more BPDs lead to shade-resilient modules [18], increasing the number of BPDs increases the number of local maximum peaks and subsequently a more accurate and complicated MPPT algorithm is required to find the GM of the module. Moreover, the efficiency losses with the BPDs are still significant [19]. One way to overcome this problem is to replace the ordinary BPD with an active BPD, which in fact is an electrical circuit consisting of Metal-Oxide-Semiconductor Field-Effect Transistors (MOSFETs).

Different configurations for module integrated electronics can be categorized in the following groups: (i) Conventional systems, consisting of three BPDs per module and a central converter to change the output voltage level; (ii) Buck converters, to which normal PV modules are connected and thus the output current of the shaded module is to be controlled; (iii) Buck-Boost converters, in which configuration both current and voltage are to be controlled; and (iv) Voltage equalizers, which are a combination of different converter or even bidirectional converters to equalize the voltage by power processing [19–21].

In the present study a generic model is developed that can be used for various module configurations and is not limited to number of cells, for instance. It uniquely combines machine learning with system architecture design. In this paper, we employ the method for mitigating the PS condition using the following smart module architecture: a certain small number of solar cells are grouped and connected to a DC-DC buck converter. Connection of the buck converters then makes up the smart module. The converter is used to control the current level and also maximizes the harvested energy from the group of cells. To investigate the shade resilience different shading patterns are modeled. The performance of the smart module is tested under these shading patterns using a full year of measured irradiation data. To have a better understanding about the smart module performance the output of this module is compared with the following PV module architectures: (i) Series connected: an ordinary module with three groups of solar cells in series, there is one BPD for each group of cells to bypass the group in case of shading; (ii) Parallel connected: a module consists of three parallel strings where each string is ended with a blocking diode; (iii) Ideal module with one converter per cell, which is assumed as the ideal reference for comparison of the output of other modules. The model is developed in MATLAB (2017a, MathWorks, Natick, MA, USA). Shading patterns in the simulations contain two different varieties of shading, random and pole shading. To have an accurate understanding about the converter efficiency a least squares support vector machine (LS-SVM, v1.8, K.U. Leuven, Leuven, Belgium) as a machine learning method is implemented to find the exact value of efficiency with respect to input and output voltage and also output current. The same machine learning method is used to calculate the maximum extracted power at different times throughout the year. In all statistical analysis simulations, real data which is extracted from the Utrecht Photovoltaic Outdoor Test (UPOT) facility is used to make the result statistically tangible [22]. Generally, in the smart module architecture a buck converter in parallel with each group of cells controls the shaded groups' current by leveling down the output voltage of the converter, which simply means that the output current flow in all converters are equal. This strategy helps the shaded groups to perform efficiently while shaded. The series architecture bypasses the shaded groups because of the implemented BPDs. In parallel architecture, the lowest rated voltage determines the voltage output of the whole array, which means wasting power. In both parallel and series architectures, a fraction of power is wasted because of mentioned reasons, but in the smart module architecture all cells, even shaded ones, are producing power efficiently and none of the cells is bypassed.

This paper is organized as follows: Section 2 discusses the principle structure of the smart module electronics including an optimization for the best value of output. Section 3 provides different shading patterns for the architectures under study. In Section 4, results from simulations in Section 3 are discussed. Finally, Section 5 summarizes the potential characteristics of the smart module in terms of shade resilience.

## 2. Smart PV Module Topology and Design

The proposed smart module architecture, as shown in Figure 1, consists of $N_G$ groups of cells with $n_g$ number of cells in each group.

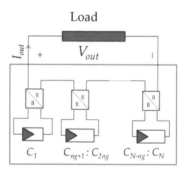

**Figure 1.** Smart module architecture with $n_g$ cells in $N_G$ groups with electronic circuits.

The total number of cells in a module, $N$, is calculated as:

$$N = N_G \times n_g \tag{1}$$

PV cells in each group are connected in series and connected to a buck converter. The DC-DC buck converter is controlled via an MPPT algorithm to ensure maximum power extraction from the group. Series connected buck converters in the output of the module is due to current control.

The test module for this simulation study consists of 60 monocrystalline silicon solar cells and we have access to each cell by two connection points for individual analysis. The characteristics at standard test conditions (STC is defined as 1000 W/m$^2$ irradiance, 25 °C cell temperature, Air mass (AM) 1.5 solar spectrum) of the cells are: open circuit voltage ($V_{OC}$) of 613 mV, short circuit current ($I_{SC}$) of 7.92 A, maximum power ($P_{max}$) of 3.7 W, and efficiency ($\eta$) of 15.4%.

In order to simulate a feasible system, many converters in the market have been studied and a comparison between the most appropriate ones is shown in Table 1. The most suitable converter chosen for the module in this work is the LTM4611 converter from Linear Technology Corporation designs (Milpitas, CA, USA), because of the following reasons: (i) its voltage and current specifications are matched to the small groups of cells in the module; (ii) it has a higher switching frequency compared with Texas Instruments converters, which leads to better performance of the MPPT algorithm; and (iii) no extra element is required for this converter besides the chip itself, which leads to higher efficiency in the complete converter circuit.

**Table 1.** Comparison of different buck converters for the smart module. The colored row indicates the chosen converter.

| Model | $V_{in_{min}}$ (V) | $V_{in_{max}}$ (V) | $V_{out_{min}}$ (V) | $V_{out_{max}}$ (V) | $I_{out_{max}}$ (A) | Switching frequency$_{max}$ (kHz) | Options | Manufacturer |
|---|---|---|---|---|---|---|---|---|
| LM2744 | 1 | 16 | 0.5 | 12.8 | 20 | 50 | Ext_Ref_Con [1] | TI [4] |
| LM2747 | 1 | 14 | 0.6 | 12 | 20 | 50 | PbStUp [2] & OpClk [3] | TI [4] |
| LM2745 | 1 | 14 | 0.6 | 12 | 20 | 250 | PbStUp [2] & OpClk [3] | TI [4] |
| LM2748 | 1 | 14 | 0.6 | 12 | 20 | 50 | PbStUp [2] | TI [4] |
| LTC3713 | 1.5 | 36 | 0.8 | 32.4 | 20 | 1500 | Ext_Ref_Con [1] | LT [5] |
| LTC3718 | 1.5 | 36 | 0.7 | 36 | 20 | 1500 | Ext_Ref_Con [1] | LT [5] |
| LTM4611 | 1.5 | 5.5 | 0.8 | 5 | 15 | 835 | Ext_Ref_Con [1] | LT [5] |

[1] Ext_Ref_Con: External reference controller, [2] PbStUp: Pre-Bias startup, [3] OpClk: optional clock, [4] TI: Texas Instruments (Dallas, TX, USA).

The converter efficiency depends on three different factors: input and output voltage, and output current. In Figure 2, the efficiency of the LTM4611 converter is depicted with these factors as parameters. The important problem is to find the optimum set of variables that lead to the maximum efficiency. To this end, the least squares support vector machine technique is implemented as a standard approach in regression analysis. This method allows to generalize the given data in the datasheet, Figure 2. Therefore, all possible combinations of variables which are necessary for designing are made available in the form of a look-up table. In the following subsection, first the aforementioned method will be introduced and then its implementation for this problem is discussed.

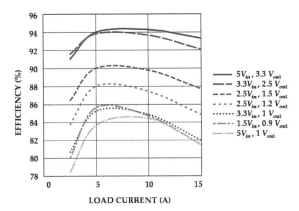

**Figure 2.** Efficiency as a function of output (load) current for the LTM4611 converter.

*LS-SVM for Efficiency Optimization*

The least squares support vector machine method is used to generalize the performance of the LTM4611 buck converter [23]. Let us assume the training set $T = \{(x_1, y_1), \ldots, (x_l, y_l)\}$ is a set of pre-determined data, where $x_j = [V_{in}(j), V_{out}(j), I_{out}(j)]^T$, $y_j = \eta(j)$, and $j = 1,\ldots,l$ is the number of elements in data set $T$. The training set is collected from the datasheet [24]. The LS-SVM uses the training data set $T$ to estimate the optimal nonlinear regression function $\hat{f}$, as shown in Equation (2):

$$\hat{f}(x_{new}) = \sum_{i=1}^{l} \beta K(x_{new}, x_i) + b \tag{2}$$

where $K$ represents a so-called kernel function and for this application, the Radial Basis Function (RBF) has been chosen as shown in Equation (3), following [23,25]:

$$K(x_{new}, x_i) = \exp\left(\frac{-\|x_{new} - x_i\|^2}{2\sigma^2}\right) \tag{3}$$

where $x_{new} = [V_{in}, V_{out}, I_{out}]^T \notin T$, and the design parameters $\beta$ and $b$ are obtained by solving the matrix-vector equation shown in Equation (4):

$$\begin{bmatrix} 0 & \begin{bmatrix} 1 & 1 & \cdots & 1 \end{bmatrix} \\ \begin{bmatrix} 1 \\ \vdots \\ 1_l \end{bmatrix} & \left[\Omega_{l\times l} + \left(\frac{1}{\gamma}\right)I_{l\times l}\right] \end{bmatrix} \begin{bmatrix} b \\ \beta \end{bmatrix} = \begin{bmatrix} 0 \\ y \end{bmatrix} \tag{4}$$

Here, $I_{l\times l}$ represents the identity matrix and $\Omega_{l\times l}$ is a $l \times l$ full matrix with computed elements from the training data as follows:

$$\Omega_{l\times l}(r,q) = \exp\left(\frac{-\|x_q - x_r\|_2^2}{2\sigma^2}\right), q,r = 1,\ldots,l \tag{5}$$

Parameters $\sigma$ and $\gamma$ in Equations (4) and (5) are tuning parameters, which can be calculated using different methods, such as k-fold cross-validation, leave one out cross-validation, etc. [23].

The next step in the analysis is to check the feasibility of grouping cells in relation to the converter specifications. Table 2 lists the options under study, showing the calculated voltage at maximum power point for different cases of grouping. From the LTM4611 specification (see Table 1), only case number 5, 6 and 7 are feasible, as the other groupings lead to voltages outside the converter specifications.

**Table 2.** Overview of cell grouping and buck converter specifications. The colored columns indicate the cases that are selected.

| Case Number | 1 | 2 | 3 | 4 | 5 | 6 | 7 | 8 | 9 | 10 |
|---|---|---|---|---|---|---|---|---|---|---|
| # Cells ($n_g$) | 60 | 30 | 20 | 15 | 10 | 6 | 4 | 3 | 2 | 1 |
| # Groups ($N_G$) | 1 | 2 | 3 | 4 | 6 | 10 | 15 | 20 | 30 | 60 |
| Vmpp (mV) * | 29,412 | 14,706 | 9804 | 7356 | 4902 | 2941.2 | 1961.6 | 1471.2 | 980.8 | 490.4 |
| Feasibility | No | No | No | No | Yes | Yes | Yes | No | No | No |

* STC is considered as the highest reference.

Using all different possible combinations of variables extracted from the buck converter datasheet [24] the efficiency is computed. For instance, Figure 3 depicts the efficiency for input voltage $V_{in}$ = 5 V, and Figure 4 shows how LS-SVM generalizes the data from Figure 3 to cover all different combinations of $V_{out}$ and $I_{out}$ in order to allow computing the efficiency. For the three different case numbers 5, 6 and 7 (Table 2), more information is presented in Table 3.

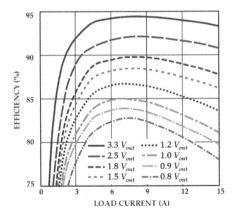

**Figure 3.** Efficiency vs Load Current at $V_{in}$ = 5 V.

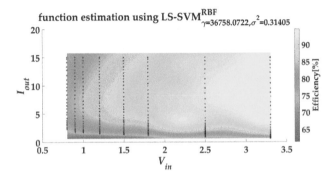

**Figure 4.** Efficiency for all values of $V_{in}$ and $I_{out}$ at $V_{out}$ = 5 V.

**Table 3.** Specifications for three cases of grouping at STC.

| Case Number | 5 | 6 | 7 |
|---|---|---|---|
| Number of Cells | 10 | 6 | 4 |
| Open Circuit Voltage (V) | 6.13 | 3.67 | 2.45 |
| Short Circuit Current (A) | 7.92 | 7.92 | 7.92 |
| Current at MPP (A) | 7.54 | 7.54 | 7.54 |
| Voltage at MPP (V) | 4.902 | 2.941 | 1.961 |

While the variation in irradiance leads to considerable linear variation in the output current at MPP, it influences the voltage at MPP only slightly ($\sim\ln(I_{sc})$). Therefore, as a simplification, the input voltage for buck converter is assumed to be constant having the value as mentioned in Table 3. Although for lower irradiation levels the PV output voltage level in case 7 may not be suitable for this converter specification, all three cases are taken into consideration for comparison. The input current according to the DC-DC buck converter basics is calculated with Equation (6):

$$P_{out} = P_{in} \times \eta \Rightarrow V_{out} \times I_{out} = V_{in} \times I_{in} \times \eta \tag{6}$$

The above-described calculations lead to three contour graphs, one for each case, that are shown in Figure 5a–c. From those, the following can be extracted:

1. Referring to the color bars, which are different per case, the highest maximum possible efficiency belongs to case 6, and the lowest maximum possible efficiency belongs to case 7.

2. The line $I_{in} = I_{MPP_{STC}}$ (where $I_{MPP_{STC}}$ is the PV output current at STC) is depicted in all figures, thus the best possible efficiency for three cases can be found at intersections of these lines with the contour graphs. Therefore, the best possible efficiencies are 97.74% for case 5, 94.51% for case 6, and 87.69% for case 7.

3. It is clear that the best output current should be in the range of 11–13 A for case number 5, where $N_G = 6$, $n_g = 10$, and it should be 8–9 A for case number 6 with $N_G = 10$, $n_g = 6$, and 7–9 A for case number 7 with $N_G = 15$, $n_g = 4$.

4. To compare all three cases thoroughly, it is assumed that irradiation level is changed such that the PV output current changes by 50% ($\Delta I_{MPP} = 50\%$). Variation of efficiency in this situation is shown regarding the aforementioned converter output current range in Figure 5d. Although the maximum possible efficiency for case number 5 is higher, the range over which efficiency varies is wider. Also the average values for efficiency (shown in green dots in Figure 5d) indicate that the highest average value belongs to case number 6. That is the reason why case number 6 is chosen for the smart module grouping in this study.

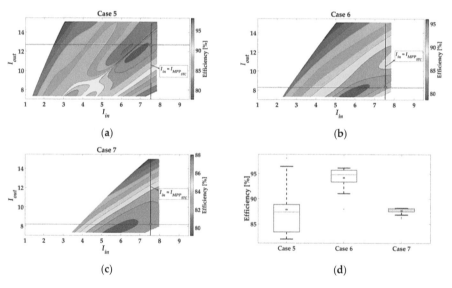

**Figure 5.** Contour plots of efficiency as a function of input and output current: (**a**) case number 5, (**b**) case number 6, (**c**) case number 7, (**d**) efficiency variation with decreasing converter input current for 50%.

## 3. Methodology

After designing the architecture and electrical topology of the smart module, let us discuss shading patterns and explain how the module is going to be modeled for this research. We consider the surface of the module to consist of 600 k pixels, which means that each cell has 10 k pixels (ignoring inter-cell distances for simplicity). The irradiation level on pixel $p$ is called $G_p$ and Equation (7) gives the value of this variable:

$$G_p = \begin{cases} G_{p,GHI} & \text{if it is not shaded} \\ G_{p,s} & \text{if it is shaded} \end{cases} \tag{7}$$

where $G_{p,GHI}$ is the global horizontal irradiance (GHI) at the pixel and $G_{p,s}$ the irradiance at the pixel under the shaded condition.

To calculate the irradiation level on each cell Equation (8) is used, the equation is extracted based on experimental results in a research study by Sinapis et al. [26]:

$$G_C = (F_{unshaded} \times G_{GHI}) + (F_{shaded} \times G_{dif}) \tag{8}$$

where $F_{unshaded} = (N_C - N_{shaded})/N_C$ is the unshaded fraction of cell C, $F_{shaded} = N_{shaded}/N_C$ is the shaded fraction of cell C, $N_C = 10000$ is the total number of pixels of cell C, $N_{shaded}$ is the number of shaded pixel for cell C, $G_{GHI}$ is the global horizontal irradiance, $G_{dif}$ is the diffuse irradiance at the cell C.

The most shaded cell in each group $N_i$ determines the output current of that group Equation (9):

$$\forall G : I_G = f(\min(G_C | C \in N_i)) \tag{9}$$

*3.1. Different Shading Patterns in This Model*

The performance of the smart PV module needs to be tested under realistic shading conditions. In this study two different shading conditions are considered: (1) Random shadow, which might result from the effect of dust, bird droppings, snow, etc.; and (2) pole shadow, which is caused by a static obstacle during daylight, and which is mostly caused by pole shapes, chimneys, dormers, or a part of the building on the roof. Also, these shading conditions can be combined. In the following both types of mentioned shading conditions and the methods for generating them in the model will be described:

1.  Random shape shading. The characteristics of this shading condition are: (i) probability of occurrence of this shading condition is equal for all surface pixels of the module; (ii) the shape of shading is arbitrary; (iii) random shadows are not necessarily made by solid objects and consequently a transparency factor ($F_{tr}$) is defined as a random function, so that the shadow intensity is randomized; (iv) blur factor ($F_{bl}$) defines how wiped out the shadow borders and edges are. This blur factor is a function of the ratio of diffuse to direct irradiation ($R_{dd}$); (v) shadow intensity is a function of both the transparency factor and irradiance, see Figure 6a,b.
2.  Pole shading with the following characteristics: (i) pole shadow position is moving depending on the time of day, which means that the angle of the pole shadow with the module's x-axis is calculated as a function of time. The length of the pole itself is assumed to be very long so that always the pole shade covers the whole module; (ii) taking into consideration that pole shading occurs only during half of the day due to the sunlight angle, the shape of shading follows the shape of pole; (iii) shading intensity may vary depending on $G_{p,GHI}$ as the pole itself is a solid object; (iv) just as for random shading, the blur factor depends on the ratio of diffuse to direct irradiation, see Figure 6c,d.

The blur factor is a function of the ratio of diffuse to direct irradiation and is determined using a 2D-Gaussian filter, Equation (10):

$$G(x,y) = \frac{1}{2\pi\sigma^2} e^{\frac{-(x^2+y^2)}{2\sigma^2}} \tag{10}$$

where $x$ and $y$ are the distances from the origin in the vertical and horizontal axis, and $\sigma$ is the standard deviation of the Gaussian distribution. In this model $\sigma$ is a function of the ratio of diffuse to direct irradiation: $\sigma = f(R_{dd})$. As an example, the output power for the ten different groups in the module is shown in Figure 6 relative to the power at STC for each iteration: $P_G(i)/P_{STC}$.

*3.2. Different Module Calculations*

Total output power is computed regarding the module architecture and topology. In this section, the total output power for four different module architectures are mathematically formulated and details are discussed. The model simulates the formulated output power for all of the modules in order to allow for comparisons.

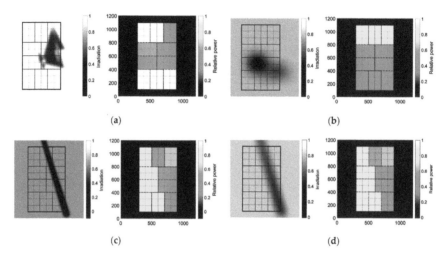

**Figure 6.** Examples of shading pattern, (**a**) Random shading for higher $R_{dd}$, (**b**) Random shading for lower $R_{dd}$, (**c**) Pole shading for higher $R_{dd}$, (**d**) Pole shading for lower $R_{dd}$.

### 3.2.1. Smart Module

As shown in Figure 1, the smart module topology consists of some groups of PV cells connected in series and connected to a DC-DC buck converter. The DC-DC buck converter is used (i) to control the operating point on the MPP using an MPPT algorithm; and (ii) to control the output current flow. The series connection of buck converters on the output-side forces the converters to work with the same current flow in their output. The MPPT on the input-side of the buck converter controls the operating point to be located at MPP. The buck converter levels down the input voltage, so that the output current could be boosted for the shaded groups with the lower current flow, following Equation (11):

$$P_{out} = P_{in} \times \eta_{Conv} \tag{11}$$

where $\eta_{Conv}$ is the converter efficiency. Total output power for the smart module is computed via

$$P_{total\_Smart} = \sum_{k=1}^{N_G} P_{MPP}^k(i) \times \eta_{Conv} \times \eta_{MPPT} \tag{12}$$

where $N_G$ is the number of groups of PV cells, $P_{MPP}^k(i)$ is the maximum group output power $k \in [1, N_G]$ at iteration $i$, and $\eta_{MPPT}$ is the MPPT algorithm efficiency. Note that in this model we assume $\eta_{MPPT}$ to be constant at 95%. The LS-SVM method is used to calculate $P_{MPP}^k(i)$ by assigning $T = \{(x_1, y_1), \ldots, (x_l, y_l)\}$ where $x_j = [G(j)]^T$, $y_j = [V_{MPP}(j), I_{MPP}(j)]$ $j = 1, \ldots, l$.

### 3.2.2. Parallel Strings with Blocking Diodes

The module with parallel-connected strings consists of three parallel strings where each string consists of 20 cells connected in series and ended with a blocking diode, see Figure 7. Normally this topology is implemented for strings of PV modules instead of PV cells. This module is controlled via a central converter with an MPPT algorithm to boost up the voltage level and control the operating point. Therefore, total output power is computed in Equation (13) [27]:

$$P_{total\_parallel} = P_{Module\_parallel\_MPP}(i) \times \eta_{Central\_Conv} \times \eta_{MPPT} \tag{13}$$

where $\eta_{Central\_Conv}$ is the efficiency of the central converter, $\eta_{MPPT}$ is the MPPT algorithm efficiency, and the maximum output power at iteration $i$, $P_{Panel\_parallel\_MPP}(i)$ is calculated in Equation (14):

$$P_{Module\_parallel\_MPP}(i) = \min(V^k(i)) \times \sum_{k=1}^{3} I^k(i) \tag{14}$$

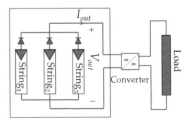

**Figure 7.** Parallel strings with blocking diode.

Let us assume that the same MPPT algorithm is implemented for all PV module architectures, which has the same efficiency $\eta_{MPPT} = 95\%$, as in the smart module architecture. The most appropriate converter topology which can be implemented for this module architecture is the boost converter to level up the voltage level, therefore for this study, a buck-boost converter, LT8390, is supposed to be implemented for both parallel and series connected architectures. Figure 8 depicts a variation of efficiency with respect to the input voltage and the output current (load current) for this very high-efficient converter [28]. The same LS-SVM method as mentioned for the LTM4611 is used to map the exact value of efficiency regarding the optimum module operating point.

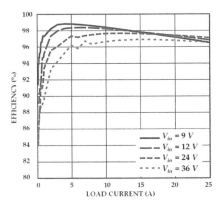

**Figure 8.** Efficiency vs load current and input voltage for LT8390.

### 3.2.3. Standard Module, Series Strings with BPD

The standard module with series-connected strings consists of three series groups of 20 cells, where each group is equipped with one BPD, is shown in Figure 9. Therefore, each group of cells which is shaded would be bypassed via the BPD for preventing cell damage and hot spots and prohibiting the cell to perform as a load instead of a source. Finally, all three groups are connected in series and a central converter may be used to control the operating point. Therefore, total output power is computed using Equation (15):

$$P_{total\_series} = P_{Module\_Series\_MPP}(i) \times \eta_{Central\_Conv} \times \eta_{MPPT} \tag{15}$$

where the maximum output power at iteration $i$, $P_{Module\_MPP}(i)$ is calculated from Equation (16):

$$P_{Module\_Series\_MPP}(i) = \max(I^k(i)) \times \left( \sum_{k=1}^{3} V_{non\_BP}^k(i) + \sum_{k=1}^{3} V_{d\_BP}^k(i) \right) \tag{16}$$

where $V_{non\_BP}^k$ is the voltage of none-bypassed group of cells and $V_{d\_BP}^k$ is the forward voltage of diodes which bypasses the group of cells with lower current.

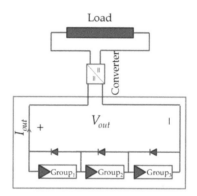

**Figure 9.** Standard Module with three BPDs.

It is assumed that the same MPPT algorithm and the same central converter are used as for the parallel-connected architecture.

3.2.4. The Ideal Module Case Study

Let us assume an ideal module, as reference for comparisons. In this ideal module for each cell a DC-DC converter is responsible to level up the current for shaded cells, thus the drop current because of shading is compensated (Figure 10).

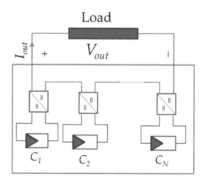

**Figure 10.** The ideal module.

The output power from the module is calculated in Equation (17), which is a summation of extracted power from all cell-converter modules:

$$P_{total\_Ideal} = \sum_{k=1}^{N} \left( P_{MPP}^k(i) \times \eta_{conv} \times \eta_{MPPT} \right) \tag{17}$$

where $P_{MPP}^k(i)$ is the maximum output power from cell $k$ at iteration $i$, $\eta_{conv}$ and $\eta_{MPPT}$ are assumed to be 100% and 95%, respectively, for the ideal module architecture.

## 4. Results and Discussion

The described model is implemented to simulate the behavior of the smart module as well as the other described architectures under different shading patterns. To understand which architecture is more shade-resilient, the harvested energy during a certain period is computed and compared for all architectures. To this end, experimental irradiance data is used as our model input, which is acquired at the Utrecht Photovoltaic Outdoor Test facility (UPOT) at Utrecht University campus in the center of the Netherlands. Irradiation measurements are done using four EKO MS-802 pyranometers (EKO Instruments, Tokyo, Japan), one EKO MS-401 pyranometer and one EKO MS-56 pyrheliometer; the measurement time is dependent on light intensity and varies from 10 milliseconds to 5 seconds. With these facilities, many variables are being measured every day like irradiation, temperature, humidity, etc. [22,29,30].

For this research available data are (i) global irradiation level; (ii) direct irradiation level; and (iii) diffuse irradiation level for four months, i.e., January, March, June and September 2016. The following steps are followed in the analysis:

1.  Figure 11 shows recorded data from UPOT at 7 September 2016. Three different time frames $t_f$ of 15 min in length are chosen to be discussed in this section and are pointed out in the figure.
2.  Generate the shading patterns: following Section 3.1, two types of shadow must be generated depending on obstacles, $R_{dd}$, $F_{bl}$, $F_{tr}$ and $G_{GHI}$. Figures 12–14 show different shading patterns and their effect on groups of PV cells for different architectures. Unlike in Figures 13 and 14, which only have the effect of pole shadow, in Figure 12 a combination of both pole and random shadows is shown. To observe all shading patterns during this day please refer to Figure S1 in the supplementary section.
3.  Analysis of the effect of shading patterns on different architectures and cell groups. In this step the effective irradiation level for each group of cells in different architecture is computed precisely.
4.  Maximum output power at each time frame is calculated using Equations (12)–(17). The output power for three time frames as shown in Figures 12–14 is given in Table 4. It is clearly shown that series connected architecture in time frame 1 performs very weak, that is the effect of BPDs in this architecture. The shade pattern in time frame 1 effects on both current and voltage significantly. The group of cells under much darker shadow are bypassed by BPD and current is very low because of the shading.
5.  Each time frame simulates 15 min of the real world with the assumption of having a constant value of irradiation variables.

**Table 4.** Output power in three time-frames.

| Architecture | Frame 1 | Frame 2 | Frame 3 |
|---|---|---|---|
| Ideal Architecture | 48.35 (W) | 84.23 (W) | 116.54 (W) |
| Smart Architecture | 18.49 (W) | 69.00 (W) | 108.85 (W) |
| Series Connected Architecture | 0.84 (W) | 30.95 (W) | 112.35 (W) |
| Parallel Connected Architecture | 4.51 (W) | 62.97 (W) | 113.42 (W) |

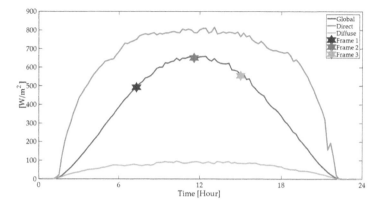

**Figure 11.** Global, Direct and Diffuse irradiation levels on 7 September 2016.

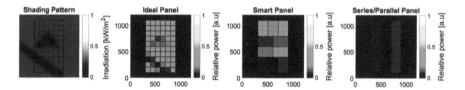

**Figure 12.** Combined pole and random shading pattern and effect of that on different architectures at time frame 1.

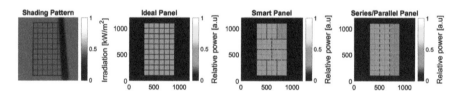

**Figure 13.** Pole shading pattern and effect of that on different architectures at time frame 2.

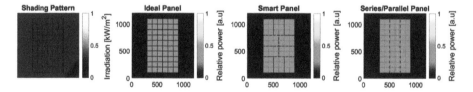

**Figure 14.** Pole shading pattern and effect of that on different architectures at time frame 3. Note that the shade is not cast on the panel.

Figure 15 depicts the output energy from different module architectures for different months of the year 2016. The output energy in January is the lowest compared to the other three months. In all three other months, it is clearly shown that the ideal module outperforms all other modules, which is the effect of both the architecture and the 100% and 95% efficiency that is considered for the converter and MPPT algorithm, respectively. For all three months of March, June, and September the second-best performing module is the smart module, followed by the parallel connected and the series connected module. Generally, the drawback of the parallel-connected module is its very low voltage compared to the series-connected module. For designing a practical PV system voltage levels need to be boosted up with a central DC-DC converter and then be controlled to be compatible with the load specifications. In contrast, the series-connected module, which performs worst of all architectures, does not need the boost up the level between load, which thus makes the whole system design easier and more cost efficient.

Figure 16 shows the ratio of output energy from the modules with respect to the output from the ideal module, Equation (18):

$$R_E(\%) = E_m / E_{ideal} \times 100, \ m \in (\text{smart, series connected, parallel connected}) \qquad (18)$$

Excluding January and the days with very low output energy the smart module performs much better compared to the rest of architectures excluding the ideal module. On the other hand, the series connected module for most of the time generates the lowest amount of energy.

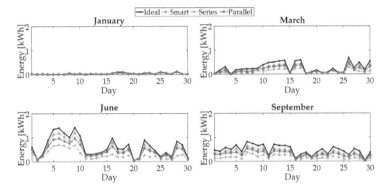

**Figure 15.** Harvested energy at four different months of the year 2016.

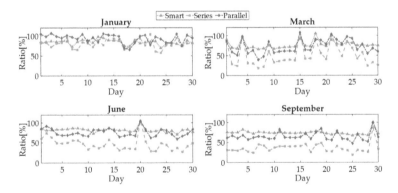

**Figure 16.** The ratio of harvested energy of different architecture compare to ideal module at four different months of the year 2016.

The summation of output energy and the average $R_E(\%)$ for the whole year 2016 are depicted in Figure 17. To sum up, the performance of the smart module outperforms all other architectures, except for the ideal module, for all months excluding winter time when the output energy in all types of architecture is almost zero. However, the series connected module as the most ordinary architecture implemented nowadays by most of the manufacturers is performing worst.

Figure 18 shows the ratio between total harvested energy from smart module, series and connected modules compared to the ideal module. It shows that the smart module harvested almost 79.5% of the energy that the ideal module harvests; the series connected harvested 42.2% and parallel connected yield 68.8% of total module capacity under the same shading patterns.

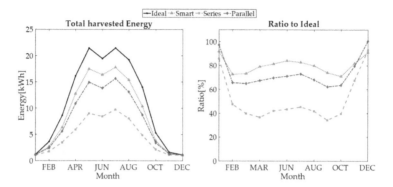

**Figure 17.** Perspective of total harvested energy during different months in 2016.

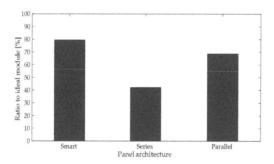

**Figure 18.** Ratio of total harvested energy from smart module, series and connected modules compared to the ideal module under shading patterns.

The method discussed and improved in this study is based on the fact that even small amounts of power which can be produced by cells should be harvested. In other feasible architectures, series and parallel, there are always some energy losses due to the electrical connections. In parallel connection, as the lowest voltage group always determines the module output voltage a fraction of power is lost. As shown in Figure 18 this energy loss is almost 31.2% of total capacity. For series connection, once the BPDs (used for safety reasons), are forward biased in anti-parallel position with shaded groups of cells those groups are bypassed and the module output voltage is decrease. This behavior of BPDs wastes some energy, and this loss is more than half of total capacity: 57.8%. For the smart module, although there is some loss in the electronic circuits due to converter efficiency the final loss compared to the other architectures is reasonably low. According to the results in Figure 18 only 20.5% of energy is lost compared to the ideal module. More importantly, the smart PV module performed 47% and 13.4% better, compared to the series connected and the parallel connected architectures, respectively. The energy loss for the smart module strongly depends on the following factors: (i) optimization of the grouping (number of cells in each group); (ii) converter choosing; and (iii) load current. It should be noted that the actual power rating of the modules, i.e., watt-peak determined under standard test conditions of 1000 W/m$^2$ actually is the same (or very similar) for all three architectures. For unshaded conditions throughout the year, also the energy rating (kWh/kWp) would be the same (or similar).

However, for shading conditions, the harvested amount of energy clearly differs. In fact, it could be recommended to develop new standards to test modules under standardized shading conditions.

The addition of electronic elements in a module will lead to additional cost, which should be offset with additional energy harvested under shading conditions. Let us consider the tradeoff between cost and harvested energy comparing both ordinary and smart modules. As the most commonly-used architecture in the market is the series connected architecture, we compare the series connected module with the proposed smart module. According to [31], a public awareness agency funded by the Dutch government, the price of buying a PV system is 1.5 €/W of which panel cost is about half. Thus, a typical panel of 220 W would costs about €170. For transforming an ordinary panel to a smart one we need to purchase 10 micro converters, a microprocessor, and some electronics elements. In this study the designed micro converter is LTM4611, which costs €16, the microprocessor is about €25 and the rest of elements costs roughly €35. To sum up, the final expenditure for a smart panel is about 1.66 times more than an ordinary panel, but as Figure 18 shows under the shading patterns, the annual harvested energy from smart module is 1.89 times more than the series connected. Thus, the economic payback time for a smart module compared to an ordinary series connected module (in shading conditions) is shorter. Note that cost of electronic components are based here on purchasing small amounts. It can be expected that large volume purchases lead to substantial limited additional costs for smart module architectures.

Finally, besides the increased resilience to shading effects, it can be expected that the occurrence of hot spots in smart modules will be limited as well. In a further experimental study, we will investigate that using Infrared (IR) thermography.

## 5. Conclusions

To mitigate partial shading effects on the performance of PV modules, a generic model is developed that is able to evaluate smart PV module architectures. In this paper the proposed architecture consisted of a number of PV cell groups, and a DC-DC buck converter for a group of PV cells is implemented to amplify the current of shaded groups. This converter was chosen after investigation of characteristics of appropriate micro-converters in the market with respect to the PV cell specifications. This resulted in the choice for the Linear Technology LTM4611 converter. The LS-SVM method was used (i) to generalize the behavior of the converter efficiency; and (ii) to optimize the group size of PV cells. In summary, the optimum grouping for the designed specifications was 10 groups of 6 cells.

After the smart module was designed, the effect of shading patterns was studied. A model for shading patterns was developed with two types of random and pole shadows based on actual measured irradiation data. Simulations demonstrated that the average amount of generated energy of the smart architecture was almost 79.5% of the energy generated by the ideal PV module. Compared with series connected and parallel connected architectures, the smart PV module performed 47% and 13.4% better, respectively. Moreover, the smart module economic payback time compared with the most commonly-used architecture in the market is expected to be shorter.

**Supplementary Materials:** The following are available online at www.mdpi.com/1996-1073/11/1/250/s1. Figure S1: Shading pattern and different architecture behaviors during the day 7 September 2016.

**Acknowledgments:** The authors gratefully acknowledge fruitful discussions with Rudi Jonkman and Robert van der Sanden (Heliox), Lenneke Sloof (ECN), and Hamed Yousefi Mesri (UMC). This work is partly financially supported by the Netherlands Enterprise Agency (RVO) within the framework of the Dutch Topsector Energy (project Scalable Shade Tolerant Modules, SSTM).

**Author Contributions:** S.Z.M.G. and A.C.d.W. conceived and designed the simulations, performed the experiments, and analyzed the data; S.Z.M.G. and W.G.J.H.M.v.S. wrote the paper; W.G.J.H.M.v.S. conceived the project.

**Conflicts of Interest:** The authors declare no conflict of interest. The funding organization had no role in the design of the study; in the collection, analyses, or interpretation of data; in the writing of the manuscript, and in the decision to publish the results.

## Nomenclature

| | |
|---|---|
| AM | Air Mass |
| BPD | Bypass diode |
| GM | Global maximum |
| Inc. Cond | Incremental conductance |
| IR | Infrared |
| LS-SVM | Least square support vector machine |
| MPPT | Maximum power point tracker |
| P&O | Perturb and observe |
| PS | Partial shading |
| PV | Photovoltaic |
| P-V | Power-Voltage |
| RBF | Radial base function |
| RES | Renewable energy sources |
| SP | Series parallel |
| STC | Standard test condition |
| TCT | Total-cross-tied |
| UPOT | Utrecht photovoltaic outdoor test facility |
| $C$ | Cell |
| $E_m$ | Harvested energy from module $m$ (kWh) |
| $F_{bl}$ | Blur factor |
| $F_{shaded}$ | Shaded fraction of cells |
| $F_{tr}$ | Transparency factor |
| $F_{unshaded}$ | Unshaded fraction of cells |
| $G_{GHI}$ | Global horizontal irradiation (W/m$^2$) |
| $Gp$ | Irradiation level of pixel p (W/m$^2$) |
| $Gs$ | Irradiation level under shadow (W/m$^2$) |
| $I_{in}$ | Converter input current (A) |
| $I_G$ | Output current of group G (A) |
| $I_{MPP\_STC}$ | Current at Maximum power point in STC (A) |
| $I_{SC}$ | Short circuit current (A) |
| $K$ | Kernel function |
| $\eta$ | Efficiency (%) |
| $N_C$ | Total number of pixels |
| $N_G$ | Number of groups of cells |
| $\eta_{Conv}$ | Efficiency of converter (%) |
| $n_g$ | Number of cells in a group |
| $\eta_{MPPT}$ | Efficiency of MPPT algorithm (%) |
| $N_{shaded}$ | Total number of shaded cells |
| $p$ | Pixel |
| $P_{max}$ | Maximum output power (W) |
| $P_{MPP}$ | Output power at maximum power point (W) |
| $P_{total}$ | Total output power (W) |
| $R_{dd}$ | Ratio of diffuse to direct irradiation (%) |
| $R_E$ | Ratio of output energy to ideal module (%) |
| $V_{d\_BP}$ | Forward voltage of diode which bypassed group of cells (V) |
| $V_{in}$ | Converter input voltage (V) |
| $V_{non\_BP}$ | Voltage of group of cells which are not bypassed (V) |
| $V_{OC}$ | Open circuit voltage (V) |
| $V_{out}$ | Output voltage (V) |
| $\sigma$ | Standard deviation |

## References

1.  Rahim, N.A.; Mekhilef, S. Implementation of three-phase grid connected inverter for photovoltaic solar power generation system. In Proceedings of the International Conference on Power System Technology, PowerCon 2002, Kunming, China, 13–17 October 2002; Volume 1, pp. 570–573.
2.  Elsaharty, M.A.; Ashour, H.A.; Rakhshani, E.; Pouresmaeil, E.; Catalão, J.P.S. A Novel DC-Bus Sensor-less MPPT Technique for Single-Stage PV Grid-Connected Inverters. *Energies* **2016**, *9*, 248. [CrossRef]
3.  Reinders, A.; van Sark, W.; Verlinden, P. Introduction. In *Photovoltaic Solar Energy*; John Wiley & Sons, Ltd.: Chichester, UK, 2017; pp. 1–12.
4.  IEA-PVPS. *Snapshot of Global Photovoltaic Markets*; International Energy Agency (IEA): Paris, France, 2016.
5.  Kreamer, N. *SEIA Annual Report 10 Compelling Stories about Solar*; Solar Energy Industries Association: Washington, DC, USA, 2016.
6.  Salas, V.; Olias, E.; Barrado, A.; Lazaro, A. Review of the maximum power point tracking algorithms for stand-alone photovoltaic systems. *Sol. Energy Mater. Sol. Cells* **2006**, *90*, 1555–1578. [CrossRef]
7.  Papathanassiou, S.A. Energy models for photovoltaic systems under partial shading conditions: A comprehensive review. *IET Renew. Power Gener.* **2015**, *9*, 340–349.
8.  Jeyaprabha, S.B.; Selvakumar, A.I. Model-Based MPPT for Shaded and Mismatched Modules of Photovoltaic Farm. *IEEE Trans. Sustain. Energy* **2017**, *8*, 1763–1771. [CrossRef]
9.  Bidram, A.; Davoudi, A.; Balog, R.S. Control and Circuit Techniques to Mitigate Partial Shading Effects in Photovoltaic Arrays. *IEEE J. Photovolt.* **2012**, *2*, 532–546. [CrossRef]
10. Mirbagheri, S.Z.; Aldeen, M.; Saha, S. A PSO-based MPPT re-initialised by incremental conductance method for a standalone PV system. In Proceedings of the 2015 23th Mediterranean Conference on Control and Automation (MED), Torremolinos, Spain, 16–19 June 2015; pp. 298–303.
11. Esram, T.; Chapman, P.L. Comparison of photovoltaic array maximum power point tracking techniques. *IEEE Trans. Energy Convers.* **2007**, *22*, 439–449. [CrossRef]
12. Simoes, M.G.; Franceschetti, N.N.; Friedhofer, M. A fuzzy logic based photovoltaic peak power tracking control. In Proceedings of the IEEE International Symposium on Industrial Electronics, ISIE '98, Pretoria, South Africa, 7–10 July 1998; Volume 1, pp. 300–305.
13. Hiyama, T.; Kouzuma, S.; Imakubo, T. Identification of optimal operating point of PV modules using neural network for real time maximum power tracking control. *IEEE Trans. Energy Convers.* **1995**, *10*, 360–367. [CrossRef]
14. Liu, Y.H.; Huang, S.C.; Huang, J.W.; Liang, W.C. A Particle Swarm Optimization-Based Maximum Power Point Tracking Algorithm for PV Systems Operating Under Partially Shaded Conditions. *IEEE Trans. Energy Convers.* **2012**, *27*, 1027–1035. [CrossRef]
15. Mirhassani, S.M.; Golroodbari, S.Z.M.; Golroodbari, S.M.M.; Mekhilef, S. An improved particle swarm optimization based maximum power point tracking strategy with variable sampling time. *Int. J. Electr. Power Energy Syst.* **2015**, *64*, 761–770. [CrossRef]
16. Karatepe, E.; Hiyama, T. Simple and high-efficiency photovoltaic system under non-uniform operating conditions. *IET Renew. Power Gener.* **2010**, *4*, 354–368. [CrossRef]
17. Serna-Garcés, S.; Bastidas-Rodríguez, J.; Ramos-Paja, C. Reconfiguration of Urban Photovoltaic Arrays Using Commercial Devices. *Energies* **2016**, *9*, 2. [CrossRef]
18. Pannebakker, B.B.; de Waal, A.C.; van Sark, W.G.J.H.M. Photovoltaics in the shade: One bypass diode per solar cell revisited. *Prog. Photovolt. Res. Appl.* **2017**, *25*, 836–849. [CrossRef]
19. Olalla, C.; Clement, D.; Rodriguez, M.; Maksimovic, D. Architectures and Control of Submodule Integrated DC–DC Converters for Photovoltaic Applications. *IEEE Trans. Power Electron.* **2013**, *28*, 2980–2997. [CrossRef]
20. Schmidt, H.; Rogalla, S.; Goeldi, B.; Burger, B. Module Integrated Electronics—An Overview. In Proceedings of the 25th European Photovoltaic Solar Energy Conference and Exhibition/5th World Conference on Photovoltaic Energy Conversion, Valencia, Spain, 6–10 September 2010; pp. 3700–3707.
21. Uno, M.; Kukita, A. Current Sensorless Equalization Strategy for a Single-Switch Voltage Equalizer Using Multistacked Buck–Boost Converters for Photovoltaic Modules Under Partial Shading. *IEEE Trans. Ind. Appl.* **2017**, *53*, 420–429. [CrossRef]

22. Van Sark, W.G.J.H.M.; Louwen, A.; de Waal, A.C.; Schropp, R.E.I. UPOT: The Utrecht Photovoltaic Outdoor Test Facility. In Proceedings of the 27th European Photovoltaic Solar Energy Conference and Exhibition, Frankfurt, Germany, 24–28 September 2012; pp. 3247–3249.

23. Burges, C.J.C. A tutorial on support vector machines for pattern recognition. *Data Min. Knowl. Discov.* **1998**, *2*, 121–167. [CrossRef]

24. Linear Technology. *LTM4611/Typical Application Ultralow VIN, 15A DC/DC μModule Regulator*; Linear Technology: Milpitas, CA, USA, 2017.

25. Suykens, J.A.; Van Gestel, T.; De Brabanter, J.; De Moor, B.; Vandewalle, J. *Least Squares Support Vector Machines*; World Scientific: River Edge, NJ, USA, 2002; Volume 4.

26. Sinapis, K.; Tzikas, C.; Litjens, G.; Van den Donker, M.; Folkerts, W.; van Sark, W.G.J.H.M.; Smets, A. A comprehensive study on partial shading response of c-Si modules and yield modeling of string inverter and module level power electronics. *Sol. Energy* **2016**, *135*, 731–741. [CrossRef]

27. Mäki, A.; Valkealahti, S. Power Losses in Long String and Parallel-Connected Short Strings of Series-Connected Silicon-Based Photovoltaic Modules Due to Partial Shading Conditions. *IEEE Trans. Energy Convers.* **2012**, *27*, 173–183. [CrossRef]

28. Linear Technology. *60V Synchronous 4-Switch Buck-Boost Controller with Spread Spectrum 98% Efficient 48W (12V 4A)*; Linear Technology: Milpitas, CA, USA, 2017.

29. UPOT—System Layout. Available online: http://upot.nl/system.html (accessed on 7 November 2017).

30. Louwen, A.; de Waal, A.C.; Schropp, R.E.I.; Faaij, A.P.C.; van Sark, W.G.J.H.M. Comprehensive characterisation and analysis of PV module performance under real operating conditions. *Prog. Photovolt. Res. Appl.* **2017**, *25*, 218–232. [CrossRef]

31. Milieu Centraal, Prijs en Opbrengst Zonnepanelen—MilieuCentraal. Available online: https://www.milieucentraal.nl/energie-besparen/zonnepanelen/zonnepanelen-kopen/kosten-en-opbrengst-zonnepanelen/ (accessed on 10 January 2018).

Article

# Organic Soiling: The Role of Pollen in PV Module Performance Degradation

Ricardo Conceição [1,2] , Hugo G. Silva [1,2,*] , José Mirão [3] and Manuel Collares-Pereira [1,2]

[1]  Renewable Energies Chair, University of Evora, 7002-554 Evora, Portugal; rfc@uevora.pt (R.C.);
     collarespereira@uevora.pt (M.C.-P.)
[2]  Institute of Earth Sciences, University of Evora, 7000-671 Evora, Portugal
[3]  Hercules Laboratory, University of Evora, 7000-089 Evora, Portugal
*    Correspondence: hgsilva@uevora.pt; Tel.: +351-967-480-736

Received: 14 December 2017; Accepted: 16 January 2018; Published: 26 January 2018

**Abstract:** Soiling is a problem for solar energy harvesting technologies, such as in photovoltaic modules technologies. This paper describes not only one complete year of Soiling Ratioindex and rates measured in a rural environment of Southern Europe, but also focuses on the seasonal variation of the type of soiling, mainly spring and summer. The Soiling Ratio index is calculated based on the maximum power output and short circuit current of two photovoltaic (PV) panels, along with Scanning Electron Microscopy and Energy Dispersive X-Ray of glass samples to provide visual and chemical inspection of the type of soiling. Mass accumulation on glass samples mounted on a "glass tree" was weekly measured with a microbalance and related with the Soiling Ratio metrics. Soiling rates were calculated to infer the degree of soiling for each season and the respective comparison made. Results show a soiling rate of 4.1%/month in April (spring), 1.9%/month in July (summer) and 1.6%/month in September (fall). Rain (the main natural cleaning agent of the photovoltaic modules) as well as aerosol optical depth (proxy for atmospheric particle concentration) were correlated with the Soiling Ratio. In-depth analysis on the type of organic soiling was performed.

**Keywords:** solar energy; photovoltaic module performance; organic soiling; Scanning Electron Microscopy (SEM)

## 1. Introduction

Soiling on solar harvesting technologies, namely in photovoltaic systems (PV), has been intensively studied in the past few decades since it induces severe performance losses on such systems by reducing the incoming radiation, through reflection, scattering and absorption [1–3]. This implies that the systems do not work at their fullest capability [4]. As a consequence, frequent cleaning of the systems is required, which represents an important slice of the kWh cost of the electricity being produced. Moreover, soiling is not only a local [5], but also a seasonal phenomenon and for that reason, if operational costs are to be reduced, a proper characterization of it is needed. This has been done extensively in the literature, but mostly focused on desert regions [6–8], which have high irradiance values and significant problems with dust deposition, namely mineral and not organic.

However, everyday, new PV plants are being deployed around the world and the tendency is to have more of them in the future. Naturally, PV also suffers from soiling, not only from local sources but also transported from remote ones [5]. One country in Europe that is certainly going to increase its use of solar technologies is Portugal, due to its high irradiance availability [9]; its southern region has, on average, an annual global horizontal irradiation of 1800 kWh/m$^2$ [10]. With this in mind, studies of soiling in this region are very important for future plants.

A first study of soiling in the region of Alentejo, Portugal, has already been published [5] and shows how Saharan dust long-range transport [11] can decrease the performance of PV systems.

However, Saharan dust is not the only source of soiling that impacts PV [12]. There is one particular season of the year, spring, where organic material, such as pollen [13], can be a problem regarding soiling. This paper describes the seasonal variation in the type of particles that adhere to the relevant surfaces from winter to summer. It also describes how rain (which is the main natural cleaning agent [14]) and aerosols, affect the Soiling Ratio index (SR) in this location (Évora—38°34'0.01" N; 7°54'0.00" W). To identify the type of deposited particles, Scanning Electron Microscopy (SEM) and Energy Dispersive Spectroscopy (EDS) were performed. For organic particle deposition, EDS cannot be used since the composition is mainly carbon based and no conclusions can be taken. For this purpose, the Portuguese Society of Aerobiology (Sociedade Portuguesa de Aerobiologia), was able to identify some of the organic material deposited using the SEM images obtained.

Environmental variables, like rain and aerosol optical depth (AOD), contribute to explaining the observed SR during the period of the measuring campaign, November 2016 to October 2017, showing how both variables shape the evolution of the SR metrics. The authors consider these variables of vital importance for soiling deposition and removal, since rain is the main natural cleaning agent and AOD is used as a proxy for particle concentration in the atmosphere. Note also that the absence of rain allows for particles to build up on top of surfaces, increasing the soiling and consequently decreasing PV module performance.

Although most soiling work considers dust to be the main soiling agent, it is restricted to mineral particles such as desert dust [15,16], this paper instead focuses mainly on the role of organic soiling on the PV module performance. Measurements were done in Alentejo, an ideal location for this type of study since this is the region with the highest pollen concentration in Portugal [17]. Pollen concentration is expected to increase in the next few decades in Europe [18], and that may imply more organic soiling in the future; thus, this is one more reason to study its impact on PV. In addition, studies of organic soiling are very scarce, making it an excellent study opportunity. An annual soiling analysis for the measurement location is also reported for the first time, as well as the soiling rates that develop during periods without rain, which are then related to the environment, through AOD, SEM imagery and image processing.

The structure of the paper is as follows: in Section 2 the experiment is explained, as well as the equipments and methodologies used to obtain the data. In Section 3 an analysis of the annual soiling, as well as soiling rates, is performed. The analysis considers rain and AOD, as the main variables to explain soiling behaviour. The effect of organic material in soiling is also highlighted in comparison to mineral material. In Section 4 the organic material deposited at the measuring location (through SEM images) is characterized in detail. Section 5 draws some conclusions.

## 2. Methodology

Mass accumulation measurements, as well as the PV performance measurements, took place in a rural location in the outskirts of Évora, in Alentejo (southern Portugal) at the Solar Colector Test Plataform (Plataforma de Ensaios de Coletores Solares, PECS) facilities from the Renewable Energies Chair (Catedra Energias Renovaveis, CER), University of Évora.

### 2.1. Mass Accumulation Experiment

A glass tree, see Figure 1, was used. Inspired by [19], it is composed of 25 glass samples, with 6 per cardinal direction, in 15° inclination steps, with one completely horizontal, as shown in Figure 1. The glass samples are from Interfloat Corporation (Ruggel, Liechtenstein), model SINA (high solar transmittance and fine micropatterned glass suitable for flat-plate solar collectors and photovoltaic modules), with 11 by 9 cm (length × width) and 3.2 mm thickness. When possible, weekly mass measurements were performed on each of the 25 samples at PECS; however, aspects such as intense rain or vacations may introduce longer periods between measurements. Monthly Scanning Electron Microscopy (SEM) and Energy Dispersive X-ray Spectroscopy (EDS) measurements were carried out on the completely horizontal sample, except for in the month of August.

**Figure 1.** Glass tree apparatus in Solar Colector Test Plataform (PECS).

Each glass sample weight was measured with a Bosch SAE 80/200 microbalance (Germany), model SAE 80/200. Mass accumulation on week $t$, $m_a(t)$, is determined by subtracting from the measured mass at that week, mass $(t)$, the initial mass of the clean glass, mass $(0)$ as in Equation (1):

$$m_a(t) = \text{mass}(t) - \text{mass}(0). \tag{1}$$

No cleaning is done to the glasses during the experiment and only environmental action (e.g., rain, dew) can act towards reducing the mass accumulated on the samples.

*2.2. PV Experiment*

The PV system used, see Figure 2, is based on the methodology developed in [20], all measurements performed manually, with an I–V curve tracer from Metrel (Horjul, Slovenia), model Eurotest PV Lite MI 3109. Another difference is that instead of calculating the mean of the PV parameters for a static position along the day, the mean of three measurements is obtained with the normal to the PV pointing towards the sun, only in clear sky days and near solar noon. However, when not being measured, the PV panels were facing south with a tilt angle of around 30°, a value close to the optimum tilt angle value for the highest annual performance in this region. The PV flat panels are mc-Si of the same model, FTS-220P, manufactured by Fluitecnik (Madrid, Spain). The clean reference PV panel is always cleaned before the measurements and the other one left to the effect of natural soiling, without any artificial cleaning.

**Figure 2.** PV apparatus in PECS.

The Soiling Ratio index (SR) can be calculated using two different metrics: based of the short-circuit current ($I_{SC}$), which is denominated ($SR_{I_{SC}}$) and the maximum power output ($P_{max}$), designated ($SR_{P_{max}}$)of the two PV panels. The main difference noted in [20] is the fact that when soiling is homogeneous, both metrics give similar results, but when the soiling is not homogeneous, calculating the soiling ratio based on the short circuit current can give either an underestimated or overestimated result, compared to what was actually lost in power output. This method works by comparing what is being currently measured, short-circuit or maximum power output, at some environment conditions with the value converted from Standard Test Conditions (STC), in a clean state, to those exact same conditions. In mathematical terms, $SR_{I_{SC}}$ and $SR_{Pmax}$ are calculated through Equations (2)–(4):

$$SR_{I_{sc}} = \frac{I_{sc}^{soil}}{I_{sc,0}^{soil}[1 + \alpha(T^{soil} - T_0)](G/G_0)},$$ (2)

where $I_{sc}^{soil}$ is the short-circuit current of the dirty PV panel, $I_{sc,0}^{soil}$ is the short-circuit current at Standard Test Conditions (STC) of the soiled panel, $\alpha$ is the short-circuit temperature correction coefficient, $T^{soil}$ is the cell temperature of the soiled panel, $T_0$ is the temperature at reference condition (25 °C), $G$ is the irradiance in the PV plane and $G_0$ the irradiance at STC conditions (1000 W/m²).

$$SR_{P_{max}} = \frac{P_{max}^{soil}}{P_{max,0}^{soil}[1 + \gamma(T^{soil} - T_0)](G/G_0)},$$ (3)

where $P_{max}^{soil}$ is the maximum power of the dirty PV panel, $P_{max,0}^{soil}$ is the maximum power at clean condition and $\gamma$ is the maximum power temperature correction coefficient. For the calculation of the irradiance in the PV plane, the clean module is used:

$$G = G_0 \frac{I_{sc}^{clean}[1 - \alpha(T^{clean} - T_0)]}{I_{sc,0}^{clean}}, \tag{4}$$

where $I_{sc}^{clean}$ is the short-circuit current of the clean PV panel, $I_{sc,0}^{clean}$ is the short-circuit current at STC of the clean panel, $\alpha$ is the short-circuit temperature correction coefficient and $T^{clean}$ is the cell temperature of the cleaned panel. As stated in [21], soiling ratios may be measured with absolute uncertainties on the order of $\pm1\%$ or better on an absolute basis, under appropriate conditions. Measurements should be restricted to the middle portion of the day to exclude high uncertainties from morning and evening hours related to lower signal amplitudes and the effects of angular alignment differences, preferably averaging data for equal periods surrounding solar noon. As stated before, it is highlighted here that the measurements were only performed at clear sky conditions (to avoid any effects on the irradiance perceived by both solar panels due to passing clouds), near to solar noon and with the panels normal to the sun. This ensures that the experiment is performed under circumstances for which the uncertainty is the lowest.

## 3. Annual Soiling Characterization

An annual series of soiling data for the region under study is reported here for the first time, with data ranging from 1 November 2016 to 31 October 2017, as presented in Figure 3. Soiling rates, $S_R \approx d(SR)/dt$, between periods without rain are shown as red dashed lines in this figure. To explain the observed soiling ratio and the respective soiling rates, rain is also added to the study, see Figure 4a and aerosol optical depth data, see Figure 4b. Rain as the main cleaning agent is very important, as well as the AOD, proxy of particle concentration in the atmosphere. Details on this analysis can be found in [5]. There are five evident cases in Figure 3: two long-range desert dust transportation events, one in February (denoted in Figure 3 as F. Event) and other in March (denoted as M. Event), with both events documented in [5], and three periods where the lack of rain, denoted as SPR (meaning spring), $S_1$ and $S_2$, which led to a linear decrease in both SR metrics, marked with the red dashed line.

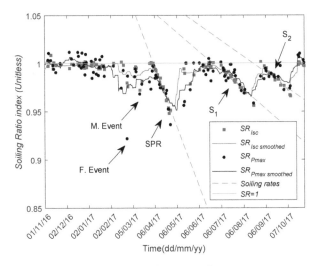

**Figure 3.** Soiling ratio and rates.

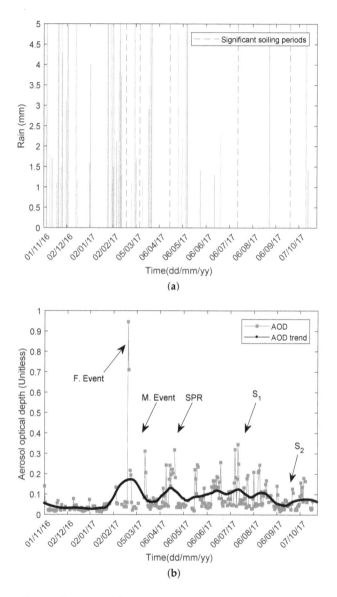

**Figure 4.** Environmental measurements: (**a**) rain; (**b**) aerosol optical depth.

During fall and winter, because of frequent rain, soiling cannot accumulate in significant amounts on the glass surfaces, yielding SR metrics close to 1 as expected. The first major soiling event was in February, the F. Event, where a major quantity of Saharan dust was transported to southern Europe, mainly Portugal and Spain [5], leading to a decrease of $\approx$8% in ($SR_{P_{max}}$). The second one was in March, the M. Event, although with less intensity led to a decrease of $\approx$3% in both SR metrics. During spring, mainly April, denoted as SPR, there was an absence in rain, which allowed for organic material (confirmed below with complementary measurements to be mostly pollen) to deposit on the PV glass surfaces, leading to a decrease of 4.1% in both metrics, see Table 1. Note that since the soiling rates are per month, that also means that at the end of the month the losses are the same as the loss rates, since at the beginning of every event the panels were clean. However, heavy rain in the first part of May, was able to remove the accumulated soiling. In June, maintenance was done to the PV tracker and because of that the PV panels were cleaned, which resulted in two weeks without data coinciding with some rain. During July, the period denoted by $S_1$, due to the scarce rain, soiling was able to settle, leading to a decrease of 1.9%. In August (vacation period), there were fewer data points at the end of the month. There was then a rainy day, which cleaned the PV panels, returning both metrics close to one. Due to the continued lack of rain during September, particles accumulated over the surface, leading to a decrease of 1.8%. After a heavy rain period in the beginning of October, the SR metrics recovered near to unity.

**Table 1.** Soiling rates and respective statistical indicators.

| Indicators | SPR | $S_1$ | $S_2$ |
|---|---|---|---|
| $S_R$ (%/month) | 4.1 | 1.9 | 1.6 |
| $r^2$ | 0.97 | 0.97 | 0.94 |
| RMSE | 0.0013 | 0.0007 | 0.0008 |

Analysis shows that, the February dust event is responsible for the highest loss registered; however, due to an increase of organic material, spring reveals itself as the second most important soiling case. Moderate March dust can be rated third, hand in hand with the soiling rates developed during summer.

The SR metrics decay was smoothed due to the noisy data derived from the reduced amount of data points [20]. In statistical terms, the values of $r^2$, for the linear decay of the SR metrics are presented on Table 1; they are all above 0.9, which represent very high correlation values, while Root Mean Square Error (RMSE) values are always bellow 0.002. This fact ensures that not only are the soiling rates close to being linear, but they also enhance confidence in the results obtained for the Soiling Ratio.

Note that in Figure 4b, AOD data is in red and the smoothed data in black, which serves the purpose of being able to visualize the trends in AOD. It can be seen that the highest AOD was in February, due to the large dust event, while lower values are in March when the event with less intensity took place. After that, a peak in April can be seen, which is due to the increase in organic material in the atmosphere, with a more or less constant trend during the summer, until it starts to decay from September on. However, as stated before, due to rain in May, maintenance in June, lack of data and high rain at the end of August, soiling was only able to develop during summer. If these exceptions are ignored, it can be concluded that the soiling rates follow the same trend as the AOD and lack of rain. As a consequence it means that the AOD is working as particle concentration proxy. Higher AOD values, means there are more particles in the atmosphere, which increases the probability of a particle to be deposited on the surface, leading to higher PV performance losses. For further proof, see Figure 5, where $m_a$ represents the mass accumulation on the 30° tilted glass sample. Note that fast dust events are not well represented by this particular glass, probably probably due to the glass tree architecture. However, for periods without rain and continuous particle deposition it relates very well to the AOD and SR metrics.

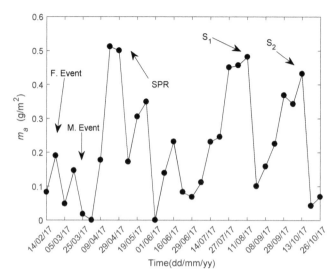

**Figure 5.** Mass accumulation on 30° tilted glass sample.

It can be seen that the mass accumulation is higher during April, then July and lastly during September. This corroborates not only the soiling rates results, but also the use of AOD to explain such phenomena. Note again that, due to maintenance, the glasses had to be clean in the beginning of June, which is the reason why the mass accumulation is zero in that period.

As seen in the data of Figure 3 and Table 1, if the rain is scarce during the spring, the organic material (namely pollen as shown below) may induce very high soiling levels, even higher than mineral material. In the next section a descriptive analysis is given on the type of soiling found in SEM images from April and July.

### 4. Organic Material Characterization and Features

Organic material has been out of the scope of most of the known soiling studies, since they are absent or in very low quantities in desert areas, where most studies were and are still being done. However, Portugal is one of the places in Europe, which is not only affected by desert dust, e.g., Saharan long dust transport, but also by high amounts of pollen, mostly during spring. More specifically, the region of Alentejo is the one with the highest pollen concentration in Portugal as in [17]. The type of pollen in higher concentrations belong to the family of *Gramineae, Oleaceae, Fagaceae, Pinaceae, Cupressaceae, Platanaeae,* and *Urticaceae,* according to [22]. It is known that the locations far from the ocean coast and situated more to the south of the country are the ones with more pollen concentration. From [22] it is also known that pollen counts can roughly go from 3000 grains/m$^3$ in January to 40,000 grains/m$^3$ in May. Assuming that most of pollen species follow the trend of low values at night to high values towards the afternoon [23], there is a potential problem, since dew that forms during the night can hold pollen during the beginning of the day, when it had not yet evaporated. However, there are some types of pollen [24] that have their maximum concentration during the night, which, following the same logic, can also cause higher deposition in PV panels, if they get trapped by the dew water.

Also from 2007 to 2009, for some of the pollen species referred to before, total pollen count was calculated, in the city of Lisbon [25] and normalized, which can be seen in Figure 6. It is assumed that the trends in Évora and Lisbon are similar, mainly because they are not far apart (around 100 km). However, Évora should have substantially higher pollen count values. It can be seen that, from January to April/May, the tendency is to have an increasing amount of pollen, which then starts to decay until the end of the year. The fact that April has the highest concentration of pollen, together with the lack of rain can lead to a harsh decrease in the PV performance, corroborates what was seen before, Figure 3.

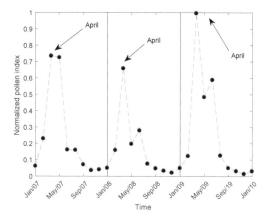

**Figure 6.** Annual pollen trends in Lisbon (Portugal) from 2007 to 2009.

In spite of the already high pollen concentration, the prediction for the future is that the concentration of some species of pollen will increase, see [26]. This could make pollen (as a soiling agent) even more important in the future. This is reinforced by NASA Earth Exchange Global Daily Downscaled Projections for 2100, predicting less frequent rain for Portugal. If it indeed becomes a reality, it may not only be possible that soiling will have more time to build up on surfaces, but also the amount could be higher (mainly during the spring), which will eventually result in higher losses than the ones shown here.

*SEM Characterization and Image Processing*

In order to have a deeper insight into what kind of organic and mineral material gets deposited on PV surface, SEM and EDS measurements were made to the central glass sample (with zero tilt angle) on the glass tree shown in Figure 1. The organic material found in the SEM was mainly pollen, which is shown in Figure 7.

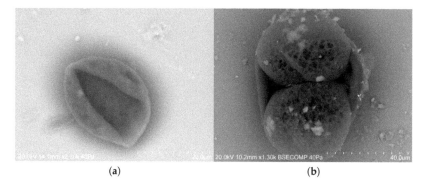

(a)                                            (b)

**Figure 7.** *Cont.*

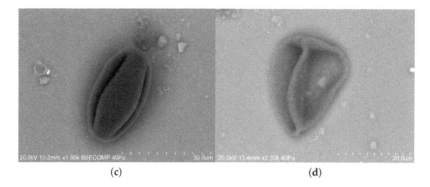

(c)                                          (d)

**Figure 7.** Pollen species from glass sample SEM: (**a**) *Arecaceae*; (**b**) *Pinaceae*; (**c**) *Quercus sp*; (**d**) *Poaceae*.

Different species were found as the *Arecaceae* in Figure 7a, *Pinaceae* in Figure 7b, *Pinaceae* in Figure 7b and *Quercus sp* in Figure 7c. The identification of the organic material in the SEM images, was made by the Portuguese Society of Aerobiology (Sociedade Portuguesa de Aerobiologia). From image observation, it can be seen that these pollen species tend to be geometrically spherical or elliptical. It can also be seen that they tend to be around 20 to 30 μm in diameter, if assumed spherical. For a better understanding of how the pollen affected PV performance during the spring and summer, the following SEM images are shown, one taken at April and other at July with the same magnification in Figure 8.

Using ImageJ, which is an open source image processing program (http://imagej.net), the following characteristics of the organic material, as well as all the material in the SEM images were found, see Table 2. The variable $O_D$ corresponds to the average diameter of the organic material (in black on the SEM images), $A_D$ is the average diameter of all the particles detected in the image, $OA_{PNR}$ is the ratio percentage between the organic and all material present in the image, in terms of particle count, while $OA_{AR}$ is the ratio in percentage (in terms of area) between the organic and all material in the image.

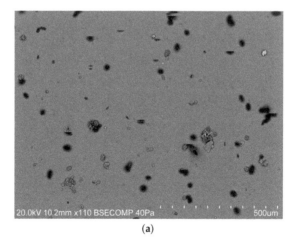

(a)

**Figure 8.** *Cont.*

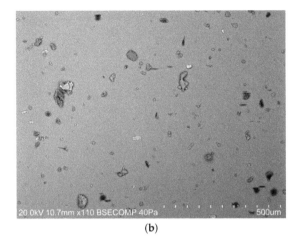

(b)

**Figure 8.** SEM images: (a) April 2017; (b) July 2017.

**Table 2.** SEM image parameter analysis.

| Parameters | April | July |
|---|---|---|
| $O_D$ (µm) | 20.5 | 19.3 |
| $A_D$ (µm) | 14.4 | 14.9 |
| $OA_{PNR}$ (%) | 33.8 | 4.1 |
| $OA_{AR}$ (%) | 40.3 | 6.5 |

The organic materials, which are essentially pollen, namely *Quercus sp*, have ≈ 20 µm diameter, which is in line with the size of that species in Figure 7c. When considering all the material in the sample, organic and mineral, the diameter drops to ≈ 14.5 µm, which is expected, since mineral soiling tends to be smaller than pollen. In April, 33.8% of the particles present are organic material, while in July this figure is only 4.1%. This illustrates the difference, considering the number of particles, between these two different seasons in terms of soiling. Regarding the percentage ratio of area occupied by organic particles, 40.3% of the area occupied by all particles is occupied by organic soiling in April, while in July only 6.5%. From this data, it can be concluded, that organic material, namely pollen, deposits on surfaces during the spring at higher concentrations and due to its higher diameter it has relatively more impact than mineral soiling (smaller particles).

## 5. Conclusions

This study reveals that organic soiling can have a significant importance on the decrease of PV performance. It provides in depth insight in terms of deposited particle characterization. It describes the contents, shape and size of particles that build up on the top of surfaces with visual and chemical analysis, which can help in future studies for soiling modelling/prediction, since these characteristics are perhaps the most important ones to characterize soiling at microscopic scales. Soiling rates have been calculated and compared for different seasons of the year and highlight how spring and periods without rain can impact solar harvesting technologies. During spring, besides the mineral soiling that can be deposited, pollen at high concentration in the atmosphere also leads to PV module performance decrease. Furthermore, the periods without rain allow for the particles to build up, increasing the soiling over time, resulting in higher losses along that period. Due to this fact, it is possible to distinguish several soiling rates along the year. Soiling rates were analysed together with rain (the main cleaning agent) and AOD as a proxy for particle concentration, showing a good agreement. SEM imagery helped in the detection of organic soiling, namely pollen, which deposited on the PV modules increasing the background soiling seen during summer, which led to a greater module performance degradation. Note that these data may be helpful to determine cleaning schedules and for modelling purposes for future PV plants in the region.

**Acknowledgments:** Ricardo Conceição acknowledges the FCT scholarship SRFH/BD/116344/2016. Hugo G. Silva is grateful to DNI-ALENTEJO, INSHIP (H2020-LCE-2016-ERA) projects and COST Action CA16235, "PEARL PV". This work is was funded by the European Union through the European Regional Development Fund, framed in COMPETE2020 (Operational Programme Competitiveness and Internationalisation) through the ICT Project (UID/GEO/04683/2013) with Reference POCI-01-0145-FEDER-007690. The authors are truly grateful to Elsa Caeiro from the Portuguese Society of Aerobiology for the information related with the pollen characterization.

**Author Contributions:** Ricardo Conceição has conducted the SEM sessions at Hercules Laboratory under the supervision and with the help of José Mirão. Ricardo Conceição and Hugo G. Silva have developed the experiments at PECS and shared the writing. The programming was done by Ricardo Conceição. Exhaustive revision of the final manuscript was done by Manuel Collares-Pereira. All authors read and approved the final manuscript.

**Conflicts of Interest:** The authors declare no conflict of interest.

## References

1.  Garg, H.P. Effect of dirt on transparent covers in flat-plate solar energy collectors. *Sol. Energy* **1974**, *15*, 299–302.
2.  El-Shobokshy, M.S.; Hussein, F.M. Effect of dust with different physical properties on the performance of photovoltaic cells. *Sol. Energy* **1993**, *51*, 505–511.
3.  Hassan, A.; Rahoma, U.A.; Elminir, H.K.; Fathy, A.M. Effect of airborne dust concentration on the performance of PV modules. *J. Astron. Soc. Egypt* **2005**, *13*, 24–38.
4.  Maghami, M.R.; Hizam, H.; Gomes, C.; Radzi, M.A.; Rezadad, M.I.; Hajighorbani, S. Power loss due to soiling on solar panel. *Renew. Sustain. Energy Rev.* **2016**, *59*, 1307–1316.
5.  Conceição, R.; Silva, H.; Mirão, J.; Gostein, M.; Fialho, L.; Narvarte, L.; Collares-Pereira, M. Saharan dust transport to Europe and its impact on photovoltaic performance: A case study of soiling in Portugal. *Sol. Energy* **2018**, *160*, 94–102.
6.  Figgis, B.; Ennaoui, A.; Ahzi, S.; Rémond, Y. Review of PV soiling particle mechanics in desert environments. *Renew. Sustain. Energy Rev.* **2017**, *76*, 872–881.
7.  Figgis, B.; Ennaoui, A.; Guo, B.; Javed, W.; Chen, E. Outdoor soiling microscope for measuring particle deposition and resuspension. *Sol. Energy* **2016**, *137*, 158–164.
8.  Alnaser, N.W.; Al Othman, M.J.; Dakhel, A.A.; Batarseh, I.; Lee, J.K.; Najmaii, S.; Alothman, A.; Al Shawaikh, H.; Alnaser, W.E. Comparison between performance of man-made and naturally cleaned PV panels in a middle of a desert. *Renew. Sustain. Energy Rev.* **2018**, *82*, 1048–1055.
9.  Šúri, M.; Huld, T.A.; Dunlop, E.D.; Ossenbrink, H.A. Potential of solar electricity generation in the European Union member states and candidate countries. *Sol. Energy* **2007**, *81*, 1295–1305.
10. Lopes, F.; Silva, H.; Salgado, R.; Cavaco, A.; Canhoto, P.; Collares-Pereira, M. Short-term ECMWF Forecasts of Solar Irradiance for Solar Energy Systems Validated in Southern Portugal. *Sol. Energy* Submitted.
11. Silva, H.G.; Lopes, F.M.; Pereira, S.; Nicoll, K.; Barbosa, S.M.; Conceição, R.; Neves, S.; Harrison, R.G.; Collares Pereira, M. Saharan dust electrification perceived by a triangle of atmospheric electricity stations in Southern Portugal. *J. Electrost.* **2016**, *84*, 106–120.
12. Stridh, B. Evaluation of economical benefit of cleaning of soiling and snow in PV plants at three European locations. In Proceedings of the IEEE Photovoltaic Specialists Conference, Austin, TX, USA, 3–8 June 2012; pp. 1448–1451.
13. Appels, R.; Lefevre, B.; Herteleer, B.; Goverde, H.; Beerten, A.; Paesen, R.; De Medts, K.; Driesen, J.; Poortmans, J. Effect of soiling on photovoltaic modules. *Sol. Energy* **2013**, *96*, 283–291.
14. Naeem, M.; Tamizhmani, G. Climatological relevance to the soiling loss of photovoltaic modules. In Proceedings of the 2015 Saudi Arabia Smart Grid, Jeddah, Saudi Arabia, 7–9 Dcember 2016; pp. 1–5.
15. Aïssa, B.; Isaifan, R.J.; Madhavan, V.E.; Abdallah, A.A. Structural and physical properties of the dust particles in Qatar and their influence on the PV panel performance. *Sci. Rep.* **2016**, *6*, 1–12.

16. Olivares, D.; Ferrada, P.; Matos, C.D.; Marzo, A.; Cabrera, E.; Portillo, C.; Llanos, J. Characterization of soiling on PV modules in the Atacama Desert. *Energy Procedia* **2017**, *124*, 547–553.

17. Caeiro, E.; Brandão, R.; Carmo, S.; Lopes, L.; de Almeida, M.M.; Gaspar, Â.; Oliveira, J.F.; Todo-Bom, A.; Leitã, T.; Nunes, C. Rede Portuguesa de Aerobiologia: Resultados da monitorização do pólen. *Rev Port Imunoalergologia* **2007**, *15*, 235–250.

18. Hamaoui-Laguel, L.; Vautard, R.; Liu, L.; Solmon, F.; Viovy, N.; Khvorostyanov, D.; Essl, F.; Chuine, I.; Colette, A.; Semenov, M.A.; et al. Effects of climate change and seed dispersal on airborne ragweed pollen loads in Europe. *Nat. Clim. Chang.* **2015**, *5*, 766–771.

19. Elminir, H.K.; Ghitas, A.E.; Hamid, R.H.; El-Hussainy, F.; Beheary, M.M.; Abdel-Moneim, K.M. Effect of dust on the transparent cover of solar collectors. *Energy Convers. Manag.* **2006**, *47*, 3192–3203.

20. Gostein, M.; Duster, T.; Thuman, C. Accurately measuring PV soiling losses with soiling station employing module power measurements. In Proceedings of the 2015 IEEE 42nd hotovoltaic Specialists Conference, New Orleans, LA, USA, 14–19 June 2015; pp. 3–7.

21. Dunn, L.; Littmann, B.; Caron, J.R.; Gostein, M. PV module soiling measurement uncertainty analysis. In Proceedings of the 2013 IEEE 39th Photovoltaic Specialists Conference, Tampa, FL, USA, 16–21 June 2013; pp. 658–663.

22. Todo-bom, A.; Brandão, R.; Nunes, C.; Caeiro, E.; Leitão, T.; Oliveira, J.F.; de Almeida, M.M. Allergenic airborne pollen in Portugal 2002–2004. *Rev Port Imunoalergologia* **2006**, *14*, 41–49.

23. Alcázar, P.; Galán, C.; Cariñanos, P.; Domínguez-Vilches, E. Diurnal variation of airborne pollen at two different heights. *J. Investig. Allergol. Clin. Immunol.* **1999**, *9*, 89–95.

24. Grewling, L.; Bogawski, P.; Smith, M. Pollen nightmare: Elevated airborne pollen levels at night. *Aerobiologia* **2016**, *32*, 725–728.

25. Marques, G.; Martins, C.; Belo, J.; Alves, C.; Paiva, M.; Caeiro, E.; Leiria-Pinto, P. Pollen Counts Influence Web Searches for Asthma and Rhinitis. *J. Investig. Allergol. Clin. Immunol.* **2016**, *26*, 192–194.

26. Albertine, J.M.; Manning, W.J.; Da Costa, M.; Stinson, K.A.; Muilenberg, M.L.; Rogers, C.A. Projected carbon dioxide to increase grass pollen and allergen exposure despite higher ozone levels. *PLoS ONE* **2014**, *9*, 1–6.

*Article*

# Prediction Model of Photovoltaic Module Temperature for Power Performance of Floating PVs

**Waithiru Charles Lawrence Kamuyu [1], Jong Rok Lim [1], Chang Sub Won [2] and Hyung Keun Ahn [1,\*]**

[1]   Konkuk University, 120 Neungdong-Ro, Gwanjin-Gu, Seoul 143-701, Korea; waithiru@gmail.com (W.C.L.K.); bangsil82@hanmail.net (J.R.L.)
[2]   LSIS R&D Campus 116 beongil 40 Anyang, Gyeonggi 431-831, Korea; cswon@lsis.com
\*   Correspondence: hkahn@konkuk.ac.kr; Tel.: +82-104-630-9972

Received: 13 December 2017; Accepted: 11 February 2018; Published: 18 February 2018

**Abstract:** Rapid reduction in the price of photovoltaic (solar PV) cells and modules has resulted in a rapid increase in solar system deployments to an annual expected capacity of 200 GW by 2020. Achieving high PV cell and module efficiency is necessary for many solar manufacturers to break even. In addition, new innovative installation methods are emerging to complement the drive to lower $/W PV system price. The floating PV (FPV) solar market space has emerged as a method for utilizing the cool ambient environment of the FPV system near the water surface based on successful FPV module (FPVM) reliability studies that showed degradation rates below 0.5% p.a. with new encapsulation material. PV module temperature analysis is another critical area, governing the efficiency performance of solar cells and module. In this paper, data collected over five-minute intervals from a PV system over a year is analyzed. We use MATLAB to derived equation coefficients of predictable environmental variables to derive FPVM's first module temperature operation models. When comparing the theoretical prediction to real field PV module operation temperature, the corresponding model errors range between 2% and 4% depending on number of equation coefficients incorporated. This study is useful in validation results of other studies that show FPV systems producing 10% more energy than other land based systems.

**Keywords:** floating PV systems (FPV); floating PV module (FPVM)

---

## 1. Introduction

A report published by International Renewable Energy Agency (IRENA) in 2016 [1] shows that the global cumulative capacity of installed solar systems was 222 GW, with China, Germany, Japan, and USA installing 43, 40, 33, and 22 GW, respectively. In many markets, we see the growing conflict between the need to convert arable land or forests to create PV installation sites installation vis-à-vis the need to protect the environment. A floating photovoltaic (PV) system installation on dams or water reservoirs is one such method that offers an installation site option with minimal interference with the environment. Additionally, it utilizes the cooling effect of water on its surface to improve the efficiency of the PV module and ultimately the performance of the PV system [2].

Extensive studies on the efficiency, power, and temperature of the conventional PV system module have been carried out by Evans [3], Duffie and Beckman [4], and many others [5]. Considering the importance of device temperature in PVM efficiency analysis, this paper proposes a model that correlates the temperature of a FPV module to the ambient temperature, solar radiation, and wind speed. A second model incorporates the influence of water temperature of the FPV installation. Well known PV module temperature models are compared under constant irradiation and constant ambient conditions as presented herein. The characteristic analysis of the FPVM temperature models shows resemblance to models proposed by Lasnier and Ang 1990 [6] and Duffie and Beckmans 2006.

Duffie and Beckmans' predictions are thus preferred for size optimization, simulation and design of solar photovoltaics. Koehl [7], Kurtz [8], and Skoplaski [9] that include wind speed in temperature predictions are also included in analysis. A simple comparison of the temperature profiles of FPVMs with the conventional land- or rooftop-based modules shows that the mean value of the yearly PV module temperature of an FPV system is 21 °C, which is 4 °C below that of land or rooftop installed PV modules [10].

The aforementioned research is important in analyzing the correlation between efficiency and temperature. Solar cells only convert a small amount of absorbed solar radiation into electrical energy with the remaining energy being dissipated as heat in the bulk region of the cell [11]. A rise in the operation temperature of a solar cell and module reduces the band gap, thus slightly increasing the short circuit current of a solar cell for a given irradiance, but largely decreasing the open circuit voltage, resulting in a lower fill factor and power output. The net effect results in a linear relation for the electrical efficiency ($\eta_c$) of a PV module as

$$\eta_c = \eta_{Tref}\left[1 - \beta_{ref}\left(T_m - T_{ref}\right)\right] \tag{1}$$

where $\eta_{Tref}$ and $\beta_{ref}$ are the electrical efficiency and temperature coefficient of the PV module, respectively. $T_m$ and $T_{ref}$ are the PV module operational temperature and reference temperature, respectively.

## 2. Floating PV System Introduction and Performance

### 2.1. Site Information of Floating PV System

Figure 1 shows the aerial views of Korea's first 100 kW and 500 kW Hapcheon Dam FPV power stations located at southern part of the country. Based on the previous research on module reliability [2], a special anti-damp proof FPVM with a unique encapsulation was certified and installed. A unique mooring system is anchored the floating system on the dam floor, aligning the FPV system to the correct azimuth. To monitor environment conditions, a portion of the floating platform is fitted with a small weather station equipped with sensors, as outlined in Table 1, and based on IEC (International Electrotechnical Commission) standard 61724-1 [12]. A low-loss cable transmitted DC power from the FPV system to dry land where an electric room housing a PV inverter and monitoring computers were installed.

**Figure 1.** Aerial view of the 100 kW (**left**) and 500 kW (**right**) floating systems on Hapcheon lake.

**Table 1.** Main sensor specifications.

| Sensor Type | Maker | Model | Accuracy | Range | Mounting |
|---|---|---|---|---|---|
| Solar Irradiance | Apogee | SP-110 | ±5% | 0–1750 Wm$^2$ | Leveling fixture |
| Anemometer (Wind Speed) | Jinyang | WM-IV-WS | ±0.15 m/s | 0–75 m/s | Pole mount |
| Accelerometer | Das | MSENS-IN360 | 0.10° | 0–360° | Pole mount |
| Humidity and Temperature Probe | Vaisala | HMP155 | ±0.176% | −80–60 °C | Protective housing |
| PVM Temperature | Taeyeon | DY-HW-7NN | ±0.20% | −5–55 °C | PVM rear surface |
| Water Temperature | Taeyeon | DY-HW-11NN | ±0.20% | −5–55 °C | Water submerged |

*2.2. Power Outputs of Floating PV Versus Rooftop-Based System*

In Table 2, we outline system specification of two FPV and a land based systems study.

**Table 2.** FPVs and rooftop PVs information.

| Project Type | Test Bed | Floating PV | Rooftop PV |
|---|---|---|---|
| Site Name | Hapcheon Dam | Hapcheon Dam | Haman |
| Site Coordinates | N 35.5°33'06" E 128°00'49" | N 35.5°33'36" E 128°02' 26" | N 35° 16'10" E 128° 24' 01" |
| Installation Capacity | 100 kW | 500 kW | 1 MW |
| Installation Year | 2011 | 2012 | 2012 |
| Module Slope | 33° | 33° | 30° |
| Module Type | c-Silicon | c-Silicon | c-Silicon |
| Mounting | Aluminum, steel | Aluminum | Aluminum |
| Mounting Type | Fixed | Fixed | Fixed |
| Water Depth | 20 m | 40 m | n/a |

Table 3 shows yearly energy results of the three PV systems. We compare the performance of the three systems after normalizing energy output with system kWp capacity (kWh/KWp). The unit (h/d) is an expression of how many hours a PV system operates at its peak power. As shown in Table 3, the *y*-axis (left) illustrates this daily monthly average energy output. For example in April 2013, average monthly output from the three PV Systems was 443, 2078, and 3976 kWh for the 100, 500, and 1000 kW PV systems, respectively. Multiplying respective monthly average but days in month, and summing monthly outputs gives 130.3, 693.2, and 1197.5 MWh respective total yearly output.

**Table 3.** General system performance and output.

| Output Energy | | | Floating PV | | Rooftop PV |
|---|---|---|---|---|---|
| | | | 100 kW | 500 kW | 1000 kW |
| Annual Output (kWh/year) | | | 130,305 | 693,219 | 1,197,547 |
| Daily Yearly Average (kWh/year/days of year) | | | 357 | 1859 | 3281 |
| Normal Power | Yearly | kWh/year/kWp (h/year) | 1303 | 1386 | 1198 |
| | Monthly | kWh/month/kWp (h/month) | Monthly details in Figure 2 | | |
| | Daily | kWh/year/days/kWp (h/d) | 3.58 | 3.80 | 3.28 |

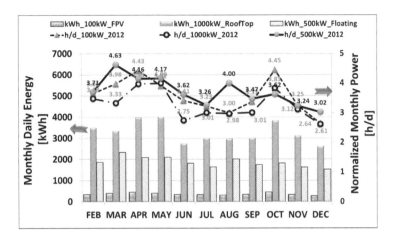

**Figure 2.** Monthly daily average energy of the month (kWh) and corresponding normalized power comparisons for FPV systems vs. 1000 kW rooftop system.

For the 100 kW FPV station, October and December are the best and worst performing months at 445 kWh and 264 kWh respectively, compared to the station's yearly average of 357 kWh. Similarly for the 500 kW FPV station, March and December are the best and worst performing months at 2316 kWh and 1512 kWh respectively, compared to the station's yearly average of 1859 kWh. Finally for the 1000 kW rooftop PV station, May and December are the best and worst performing months at 3998 Wh and 2612 kWh respectively, compared to the station's yearly average of 3281 kWh. Whereas the rooftop produces more power quantitatively, the FPV systems are more efficient in qualitative power delivery. Output energy (kWh) normalization to name plate peak power (kWp) results in hours per day (h/day) unites. Table 3 shows an average of all monthly values gives yearly normalized output of 3.58, 3.80, and 3.28 h/day, respectively, as shown in Table 3. Analysis of the latter values proves the two FPV systems are outperforming the rooftop systems by 9% and 16%, warranting investigation into temperature performance.

## 3. Methodology of Floating PV Temperature Model

In this section, we formulate a multiple linear equation for the dependent PVM variable ($y$; FPV module operation temperature) using four independent linear variables $x_1$, $x_2$, $x_3$ and $x_4$ representing solar irradiance ($G_T$), ambient temperature ($T_a$), wind speed ($V_w$), and water temperature ($T_w$), respectively.

The equation is linear for unknown parameters $\beta_0 - \beta_{k-1}$, and is of the form given in Equation (2)

$$y_i = \beta_0 + \beta_1 x_{i1} + \beta_2 x_{i2} + \cdots + \beta_{k-1} x_{ik-1} + \epsilon_i \qquad (2)$$

where $y_i$ is the predicted value of module temperature $y$ and assumes $i$th independent error $\epsilon_i \sim N(0, \sigma^2)$ following a normal distribution with independent mean and variance squared. The matrix can be expressed as

$$Y = X\beta + \epsilon \qquad (3)$$

where
$Y = \begin{pmatrix} Y_1 \\ Y_2 \\ \vdots \\ Y_n \end{pmatrix}$,
$\beta = \begin{pmatrix} \beta_1 \\ \beta_2 \\ \vdots \\ \beta_n \end{pmatrix}$,
$\epsilon = \begin{pmatrix} \epsilon_1 \\ \epsilon_2 \\ \vdots \\ \epsilon_n \end{pmatrix}$
and
$X = \begin{pmatrix} 1 & x_{11} & x_{12} & \cdots & x_{1k-1} \\ 1 & x_{21} & x_{22} & \cdots & x_{2k-1} \\ \vdots & \vdots & \vdots & \cdots & \vdots \\ 1 & x_{n1} & x_{n2} & \cdots & x_{nk-1} \end{pmatrix}$.

The multiple linear regression form is expressed in Equation (3) with $Y$, $\beta$, $\epsilon$, and $X$ representing $y$ observations, vector of parameters, error, and $n \times k$ matrix vectors, respectively. The field data of PV system data is given in the forms of $Y_i$, $x_{i1}$, $x_{i2}$, $x_{i3}$, and $x_{i4}$, for $T_m$, $T_a$, $G_T$, $V_w$, and $T_w$, respectively.

We use the standard least-squares minimization to determine the aforementioned model parameters by minimizing the sum of squares of residuals ($SS_{Res}$) as shown in a matrix form in Equation (4)

$$SS_{Res} = \sum_{i=1}^{n} e_i^2 \tag{4}$$

where $e = (Y - \overline{Y})$, $e^2 = e'e = (Y - \overline{Y})'(Y - \overline{Y})$ and $\overline{Y} = \overline{\beta}_0 + \overline{\beta}_{x_1} + \overline{\beta}_{x_2} + \cdots + \overline{\beta}_{k-1}x_{k-1}$.

Substituting the former into Equation (4) leads to the definition of $SS_{Res}$ in terms of the unknown parameters in Equation (5).

$$SS_{Res} = \sum_{i=1}^{n} (Y - \overline{Y})_i^2 = (Y - \overline{Y})'(Y - \overline{Y}) = Y'Y - 2\overline{\beta}'X'Y' + \overline{\beta}'X'X\overline{\beta}' \tag{5}$$

Equation is expanded using $\overline{Y} = X\overline{\beta}'$. Integrating $SS_{Res}$ with respect to $\overline{\beta}'$ results in normal equations, which have to be solved for unknown equations in Equation (6). For easy computation, an alternative matrix equation is presented for solving the coefficients.

$$\frac{\partial SS_{Res}}{\partial \overline{\beta}'} = \frac{\partial(Y'Y - 2\overline{\beta}'X'Y' + \overline{\beta}'X'X\overline{\beta}')}{\partial \overline{\beta}'} \tag{6}$$

$$\overline{\beta}' = (X'X)^{-1} + X'Y \tag{7}$$

where $X'$ is the inverse $X$ matrix of predictor variables listed on Table 4.

**Table 4.** Multiple regression variables for FPVM temperature (symbol; $T_m$).

| Term | Predictor Variables | Symbol | Unit |
|------|--------------------|--------|------|
| $x_1$ | Ambient Temperature | $T_a$ | °C |
| $x_2$ | Solar Irradiance | $G_T$ | W/m$^2$ |
| $x_3$ | Wind Speed | $V_w$ | m/s |
| $x_4$ | Water Temp. | $T_w$ | °C |

## 4. FPV Temperature Model Results and Comparisons

### 4.1. Model Results

With data collected from the floating PV site, we formulate the X matrix containing FPVM data points, as is plotted in Figure 3, and corresponds to five minutes of site data. The Y matrix corresponds to the measured module temperature. The coefficients of models 1 and models 2 in Equation (4) for the FPVM suggested in this paper are expressed as

$$T_{m_1} = 2.0458 + 0.9458T_a + 0.0215G_T - 1.2376V_w \tag{8}$$

$$T_{m_2} = 1.8081 + 0.9282T_a + 0.021G_T - 1.2210V_w + 0.0246T_w \tag{9}$$

$T_{m_1}$ and $T_{m_2}$ explains the operation temperature behavior of the FPV module with seasonal variables $T_a$, $G_T$, $V_w$, and $T_w$.

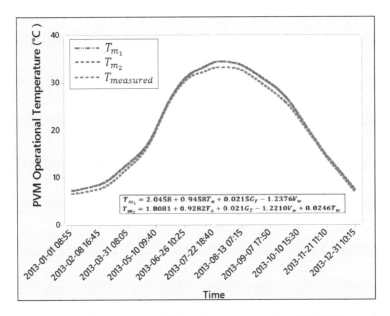

**Figure 3.** Predicted versus measured PV module temperature data (100 kW PV system).

Based on the two predictions of the temperature of an FPVM proposed herein, we propose two corresponding modifications to Equation (1) based on the input parameters of $T_m$ as

$$\eta_{c,FPV_1} = \eta_{Tref}\left[1 - \beta_{ref}\left(2.0458 + 0.9458T_a + 0.0215G_T - 1.2376V_w - T_{ref}\right)\right] \quad (10)$$

$$\eta_{c,FPV_2} = \eta_{Tref}\left[-\beta_{ref}\left(1.8081 + 0.9282T_a + 0.021G_T - 1.2210V_w + 0.0246T_w - T_{ref}\right)\right] \quad (11)$$

where $\eta_{c,FPV}$, $T_a$, $G_T$, and $V_w$ represent average values of corresponding model PV module efficiency, ambient temperature, solar irradiation, and wind speed respectively. Equation (9) includes an additional variable, i.e., water temperature ($T_w$).

Figure 3 is a time series plot of predicted and measured module temperatures in 2013. From the graph, predicted PVM temperatures are almost always higher than measured except during third quarter where $T_{measured} > T_{m_1}, T_{m_2}$. Coincidentally, wind speeds ($T_W$) are also low during same period, implying the dominance of $T_W$ in the two models Equations (8) and (9).

Equation (12) introduces the average error, as the difference in area of the curves of measured versus modeled temperatures in Figure 3 of FPV models, by showing the average difference between predicted and measured values. $T_{Error}$ ranges between 2.06% and 4.40%, for respective FPV models.

$$T_{Error} = \int_{n=1}^{n=k} (T_{measured} - T_m)_n \quad (12)$$

In Equation (12), $k$ is total data points. Inclusion of additional parameter $T_w$ in Equation (11) increases $T_{Error}$ by 2% to 4%.

*4.2. Comparison with Land-Based PV System*

In Figure 4, we compare the floating PV and rooftop system yearly energy profile based on operation module temperature. As can be seen, the FPV system produces a larger portion of energy at lower temperatures [13] compared to the rooftop system.

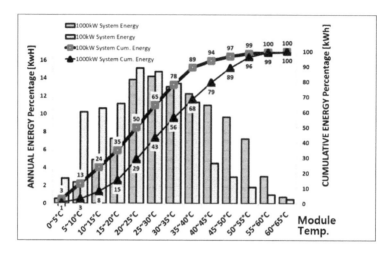

**Figure 4.** Annual energy generated over module temperature of floating PV temperatures.

Cumulative PV energy produced throughout the year, is plotted against corresponding module temperature. As indicated, 89% of annual energy produced by the FPV system, and 68% of the rooftop system's energy is produced when module temperatures of both systems are below 40 °C. Energy produced beyond the 25 °C is subject to power loss due to loss of open circuit ($V_{oc}$) and fill factor (F.F.).

### 4.3. Comparison with Selected Temperature Models

A select group of PV temperature models [5] is presented in Table 5 for comparison. The models incorporate a reference, stating for example air temperature ($T_a$), and the corresponding values of relevant variables ($G_T$, $V_w$, etc.). Owing to the complexities involved, some authors presented explicit correlation in addition to implicit relations requiring iterations.

**Table 5.** PV module models.

| Model | Empirical Models |
|---|---|
| Ross (1976) [14] | $T_c = T_a + kG_T$ where $k = \Delta(T_c - T_a)/\Delta G_T$ |
| Rauschenbach (1980) [15] | $T_c = T_a + (G_T/G_{T,NOCT})(T_{c,NOCT} - T_{c,NOCT})(1 - n_m/\gamma)$ |
| Risser & Fuentes (1983) [16] | $T_c = 3.81 + 0.0282 \times G_T 1.31 \times T_a - 165V_w$ |
| Schott (1985) [17] | $T_c = T_a + 0.028 \times G_T - 1$ |
| Ross & Smokler (1986) [14] | $T_c = T_a + 0.035 \times G_T$ |
| Mondol et al. (2005, 2007) | $T_c = T_a + 0.031G_T$ <br> $T_c = T_a + 0.031G_T - 0.058$ |
| Lasnier & Ang (1990) [5] | $T_c = 30.006 + 0.0175(G_T - 300) + 1.14(T_a - 25)$ |
| Servant (1985) | $T_c = T_a + \alpha G_T(1 + \beta Ta)(1 - \gamma V_w)\left(1 - 1.053 n_{m,ref}\right)$ |
| Duffie & Beckman [3] | $T_c = T_a + (G_T/G_{NOCT})(9.5/5.7 \times 3.8V_w)(T_{NOCT} - T_{a,NOCT})(1 - n_m)$ |
| Koehl (2011) [6] | $T_c = T_a + G_T/(U_0 + U_1 \times V_w)$ |
| Kurtz S (2009) [7] | $T_c = T_a + G_T \times e^{-3.473 - 0.0594 \times V_w}$ |
| Skoplaki (2009) [8] | $T_c = T_a + (G_T/G_{NOCT}) \times (T_{NOCT} - T_{a,NOCT}) \times h_{w,NOCT}/h_w$ <br> $\times [1 - \eta_{STC}/\tau \times \alpha(-\beta_{STC}T_{STC})]$ |

In Figure 5 below, temperature models listed above are plotted against both ambient ($T_a$) temperature and solar radiation ($G_T$).

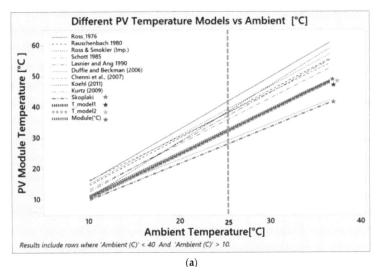

(a)

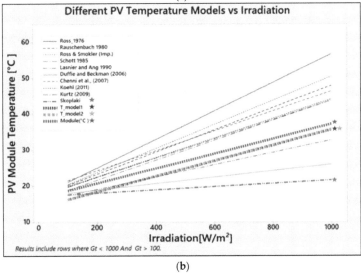

(b)

**Figure 5.** PV predicted cell/module temperature verses ambient temperature (a) and irradiance (b).

As can be seen, all models vary linearly with both $T_a$ and $G_T$ with varying specific gradients. The analytical implication is different model interpret heat dissipation by the PV module differently when exposed to the elements. For example, Ross [14] and Rauschenbach [15] model display the highest PV operation temperatures when exposed to $G_T$ at constant $T_a$. Koehl, Kurtz, and Skoplaki's research incorporates wind ($V_w$) data in temperature prediction.

In Figure 5a, Duffie & Beckman [4] and Skoplaki [9] recorded low temperatures with increasing $T_a$, suggesting adequate heat dissipation by the modules due to incorporation of wind data. To the contrary, Ross and Rauschenbach show high temperatures near 60 °C suggesting the PVM retains heat.

Model 1 ($T_{m1}$) and Model 2 ($T_{m2}$) plots vary slightly with real PV module data, and operate at much lower temperatures when compared to all other models. Lower operation temperatures suggests heat dissipation to FPV ambient environment

In Figure 5b, Skoplaki has lowest operation temperature with increasing $G_T$, while Ross has highest temperature values because of PVM heat retention. Skoplaki model reacts very slowly to rising $G_T$ due to quick heat dissipation by the $V_w$ factor. A slight deviation from real temperature by Model 1 ($T_{m1}$) and Model 2 ($T_{m2}$) is noted with increasing $G_T$.

Based on the two graphs in Figure 5, we conclude that our two FPV models operating temperatures are significantly lower than conventional PV module ranges.

### 4.4. Comparison of Models with Minitab Model

MINITAB [18] is an advanced statistics program has well-defined algorithms that describe the change of any dependent variable $y$ with the interaction between the respective independent variables $x_i$. Minitab generates an equation that shows the interaction between the dependent variable (module temperature) and independent variables. Equation (13) derived by MINITAB is highly accurate (0.1%) but incurs the risk of equation complexity due to over-fitting.

$$
\begin{aligned}
\text{Module} = \ & -1.9034 + 1.12322\, x_1 + 0.028655\, x_2 - 0.6517\, x_3 - 0.09362\, x_4 \\
& -0.001328\, x_1^2 - 0.000014\, x_2^2 + 0.08382\, x_3^2 - 0.000604\, x_1 \times x_2 \\
& -0.031334\, x_1 \times x_3 + 0.001389\, x_1 \times x_4 - 0.000981\, x_2 \times x_3 \\
& +0.000545\, x_2 \times x_4 + 0.039145\, x_3 \times x_4
\end{aligned}
\tag{13}
$$

for $x_1$, $x_2$, $x_3$, $x_4$, representing $T_a$, $G_T$, $V_w$, $T_w$.

In Figure 6, four histograms compare the normal distribution of real FPV module temperature data (lower right) to Model 1, Model 2, MINITAB's prediction.

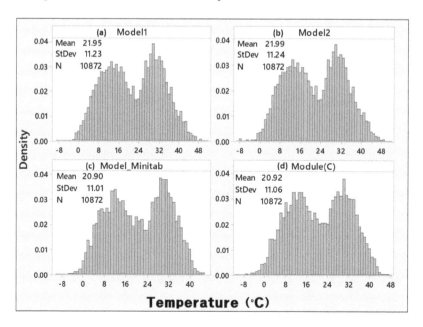

**Figure 6.** Histograms comparison of FPV Model 1 (**a**), Model 2 (**b**), Minitab data fitting model (**c**), and module field data (**d**).

The sub figures *x*-axis is operational temperature from $-10\ °C$ to over $50\ °C$. The *y*-axis plots the density of respective temperature range throughout the year. All model distributions show a bimodal shape, with two peaks temperatures at $10\ °C$ and $30\ °C$. Analysis with fewer data points shows a normal distribution curve.

## 5. FPV Module Efficiency and Power Prediction

Operating a PV system on the water surface has the added benefit of increasing conversion efficiency due to the cooling effect on water's surface.

Figure 7 is a 3D plot of FPV module efficiency/$T_a$/$G_T$ . In the plot, a decrease in ambient temperature ($T_a$) has a positive effect of increasing efficiency between 1–2% points. The plot shows the importance $T_a$ in defining PVM operation temperature and, ultimately, conversion efficiency. Radiation ($G_T$) plays a secondary role given the minimal impact on efficiency. It can be observed that, at higher radiation levels, $G_T$ varies more frequently, and this impacts power stability.

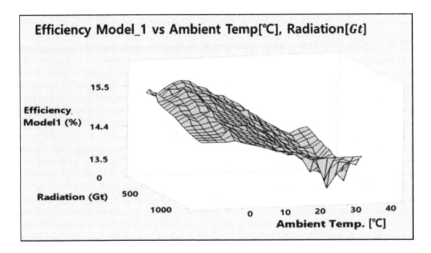

**Figure 7.** 3D surface plots of Model 1 efficiency/$T_a$/$G_t$.

In summary, as observed from the FPV data, low ambient conditions are ideal for higher system efficiency and power performance as shown by seasonal variation in efficiency in Figure 8. In June through August when ambient temperature are high, PVM efficiency drops between 1–2% points. For a land based system, a more severe dip is expected. During fall and winter when temperatures drop, we notice a step rise in efficiencies to the mid-15% level. Based on graphical description on Figure 8, for FPV module temperature models 1, we predict that a 1 °C increase in $T_m$ results in a 0.058% decrease in $\eta_{c,FPV_1}$, model 1 PVM efficiency, as shown for in Equation (14).

$$\eta_{c,FPV_1} = 15.96 - 0.058T_{m_1} \tag{14}$$

PV module operational temperature has an important role in the energy conversion process [19]. From Figure 8, the electrical efficiency of the PV module varies linearly with temperature as shown. From the graph above for Model 1, Latifa [20] has done important work (2014) comparing crystalline (c-Si) and amorphous silicon (a-Si) coefficients per °C. This work shows coefficient values for a-Si closely identical to FPV.

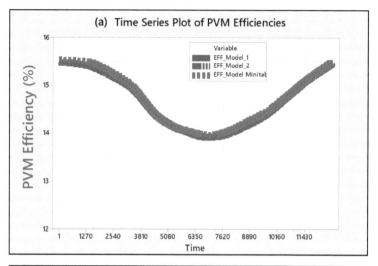

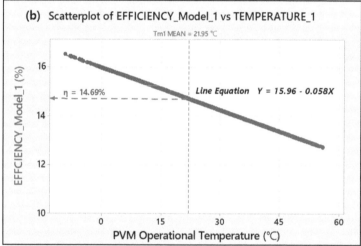

**Figure 8.** (a) Time series plot of module efficiency (Model 1); (b) efficiency over module temperature.

## 6. Conclusions

Two prediction models of the FPV module operation temperature are suggested for the analysis of performance of the FPV module and system. Model 1 includes the effects the independent variables, i.e., ambient temperature ($T_a$), solar irradiance ($G_t$), and wind speeds ($V_w$). When compared to the measured FPV module temperature over entire year, the error of model 1 is 2%. Model 2 includes the three aforementioned independent variables in addition to water temperature ($T_w$). Although the error of Model 2 increases slightly to 4%, the results are within the reasonable range of error. Fitting of the experimental data is reproduced with a minor error.

Through this research, a correlation between the temperature of the floating PV operating environment and system efficiency is derived. Beyond solar irradiation of 100 W/m², the floating system records an ideal efficiency averaging more than 14.69% based on a yearly mean PVM

*Energies* **2018**, *11*, 447

temperature of 21.95 °C. It was also observed that approximately two-thirds of the annual yield produced by the FPV system occurs when the module operational temperature was below 40 °C.

**Acknowledgments:** This work was supported by the New and Renewable Energy Technology Program of the Korea Institute of Energy Technology Evaluation and Planning (KETEP) and granted financial resources from the Ministry of Trade, Industry and Energy, Republic of Korea (NO. 20153010012060). This paper was also written as part of Konkuk University's research support program for its faculty on sabbatical leave in 2017.

**Author Contributions:** Waithiru Charles Lawrence Kamuyu developed his own primary model for the power with insolation environments, especially focusing on the photovoltaic module temperature. Jong Rok Lim added experimental data analysis with different seasons in two different areas. Chang Sub Won and Hyung Keun Ahn integrated the power tracing from the respect of environmental conditions to provide reliable temperature dependent tracking data for floating photovoltaic power station which is one of hottest systems these days for the micro-grid and land saving power network.

**Conflicts of Interest:** The authors declare no conflict of interest.

## References

1. International Renewable Energy Agency (IRENA). *Solar PV in Africa: Costs and Markets*; IRENA: Abu Dhabi, UAE, 2016; Chapter 3, p. 29.
2. Won, C.; Waithiru, L.; Kim, D.; Kang, B.; Kim, K.; Lee, G. Floating PV Power System Evaluation Over Five Years (2011–2016 (EU PVSEC 2016)). In Proceedings of the 32nd European Photovoltaic Solar Energy Conference and Exhibition, Munich, Germany, 20–24 June 2016.
3. Evans, D.L. Simplified method for predicting photovoltaic array output. *Sol. Energy* **1981**, *27*, 555–560. [CrossRef]
4. Duffie, J.A.; Beckman, W.A. *Solar Energy of Processes*, 3rd ed.; Wiley: Hoboken, NJ, USA, 2006; p. 3.
5. Jakhrani, A.Q. Comparison of Solar Photovoltaic Module Temperature Models. *World Appl. Sci. J. Spec. Issue Food Environ.* **2011**, *14*, 1–8.
6. Lasnier, F.; Ang, T.G. *Photovoltaic Engineering Handbook*; Adam Hilger: New York, NY, USA, 1990; p. 80.
7. Koehl, M.; Heck, M.; Wiesmeier, S.; Wirth, J. Modeling of the nominal operating cell temperature based on outdoor weathering. *Sol. Energy Mater. Sol. C* **2011**, *95*, 1638–1646. [CrossRef]
8. Kurtz, S.; Whitfield, K.; Miller, D.; Joyce, J.; Wohlgemuth, J.; Kempe, M.; Dhere, N.; Bosco, N.; Zgonena, T. Evaluation of high-temperature exposure of rack mounted photovoltaic modules. In Proceedings of the 34th IEEE Photovoltaic Specialists Conference (PVSC), Philadelphia, PA, USA, 7–12 June 2009; pp. 2399–2404.
9. Skoplaki, E.; Palyvos, J.A. On the temperature dependence of photovoltaic module electrical performance: A review of efficiency/power correlations. *Sol. Energy* **2009**, *83*, 614–624. [CrossRef]
10. Young, K.C.; Nam, H.L.; Kern, J.K. Empirical Research on the efficiency of Floating PV systems compare with Overland Systems. In Proceedings of the 3rd International Conference on Circuits, Control, Communication, Electricity, Electronics, Energy, System, Signal and Simulation (CES-CUBE 2013), Tamuning, GU, USA, 18–20 July 2013; Volume 25, pp. 284–289.
11. Green, M.A.; Blakers, A.W.; Shi, J.; Keller, E.M.; Wenham, S.R. High-Efficiency silicon solar cells. *IEEE Trans. Electron Devices* **1984**, *31*, 679–683. [CrossRef]
12. IEC Standard 61724-1:2017, Photovoltaic System Performance—Part 1: Monitoring. Available online: https://webstore.iec.ch/publication/33622 (accessed on 1 July 2017).
13. Malik, A.Q.; Ming, L.C.; Sheng, T.K.; Blundell, M. Influence of Temperature on the Performance of Photovoltaic. *ASEAN J. Sci. Technol. Dev.* **2010**, *26*, 61–72. [CrossRef]
14. Ross, R.G. Interface design considerations for terrestrial solar cells modules. In Proceedings of the 12th IEEE Photovoltaic Specialist's Conference, Baton Rouge, LA, USA, 7–10 January 1976; pp. 801–806.
15. Rauschenbach, H.S. *Solar Cell Array Design Handbook*; Van Nosstrand Reinhold: New York, NY, USA, 1980; pp. 390–391.
16. Risser, V.V.; Fuentes, M.K. Linear regression analysis of flat-plate photovoltaic system performance data. In Proceedings of the 5th Photovoltaic Solar Energy Conference, Athens, Greece, 17–21 October 1983.
17. Schott, T. Operation temperatures of PV modules: A theoretical and experimental approach. In Proceedings of the Sixth EC Photovoltaic Solar Energy Conference, London, UK, 15–19 April 1985.

18. MINITAB 17 Statistical Software 2010. Computer Software. Konkuk University Seoul. Minitab, Inc. Available online: www.minitab.com (accessed on 1 July 2017).

19. Martineac, C.; Hopîrtean, M.; De Mey, G.; Ţopa, V.; Ştefănescu, S. Temperature Influence on Conversion Efficiency in the Case of Photovoltaic Cells. In Proceedings of the 10th International Conference on Development and Application Systems, Suceava, Romania, 27–29 May 2010; pp. 1–4.

20. Sabri, L.; Benzirar, M. Effect of Ambient Conditions on Thermal Properties of Photovoltaic Cells: Crystalline and Amorphous Silicon. *Int. J. Innov. Res. Sci. Eng. Technol.* **2010**, *3*, 17815–17821. [CrossRef]

Article

# Hypothesis Tests-Based Analysis for Anomaly Detection in Photovoltaic Systems in the Absence of Environmental Parameters

Silvano Vergura

Department of Electrical and Information Engineering, Polytechnic University of Bari, st. E. Orabona 4, I-70125 Bari, Italy; silvano.vergura@poliba.it; Tel.: +39-080-5963590; Fax: +39-080-5963410

Received: 28 January 2018; Accepted: 22 February 2018; Published: 25 February 2018

**Abstract:** This paper deals with the monitoring of the performance of a photovoltaic plant, without using the environmental parameters such as the solar radiation and the temperature. The main idea is to statistically compare the energy performances of the arrays constituting the PV plant. In fact, the environmental conditions affect equally all the arrays of a small-medium-size PV plant, because the extension of the plant is limited, so any comparison between the energy distributions of identical arrays is independent of the solar radiation and the cell temperature, making the proposed methodology very effective for PV plants not equipped with a weather station, as it often happens for the PV plants located in urban contexts and having a nominal peak power in the 3÷50 kWp range, typically installed on the roof of a residential or industrial building. In this case, the costs of an advanced monitoring system based on the environmental data are not justified, consequently, the weather station is often also omitted. The proposed procedure guides the user through several inferential statistical tools that allow verifying whether the arrays have produced the same amount of energy or, alternatively, which is the worst array. The procedure is effective in detecting and locating abnormal operating conditions, before they become failures.

**Keywords:** ANOVA; Bartlett's test; Hartigan's dip test; Jarque-Bera's test; Kruskal-Wallis' test; Mood's Median test; residential buildings; Tukey's test; urban context

## 1. Introduction

The random variability of atmospheric phenomena affects the available irradiance intensity for photovoltaic (PV) generators. During the clear days an analytic expression for the solar irradiance can be defined, whereas this is not possible for cloudy days. The effects of the environmental conditions are studied in [1–5]. After the installation of a PV plant, a system for monitoring the energy performance in every environmental condition is needed. As the modules are main components of a PV plant, deep attention is focused on their state of health [6]. For this reason, techniques commonly used to verify the presence of typical defects in PV modules are based on infrared analysis [7,8], eventually supported by unmanned aerial vehicles [9], on luminescence imaging [10], or on their combination [11], while automatic procedures to extract information using thermograms are proposed in [12,13]. Nevertheless, these approaches regard single modules of the PV plants. When there is no information about the general operation of the PV plant, other techniques can be considered to prevent failures and to enhance the energy performance of the PV system, such as artificial neural networks [3,14], statistics [15–17], and checking the electrical variables [18–20]. More in detail, some of the PV fault detection algorithms are based on electrical circuit simulation of the PV generator [21,22], while other researchers use approaches based on the electrical signals [23,24]. Moreover, predictive model approaches for PV system power production based on the comparison between measured and modeled PV system outputs are discussed in [15,25–27]. Standard benchmarks [28], called *final PV system yield, reference yield* and *Performance Ratio (PR)*, are currently used to assess the overall system performance in terms of energy production, solar resource, and system losses. These benchmarks have been recently used to review the energy performance of 993 residential PV systems in Belgium [29] and 6868 PV installations in France [30]. Unfortunately, these indices have two drawbacks: they only supply rough information about the performance of the overall PV plant and they do not allow any assessment of the behavior of single PV plant parts. Moreover, when important faults such as short circuits or islanding occur, the electrical variables and the produced energy have fast and not negligible variations, so they are easily detected. These events produce drastic changes and can be classified as high-intensity anomalies. On the other hand, low-intensity anomalies such as the ageing of the components or minimal partial shading produce minimal variations on the electrical variables and on the produced energy, so they are not easily detectable. Moreover, these minor anomalies can evolve into failures or faults, so their timely identification can avoid more serious failures and limit the occurrence of out of order states. With respect to the configuration defined in the design stage, any PV plant can be single-array or multi-array, being an array a set of connected PV modules, for which the electrical variables and the produced energy are measured. Moreover, PV plants with only two arrays are not common: the alternatives are between one-array PV plant—this is the case of small nominal peak power PV plant—and multiple-array PV plant for higher nominal peak power PV plant. The multiple-array solution is very common, because it has several advantages, thanks to the partition of the produced energy: lower current for each array (thus reduced section of the solar cables), high flexibility in the choice of the components (inverter, switches, electrical boxes, etc.), O&M services on each single array, avoiding situations where the whole plant is out of order, and so on. Moreover, the large PV plants, having a nominal peak power higher than 100 kWp are usually equipped with a weather station, able to measure and store the environmental parameters, which affect the energy production, i.e., the solar irradiance, temperature and wind. Frequently, the large PV plants with nominal peak power higher than 1 MWp are equipped with more than one weather station, because of the large occupied area, typically about 2 ha/MWp. Obviously, these last ones are solar farms and are located in extra-urban territory. Instead, the PV plants usually installed in a urban context have a nominal peak power that ranges between 3÷50 kWp; the minimum value refers, for example, to a PV plant on the roof of a residential building, while the maximum value corresponds to a PV plant of a small company that locates it on the roof of its industrial building or in a free private area. These PV plants are usually multi-array and are not equipped with a weather station, because the costs of an advanced monitoring system are not negligible with respect to the initial investment as well as to the costs of a yearly O&M service, so these medium size PV plants are usually equipped with a simplified monitoring system, which stores the total produced energy, the electrical variables on the AC and DC side, having also the possibility to send alerts to the owner via SMS or email. This system does not perform any analysis of the produced energy, so it cannot detect any anomaly before it becomes a failure, and it can only alert when the failure is already happened. In these cases, valid support is provided by the PhotoVoltaic Geographical Information System (PVGIS) [31] of the European Commission Joint Research Centre (EC-JRC) that is based on the historical data of the solar irradiance. Figure 1 is a screenshot of the website. On the left hand-side, a colored map with the solar radiation is reported and the user can select the location of the PV plant, whereas, on the right hand-side, the user can insert the information on the typology of the PV plant (off-grid, grid-connected, tracking-based), the specifications of the PV plant (module technology, slope, rated power, etc.), and the required energy production data (monthly, daily, hourly). In this way, it is possible to estimate the productivity of the PV plant under investigation and to compare it with the real energy production. This can represent a preliminary check of the operation of the PV plant and will be used later, in the Sections 3 and 4. Nevertheless, it is extremely important to prevent a failure, detecting any anomaly in a timely way for two reasons. Firstly, when an anomaly is present, the energy production is already lower than the expected one and this implies an economic loss. Secondly, a timely action of the O&M service allows restoring the damaged parts of the PV systems with minimum costs and minimum time out of order, reducing either the Mean Time To

Repair (MTTR)—because the damage is limited—or the Mean Down Time (MDT), because the restore action is planned while the PV plant is still operating. This strategy, evidently, greatly increases the availability of the PV plant and its yearly energy performance.

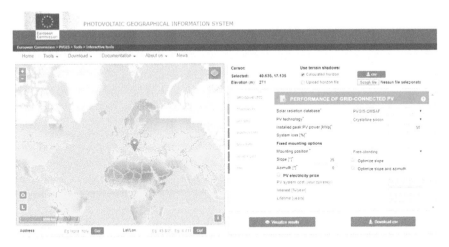

**Figure 1.** PVGIS of the European Commission Joint Research Centre (EC-JRC).

With this in mind, this paper proposes a methodology to detect an anomaly in the operation of a PV system; this methodology can be easily applied to any multi-array PV plant, but it is particularly useful for PV plants not equipped with a weather station. This is often the situation of the PV plants in urban contexts, as previously explained. The proposed methodology, in fact, compares the energy distributions of the arrays with each other, on the basis of a statistical algorithm that does not consider the environmental parameters as inputs. This is possible because the area occupied by the PV modules of a PV plant in the urban context is limited and then the average environmental conditions can be considered to affect the identical arrays in the same way. The proposed procedure is completely based on several hypothesis tests and is a cheap and fast approach to monitor the energy performance of a PV system, because no additional hardware is required. The procedure also allows continuous monitoring because it is cumulative and new data can be added to the initial dataset, as they are acquired by the measurement system. The methodology is based on an algorithm, which suggests the user, step by step, the suitable statistical tool to use. The first one is the Hartigan's Dip Test (HDT) that is able to discriminate an unimodal distribution from a multimodal one. The verification of the unimodality can be also done on the basis of a relationship between the values of skewness and kurtosis [32,33]; nevertheless, in this paper only HDT will be used, because it is usually more sensitive than other methods. The check on the unimodality is very important to decide whether a parametric test can be used to compare the energy distributions of the arrays or not, because the parametric tests, being based on known distributions, are more performing than the nonparametric ones. Nevertheless, the parametric tests can be applied, only if specific assumptions are satisfied. A powerful parametric test to compare more than two statistical distributions is the well-known ANalysis Of VAriance (ANOVA) [34] that is based on three assumptions. The proposed algorithm suggests using the Jarque-Bera's test and the Bartlett's test to verify the assumptions. In the negative case, the procedure suggests to use the Kruskal-Wallis test or the Mood's median test, in absence or in presence of outliers in the dataset, respectively. As a last step, Tukey's test is run to do a multi-comparison one-by-one between the mean values of the distributions, in order to determine which estimates are significantly different.

A case-study is discussed in the paper. The algorithm is applied to a real operating PV plant and the methodology is run four times: the first one, based on the energy dataset of one month; the second one, based on the energy dataset of three months; the third one, based on the energy dataset of six months; the last one, based on the energy dataset of the whole year. The paper is structured as follows: Section 2 describes the proposed algorithm, Section 3 describes the PV system under examination, Section 4 discusses the results, and the Conclusions end the paper.

## 2. Statistical Methodology

In this paper, it is consider that the PV plant is composed of $A$ identical arrays, with $A > 2$ for the reasons already explained in the Introduction. This constraint is mandatory for the proposed methodology, because it is based on the comparison among the energy distributions of the arrays constituting the PV system. Each array is usually equipped with a measurement system that measures the values of the produced energy in AC, other than the values of voltage and current of both the DC and AC sides of the inverter, with a fixed sampling time, $\Delta t$. At the generic time-instant $t = q\Delta t$ of the $j$-th day, the $q$-th sample vector of the $k$-th array is defined as $x_{j,k}(q) = \left[ E_{j,k}(q)\, v_{j,k,DC}(q)\, i_{j,k,DC}(q)\, v_{j,k,AC}(q)\, i_{j,k,AC}(q) \right]$, for $k = 1, \ldots, A$, $j = 1, \ldots, D$ (being $D$ the number of the investigated days), $q = 1, \ldots, Q$, where $q = 1$ characterizes the first daily sample at the analysis time $t = \Delta t$ and $q = Q$ defines the last daily sample, acquired at the time $t = Q \cdot \Delta t$. For our aims, let us consider only the dataset constituted by the energy values $E_{j,k}(q)$. Thus, the proposed methodology can be applied to any PV plant, having a measurement system that measures at least the produced energy, no matter which are the other measured variables. The $k$-th array, at the end of the $j$-th day, has produced the energy $E_{j,k} = \sum_{q=1}^{Q} E_{j,k}(q)$, therefore the complete dataset of the energy produced by the PV plant in a fixed investigated period can be represented in a matrix form:

$$E = \begin{pmatrix} E_{1,1} & \cdots & E_{1,A} \\ \vdots & \ddots & \vdots \\ E_{D,1} & \cdots & E_{D,A} \end{pmatrix} \tag{1}$$

The columns of the matrix (1) are independent each other, because the values of each array are acquired by devoted acquisition units, so no inter-dependence exists among the values of the columns, which can be considered as separate statistical distributions. The flow chart in Figure 2 proposes the methodology to detect and locate any anomaly, before it becomes a fault.

It is based on the mutual comparison among the energy distributions of the arrays; therefore the environmental data are not necessary. Obviously, this approach is valid only if the arrays are identical (same PV modules, same number of modules for each array, same slope, same tilt, same inverter, and so on); in fact, under this assumption, the energy produced by any array must be almost equal to the energy produced by any other array of the same PV plant, in each period as well as in the whole year (the changing environmental conditions affect the arrays in the same way, if they are installed next to each other without any specific obstacle).

Thus, the comparative and cumulative monitoring of the energy performance of identical arrays allows one to determine, within the uncertainty defined by the value of the significance level $\alpha$, if the arrays are producing the same energy or not. The first step of Figure 2 is the pre-processing of the energy dataset collected as previously explained, in order to check if outliers are present; the information about the presence or not of the outliers will be also useful later (green block). By default, an outlier is a value that is more than three scaled Median Absolute Deviations (MAD) away from the median. For a random dataset $X = [X_1, X_2, \ldots, X_D]$, the scaled MAD is defined as:

$$MAD = F \times median\left(\left|X_j - median(X)\right|\right) \text{ for } j = 1, \ldots, D \tag{2}$$

where $F$ is the scaling factor and is approximately 1.4826 for a normal distribution.

After the data pre-processing, it is necessary to verify if the arrays have produced the same amount of energy. This goal can be pursued by using parametric tests or non-parametric tests. As the parametric tests are based on a known distribution of the dataset, they are more reliable than the non-parametric ones, which are, instead, distribution-free. For this reason, it is advisable to use always the parametric tests, provided that all the needed assumptions are satisfied.

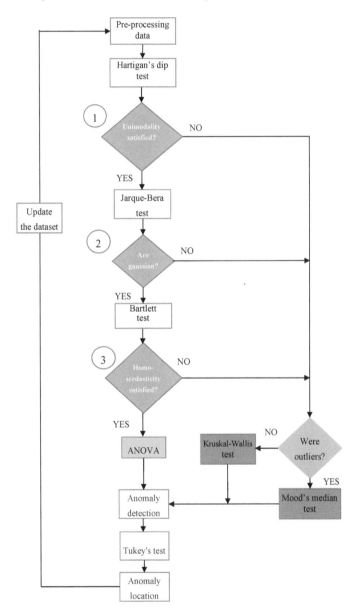

**Figure 2.** Statistical methodology.

In particular, the parametric test known as ANOVA calculates the ratio between the variance *among* the arrays' distributions (divided by the freedom degree) and the variance *within* each array distribution (divided by the freedom degree). In other words, ANOVA evaluates whether the differences of the mean values of the different groups are statistically significant or not. For this aim, ANOVA calculates the following Fisher's statistic, $F$, [35]:

$$F = \frac{\frac{D}{A-1} \cdot \sum_{k=1}^{A} \left( \overline{x}_k - \overline{X} \right)^2}{\frac{1}{A(D-1)} \cdot \sum_{k=1}^{A} \sum_{j=1}^{D} \left( x_{kj} - \overline{x}_k \right)^2} \tag{3}$$

where $\overline{x}_k$ is the mean value of the $k$-th distribution, $\overline{X}$ the global mean, $x_{kj}$ the $j$-th occurrence of the $k$-th distribution. The cumulative distribution function F allows to determine a $p$-value, which has to be compared with the significance level $\alpha$, as later explained.

ANOVA is based on the null hypothesis $H_0$ (Equation (4)) that the means of the distributions, $\mu_k$, are equal:

$$H_0 : \ \mu_1 = \mu_2 = \mu_3 = \cdots = \mu_A \tag{4}$$

versus the alternative hypothesis that the mean value of at least one distribution is different from the others. The output of the ANOVA test, as any other hypothesis test, is the $p$-value, which has to be compared with the pre-fixed significance value $\alpha$. Usually, $\alpha = 0.05$, so, if $p$-value $< \alpha$ then the null hypothesis is rejected, considering acceptable to have a 5% probability of incorrectly rejecting the null hypothesis (this is known as type I error).

Smaller values of $\alpha$ are not advisable to study the data of a medium-large PV plant, because the complexity of the whole system requires a larger uncertainty to be accepted. Nevertheless, ANOVA can be used only under the following assumptions:

(a) all the observations are mutually independent;
(b) all the distributions are normally distributed;
(c) all the distributions have equal variance.

Finally, ANOVA can be applied also for limited violations of the assumptions (b) and (c), whereas the assumption (a) is always verified, if the measures come from independent local measurement units. So, before applying ANOVA test, several verifications are needed and they are represented by the three blue blocks of Figure 2. The first check (blue block 1) regards the unimodality of the dataset of each array, because a multimodality distribution, e.g., the bimodal distribution in Figure 3, is surely not Gaussian and violates the condition (b). Moreover, the daily-based energy distribution of an array of a well-working PV system is unimodal, because the daily solar radiation has the typical Gaussian waveform, which is unimodal; therefore, the multimodality of a daily-based energy distribution is a clear alert of a high-intensity anomaly. The Hartigan's Dip Test (HDT) is able to check the unimodality [36] and is based on the null hypothesis that the distribution is unimodal versus the alternative one that it is at least bi-modal. The HDT is a non-parametric test, so it is distribution-free. HDT return a $p$-value$_{\text{HDT}}$. By fixing the significance value $\alpha = 0.05$, if $p$-value$_{\text{HDT}} < \alpha$ is satisfied, the null hypothesis of the unimodality is rejected, the distribution is surely not Gaussian, ANOVA cannot be applied and a nonparametric test has to be used.

In the general case of $A$ arrays, with $\underline{A > 2}$, the nonparametric test has to be chosen between Kruskal-Wallis test (K-W) [37,38] and Mood's Median test (MM), under the constraint of the green block; both K-W and MM do not require that the distributions are Gaussian, but only that the distributions are continuous.

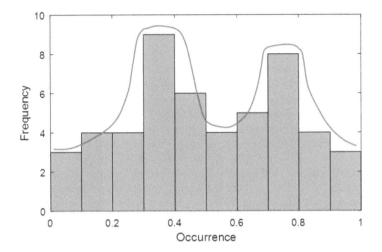

**Figure 3.** Example of histogram of a bimodal distribution.

In the presence of outliers (detected, if present, in the first block), MM performs better than K-W, otherwise K-W is a good choice. Both K-W and MM are based on the null hypothesis that the median values of all the distributions are equal versus the alternative one that at least one distribution has a median value different from the others. As K-W as MM returns a $p$-value$_{\text{K-W(MM)}}$ that has to be compared with the significance value $\alpha = 0.05$. If $p$-value$_{\text{K-W(MM)}} < \alpha$ is satisfied, the null hypothesis is rejected and the arrays have not produced the same energy; otherwise, they have. Instead, if the unimodality is satisfied, other checks are needed, before deciding whether ANOVA can be applied. In fact, it is needed to verify the previous assumptions (b) and (c). Only if both of them are satisfied (blue blocks 2 and 3, respectively), ANOVA can be applied.

To check the condition (b), an effective statistical tool is the Jarque-Bera's Test (JBT). The JBT is distribution-free and based on independent random variable. It is a hypothesis test, whose null hypothesis is that the distribution is gaussian. Then, it calculates a statistical parameter, called JB, and returns a $p$-value$_{\text{JBT}}$. By fixing the significance value $\alpha = 0.05$, if $p$-value$_{\text{JBT}} < \alpha$ is satisfied, the null hypothesis is rejected and the distribution is not gaussian, otherwise it is. It results:

$$JB = \frac{D}{6} \cdot \left[ \sigma_k{}^2 + \frac{k_u{}^2}{4} \right]$$ (5)

Being $D$ the sample size, $\sigma_k$ the skewness and $k_u$ the Pearson's kurtosis less 3 (also known as excess kurtosis). The skewness is defined as:

$$\sigma_k = \frac{1}{D} \frac{\sum_{j=1}^{D} (x_j - \overline{x})^3}{(\hat{\sigma}^2)^{\frac{3}{2}}}$$ (6)

Being D the sample size, $\overline{x} = \frac{1}{D} \sum_{j=1}^{D} x_j$ the mean value, and $\hat{\sigma}^2 = \frac{1}{D} \sum_{j=1}^{D} (x_j - \overline{x})^2$ the variance. The skewness is the third standardized moment and measures the asymmetry of the data around the mean value. Only for $\sigma_k = 0$ the distribution is symmetric; this is a necessary but not sufficient condition for a gaussian distribution. In fact, while the Gaussian distribution is surely symmetric, nevertheless there exist also symmetric but not gaussian distributions.

The excess kurtosis, instead, is defined as:

$$k_u = \frac{1}{D} \frac{\sum_{j=1}^{D}(x_j - \bar{x})^4}{(\hat{\sigma}^2)^2} - 3 \tag{7}$$

with the previous meaning of the parameters. The kurtosis is the fourth standardized moment and measures the tailedness of the distribution. Only for $k_u = 0$, the distribution is mesokurtic, which is the necessary but not sufficient condition for a Gaussian distribution. If the check of the blue block 2 is not passed, a non-parametric test (K-W or MM) has to be used, in accordance with the green block. Instead, if this verification is passed, it needs to test the assumption (c) of ANOVA, i.e., the homoscedasticity (blue block 3). This assumption can be verified by means of the Bartlett's Test (BT), which is again a hypothesis test that returns a $p$-value$_{BT}$. The BT is effective for Gaussian distributions; in fact, in the flow-chart of Figure 2 it is used only if the distributions are Gaussian. Also in this case, it is possible to fix the common significance value $\alpha = 0.05$ and to compare it with the $p$-value$_{BT}$. If the inequality $p$-value$_{JBT} < \alpha$ is satisfied, the null hypothesis is rejected and the variances of the distributions of the arrays are different, then the condition (c) is violated, and ANOVA cannot be applied. In this case, it is necessary to use K-W or MM, in accordance with the green block. Otherwise, ANOVA can be applied and it return another $p$-value$_{AN}$ that must be compared with the significance level $\alpha = 0.05$. If the inequality $p$-value$_{AN} < \alpha = 0.05$ is satisfied, then the null hypothesis ($H_0 : \mu_1 = \mu_2 = \mu_3 = \cdots = \mu_A$) is rejected and the conclusion is that the identical arrays have not produced the same amount of energy; so, a low-intensity anomaly is present and it is located in the array that has the mean value different from the other ones. To detect it, a multi-comparison analysis—one-to-one—between the distributions is done by means of the Tukey's Test (TT), which is a modified version of the well-known $t$-test and returns a $p$-value$_{TT}$, which states whether the means between two distributions are equal or not. For a small sample size (about 20 samples) the TT is reliable only for normal distribution, instead, for a lager sample size it is valid also for not normal distributions, because of the central limit theorem. Otherwise, no criticality is present and the dataset can be updated with new data to continue the monitoring of the PV plant. As the energy dataset increases, the monitoring becomes more accurate.

## 3. Description of the PV Plant under Investigation

The system under examination is a real operating 49.5 kWp grid-connected PV plant, installed in the south of Italy on the roof of the industrial building of a company. The PV plant has been designed and installed under the scheme of the feed-in tariff, financed by the Government. The 330 modules of the PV plant are equally divided in five arrays, each of them constituted by 66 PV modules. The nominal peak power of a single module is 150 W, and then the nominal peak power of a single array is 9.9 kWp. Each array is connected to the grid via a 10 kW inverter. The system faces the south and the slop is about 30°. By inserting these values in the previously mentioned PVGIS [31] of the EC-JRC, it results that the estimated yearly energy production is about 64,724 kWh, corresponding to about 1307 kWh/kWp per year. Moreover, the website provides also the estimated monthly energy production, which will be used in the next Section 4.1, Section 4.2, and Section 4.3. The PV plant is equipped with a datalogger that stores the data from the five arrays. The datalogger has a sample time of 2 s. After 10 min, the measured samples are equal to 30 (samples/min) × 10 (min) = 300 samples; an internal software calculates the average value of these 300 measured samples, whereas the energy produced in this time-slot of 10 min is calculated as $P_{average} \cdot \frac{10}{60} [kWh]$. This value is stored into the datalogger. So, the sampling time of the energy is 10 min, therefore there are 6 samples/hour, hence 144 samples/day, that are summed in the proposed procedure. Thus, the unique value of a day is not an average value, but a cumulative data that takes into account the variability of the environmental conditions happened during the whole day.

The measured variables are the power in AC, the energy in AC, and the voltage $V_{dc}$ of each inverter; moreover, the number of the operating hours is stored. The default monitoring system of the PV plant uses the power in AC and the voltage $V_{dc}$, the daily and total energy produced by each inverter, and the number of the operating hours. It is worth noting that the monitoring system is an internal software of the datalogger. As the operation of the monitoring system occupies the internal memory, for default the internal monitoring system does not utilize all the data available into the datalogger, in order not to occupy the internal memory quickly. This approach allows to monitor the PV plant for a longer time, but only the high-intensity anomalies can be detected. Instead, to detect even the low-intensity anomalies, it is necessary to use the methodology described in Figure 2 and all the data stored into the datalogger. Moreover, even if the measurement system of this PV plant does not measure all the variables mentioned in the Section 2 (the produced energy, other than the voltage and current in both the DC and AC side), nevertheless it acquires the produced energy that is the unique variable necessary for the proposed methodology; so it can be applied. The observation period refers to a full year during which the plant has shown some malfunctions, whereas in the previous years the PV plant has not shown any malfunctions, therefore the results of the previous years are not reported in the paper.

## 4. Cumulative Statistical Analysis

The energy performance of the PV plant described in Section 3 has been studied by means of the statistical methodology proposed in Section 2. Statistical data analysis has been carried out in Matlab R2017 environment by using the standard routines of the Statistics toolbox and by implementing the flow chart of Figure 2. In particular, a Matlab routine that implements just the procedure of Figure 2 has been written and run for each analysis discussed later. As some tests (ANOVA, K-W, JBT) are implemented in the *Statistics Toolbox* of Matlab, these native-routines are recalled from the main routine, when necessary. As already explained, each array is equipped by a devoted measurement system, then the five distributions are mutually independent.

Four analyses are discussed, based on the dataset of the energy produced by each array:

- one-month analysis (January);
- three-months analysis (January–March);
- six-months analysis (January–June);
- one-year analysis (January–December).

The increase of the time window, updating the dataset as described in Figure 2, allows understanding how some characteristic benchmarks of the PV plants vary during the year, as new data are acquired. The following results will be reported for each analysis: the $p$-value$_{HDT}$ of each distribution to test the unimodality; the $p$-value$_{JBT}$ of each distribution to test whether each one of them is gaussian; the $p$-value$_{BT}$ to test the homoscedasticity among the distributions; the $p$-value$_{AN}$ to test whether all the distributions have the same mean value; the $p$-value$_{K-W(MM)}$ of the non-parametric test (when ANOVA cannot be applied) to check whether all the distributions have the same median value; the box plot of the ANOVA test or of the non-parametric test; the mean value of each distribution and its spread with respect to the global mean of the PV plant.

### 4.1. One-Month Analysis (January)

Table 1 reports the main numerical values of the parameters calculated by applying the procedure in Figure 2.

This dataset is constituted by 31 cumulative samples/array, each sample being the sum of 144 samples/day. The energy dataset of the first month does not contain outliers. The $p$-value$_{HDT} > \alpha = 0.05$ for each distribution, so all the distributions are unimodal. To apply ANOVA, conditions (b) and (c) have to be verified.

**Table 1.** *p*-Value of HDT, JBT, BT, ANOVA for the energy distribution of the arrays, and mean in kWh and spread with respect to the global mean for one-month analysis (January).

| Array Number | | 1 | 2 | 3 | 4 | 5 |
|---|---|---|---|---|---|---|
| *p*-value$_{HDT}$ | | 0.628 | 0.364 | 0.674 | 0.658 | 0.670 |
| JBT | JB | 0.699 | 0.676 | 0.674 | 0.718 | 0.638 |
| | *p*-value$_{JBT}$ | 0.500 | 0.500 | 0.500 | 0.500 | 0.500 |
| *p*-value$_{BT}$ | | | | 0.999 | | |
| *p*-value$_{AN}$ | | | | 0.999 | | |
| Mean (kWh) | | 19.59 | 19.90 | 19.91 | 19.46 | 19.84 |
| Global mean | | | | 19.74 | | |
| Spread % | | −0.76 | 0.80 | 0.85 | −1.40 | 0.51 |

Table 1 reports the JB values and the related *p*-value$_{JBT}$; as *p*-value$_{JBT}$ > $\alpha$ = 0.05, all the distributions are Gaussian, so condition (b) of ANOVA is satisfied. Condition (c) about the homoscedasticity has to be verified by means of BT (see Figure 2). The *p*-value$_{BT}$ = 0.999 in Table 1 (again higher than $\alpha$ = 0.05) says that the homoscedasticity is verified, then all the variances are equal. Therefore, the main conditions of the flow chart in Figure 2 (blocks 1,2,3) are satisfied and ANOVA can be applied. The *p*-value$_{AN}$ = 0.999 in Table 1 says that the distributions have the same mean values, so all the arrays have produced the same energy in this month. Figure 4 is the box plot of ANOVA. For each box, the central red mark indicates the median, and the bottom and top edges of the box indicate the 25th and 75th percentiles, respectively. The whiskers extend to the most extreme data points. Figure 4 highlights that the five distributions have produced almost the same energy, both with respect to the median value (in red color) and to the first and third inter-quartiles; moreover, outliers are absent. Therefore, no anomaly is present in the PV plant. Particularly, from PVGIS [31], it results that the estimated average energy of the PV plant in January should be about 3173 kWh, corresponding to a daily average energy for each array of about 3173/(31 × 5) = 20.5 kWh, that is almost equal to the global mean value 19.74 kWh of Table 1. Figure 5 diagrams the mean value and confidence interval at 95% of each distribution; the values are very similar each other, as it results also from Table 2 that reports the one-to-one comparisons of the mean values. In particular, the high *p*-values confirm that the differences are not significant.

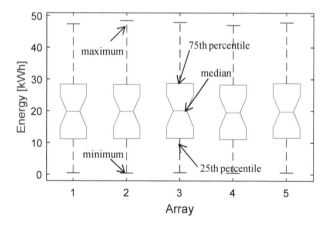

**Figure 4.** Box plot of ANOVA test of the five distributions for the one-month analysis (January).

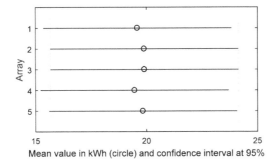

**Figure 5.** Mean value of the energy produced by each array for the one-month analysis (January).

**Table 2.** One-to-one comparison of the means for the one-month analysis (January).

| Comparison between Samples | | LowerBound | DifferenceEstimate | UpperBound | $p\text{-Value}_{TT}$ |
|---|---|---|---|---|---|
| 1 | 2 | −8.77 | −0.31 | 8.16 | 0.999 |
| 1 | 3 | −8.78 | −0.32 | 8.15 | 0.999 |
| 1 | 4 | −8.34 | 0.12 | 8.59 | 0.999 |
| 1 | 5 | −8.71 | −0.25 | 8.21 | 0.999 |
| 2 | 3 | −8.47 | −0.01 | 8.45 | 1 |
| 2 | 4 | −8.03 | 0.43 | 8.90 | 0.999 |
| 2 | 5 | −8.41 | 0.06 | 8.52 | 1 |
| 3 | 4 | −8.02 | 0.44 | 8.91 | 0.999 |
| 3 | 5 | −8.40 | 0.06 | 8.53 | 1 |
| 4 | 5 | −8.84 | −0.38 | 8.09 | 0.999 |

*4.2. Three-Months Analysis (January–March)*

Table 3 reports the main numerical values of the parameters calculated by following the algorithm of Figure 2 for the energy dataset of three months, including the first month already considered in the previous analysis. This dataset is constituted by 90 cumulative samples/array, each sample being the sum of 144 samples/day. The $p\text{-value}_{HDT} > \alpha = 0.05$ for each distribution, so all the distributions are still unimodal. The $p\text{-value}_{JBT} > \alpha = 0.05$, so all the distributions are gaussian and the condition (b) of ANOVA is satisfied. The homoscedasticity is also satisfied ($p\text{-value}_{BT} > \alpha = 0.05$). Therefore, the main conditions of the flow chart in Figure 2 (blocks 1,2,3) are satisfied and ANOVA can be newly applied. The $p\text{-value}_{AN} = 0.998$ affirms that the distributions have the same mean values, so all the arrays have produced the same energy also in these three months. Figure 6 is the box plot of ANOVA and it highlights that the five distributions have produced almost the same energy, both with respect to the median value (in red color) and to the first and third inter-quartiles; moreover, outliers are not present. Therefore, no anomaly is present in the PV plant in these three months. Particularly, from PVGIS [31], it results that the estimated average energy of the PV plant in the period January–June should be about 11,695 kWh, corresponding to a daily average energy for each array of about 11,695/(90 × 5) = 25.99 kWh, that is almost equal to the global mean value 25.95 kWh of Table 2.

Figure 7 plots the mean value and confidence interval at 95% of each distribution; the values are very similar each other, as it results also from Table 4 that reports the one-to-one comparisons of the mean values. In particular, the high $p$-values confirm that the differences are not significant.

**Table 3.** *p*-Value of HDT, JBT, BT, ANOVA for the energy distribution of the arrays, and mean in kWh and spread with respect to the global mean for three-month analysis (January–March).

| Array Number | | 1 | 2 | 3 | 4 | 5 |
|---|---|---|---|---|---|---|
| *p*-value$_{HDT}$ | | 0.776 | 0.892 | 0.818 | 0.856 | 0.830 |
| JBT | JB | 5.127 | 5.198 | 5.114 | 5.177 | 5.090 |
| | *p*-value$_{JBT}$ | 0.054 | 0.053 | 0.054 | 0.053 | 0.055 |
| *p*-value$_{BT}$ | | | | 0.999 | | |
| *p*-value$_{AN}$ | | | | 0.998 | | |
| Mean (kWh) | | 25.84 | 25.82 | 26.23 | 25.62 | 26.39 |
| Global mean | | | | 25.95 | | |
| Spread % | | −0.54 | −0.62 | 0.98 | −1.36 | 1.58 |

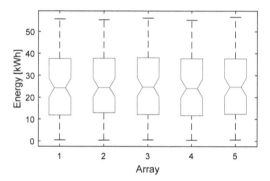

**Figure 6.** Box plot of K-W test of the five distributions for the three-months analysis (January–March).

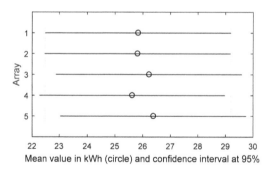

**Figure 7.** Mean value of the energy produced by each array for the three-months analysis (January–March).

**Table 4.** One-to-one comparison of the means for the three-months analysis (January–March).

| Comparison between Samples | | LowerBound | DifferenceEstimate | UpperBound | *p*-Value$_{TT}$ |
|---|---|---|---|---|---|
| 1 | 2 | −6.68 | 0.02 | 6.73 | 1 |
| 1 | 3 | −7.10 | −0.39 | 6.31 | 0.999 |
| 1 | 4 | −6.49 | 0.21 | 6.92 | 0.999 |
| 1 | 5 | −7.25 | −0.55 | 6.15 | 0.999 |
| 2 | 3 | −7.12 | −0.41 | 6.29 | 0.999 |
| 2 | 4 | −6.51 | 0.19 | 6.90 | 0.999 |
| 2 | 5 | −7.27 | −0.57 | 6.13 | 0.999 |
| 3 | 4 | −6.09 | 0.60 | 7.31 | 0.999 |
| 3 | 5 | −6.86 | −0.15 | 6.55 | 0.999 |
| 4 | 5 | −7.47 | −0.76 | 5.94 | 0.997 |

### 4.3. Six-Months Analysis (January–June)

Table 5 displays the main numerical values of the parameters obtained by the application of the procedure in Figure 2, after updating the previous energy dataset (used for the January–March analysis), by adding the data of the successive three months. This dataset is constituted by 181 cumulative samples/array, each sample being the sum of 144 samples/day. The pre-processing of the new dataset has excluded the presence of outliers. The $p$-value$_{HDT} > \alpha = 0.05$ for each distribution, so all the distributions are still unimodal. As $p$-value$_{JBT} < \alpha = 0.05$ for each distribution, then the null hypothesis is rejected and the constraint of the block 2 (corresponding to the condition (b) of ANOVA) is not satisfied: the distributions are not Gaussian. Therefore, it has no sense to verify the homoscedasticity, because it is mandatory to use a nonparametric test. As no outlier is present, it is advisable to use K-W, as suggested by the green block. The $p$-value$_{K-W} = 0.861$ affirms that the distributions have the same median values, so all the arrays have produced the same energy also in these six months, even if the distributions are no longer Gaussian. Figure 8 is the box plot of K-W and it highlights that the five distributions have produced almost the same energy, both with respect to the median value (in red color) and to the first and third inter-quartiles; moreover, it is confirmed that outliers are not present. Therefore, no anomaly is present in the PV plant in these six months. Particularly, from PVGIS [31], it results that the estimated average energy of the PV plant in the period January–June should be about 32,285 kWh, corresponding to a daily average energy for each array of about $32,285/(181 \times 5) = 35.67$ kWh, that is almost equal to the global mean value 36.23 kWh of Table 3. Figure 9 illustrates the mean value and confidence interval at 95% of each distribution; the values are very similar each other, as it results also from Table 6 that reports the one-to-one comparisons of the mean values. In particular, the high $p$-values confirm that the differences are not significant.

**Table 5.** $p$-Value of HDT, JBT, BT, (KW) for the energy distribution of the arrays, and mean in kWh and spread with respect to the global mean for six-months analysis (January–June).

| Array Number | | 1 | 2 | 3 | 4 | 5 |
|---|---|---|---|---|---|---|
| $p$-value$_{HDT}$ | | 0.794 | 0.722 | 0.782 | 0.808 | 0.842 |
| JBT | JB | 13.95 | 14.02 | 13.92 | 13.94 | 13.85 |
| | $p$-value$_{JBT}$ | 0.007 | 0.007 | 0.007 | 0.007 | 0.007 |
| $p$-value$_{BT}$ | | | | | | |
| $p$-value$_{K-W}$ | | | | 0.861 | | |
| Mean (kWh) | | 36.01 | 35.78 | 36.60 | 35.74 | 37.04 |
| Global mean | | | | 36.23 | | |
| Spread % | | −0.60 | −1.23 | 1.02 | −1.35 | 2.23 |

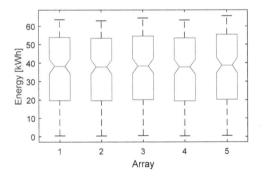

**Figure 8.** Box plot of K-W test of the five distributions for the six-months analysis (January–June).

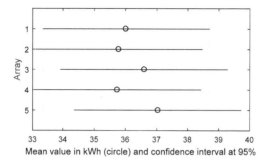

**Figure 9.** Mean value of the energy produced by each array for the three-months analysis (January–June).

**Table 6.** One-to-one comparison of the means for the three-months analysis (January–June).

| Comparison between Samples | | LowerBound | DifferenceEstimate | UpperBound | $p$-Value$_{TT}$ |
|:---:|:---:|:---:|:---:|:---:|:---:|
| 1 | 2 | −5.14 | 0.22 | 5.60 | 0.999 |
| 1 | 3 | −5.96 | −0.58 | 4.78 | 0.999 |
| 1 | 4 | −5.10 | 0.27 | 5.64 | 0.999 |
| 1 | 5 | −6.40 | −1.02 | 4.34 | 0.985 |
| 2 | 3 | −6.18 | −0.81 | 4.55 | 0.993 |
| 2 | 4 | −5.33 | 0.04 | 5.41 | 1 |
| 2 | 5 | −6.62 | −1.25 | 4.11 | 0.968 |
| 3 | 4 | −4.51 | 0.85 | 6.23 | 0.992 |
| 3 | 5 | −5.81 | −0.43 | 4.93 | 0.999 |
| 4 | 5 | −6.67 | −1.29 | 4.07 | 0.964 |

*4.4. One-Year Analysis (January–December)*

Table 7 displays the main numerical values of the parameters obtained by the application of the procedure in Figure 2, after updating the previous energy dataset (used for the January–June analysis), by adding the data of the successive six months. This dataset is constituted by 365 cumulative samples/array, each sample being the sum of 144 samples/day. The pre-processing of the new dataset has excluded the presence of outliers. The $p$-value$_{HDT} > \alpha = 0.05$ for each distribution, except for the distribution n. 4, for which $p$-value$_{HDT}(4) = 0.006 < \alpha = 0.05$; therefore, the condition of the block 1 about the unimodality is not satisfied for all the distributions and ANOVA cannot be applied. Consequently, it is mandatory to use a nonparametric test. As no outlier is present, it is advisable to apply K-W, as suggested by the green block. As $p$-value$_{K-W} = 0.009 < \alpha = 0.05$, the null hypothesis is rejected, so the distributions have different median values. This implies that the arrays have not produced the same energy in the complete year, even if they had produced the same energy for the first six months. Figure 10 is the box plot of K-W and it highlights that the median value of the distribution n. 4 is significantly different from the others. It is also confirmed that outliers are not present. Particularly, from PVGIS [31], it results that the estimated average energy of the PV plant in the period January–December should be about 64,724 kWh, corresponding to a daily average energy for each array of about $64{,}724/(365 \times 5) = 35.46$ kWh, that is almost equal to the global mean value 35.13 kWh of Table 4. Therefore, high-intensity anomaly is not present, but a low-intensity anomaly is detected in the array n. 4, as confirmed also by the spreads of the mean values reported in Table 4. It can be observed that the array n. 4 produced 6.54% less than the average energy of the PV plant. Figure 11 shows the mean value and confidence interval at 95% of each distribution. It can be observed that the array n. 4 is very different from the other ones, as it results also from Table 8 that reports the one-to-one comparisons of the mean values. In particular, the $p$-value $0.043 < \alpha = 0.05$ rejects the hypothesis that the distribution 4 and 5 have the same mean value.

**Table 7.** $p$-Value of HDT and K-W for the energy distribution of the arrays, and mean in kWh and spread with respect to the global mean for one-year analysis.

| | Array Number | | | | |
|---|---|---|---|---|---|
| | **1** | **2** | **3** | **4** | **5** |
| $p$-value$_{HDT}$ | 0.820 | 0.846 | 0.898 | 0.006 | 0.636 |
| $p$-value$_{JBT}$ | - | - | - | - | - |
| $p$-value$_{BT}$ | - | - | - | - | - |
| $p$-value$_{K\text{-}W}$ | | | 0.009 | | |
| Mean (hWh) | 35.37 | 35.24 | 35.92 | 32.83 | 36.31 |
| Global mean | | | 35.13 | | |
| Spread % | 0.68 | 0.32 | 2.25 | −6.54 | 3.35 |

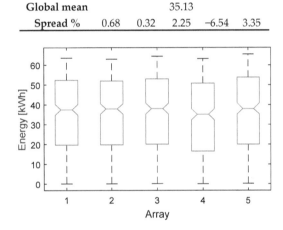

| Global mean | | 35.13 | | | |
|---|---|---|---|---|---|
| Spread % | 0.68 | 0.32 | 2.25 | −6.54 | 3.35 |

**Figure 10.** Box plot of K-W test of the five distributions for the one-year analysis.

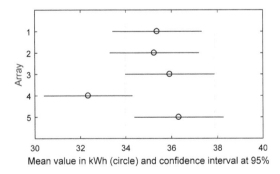

**Figure 11.** Mean value of the energy produced by each array for the one-year analysis.

**Table 8.** One-to-one comparison of the means for the one-year analysis.

| Comparison between Samples | | LowerBound | DifferenceEstimate | UpperBound | $p$-Value$_{TT}$ |
|:---:|:---:|:---:|:---:|:---:|:---:|
| 1 | 2 | −3.76 | 0.12 | 4.01 | 0.999 |
| 1 | 3 | −4.44 | −0.55 | 3.33 | 0.995 |
| 1 | 4 | −0.86 | 3.02 | 6.91 | 0.210 |
| 1 | 5 | −4.83 | −0.94 | 2.95 | 0.964 |
| 2 | 3 | −4.56 | −0.67 | 3.21 | 0.989 |
| 2 | 4 | −0.98 | 2.90 | 6.79 | 0.249 |
| 2 | 5 | −4.95 | −1.06 | 2.82 | 0.945 |
| 3 | 4 | −0.31 | 3.57 | 7.47 | 0.088 |
| 3 | 5 | −4.28 | −0.38 | 3.50 | 0.998 |
| 4 | 5 | −7.85 | −3.96 | −0.07 | 0.043 |

## 5. Conclusions

The paper proposes a statistical algorithm to monitor the energy performance of PV plants and detect anomalies. The procedure is cumulative and the algorithm can be iterated, as new data are acquired by the measurement system, in order to follow the most important benchmarks. The case study, referred to a real operating PV system, has shown the results of four cumulative analyses, starting from the dataset of only one month and finishing with a yearly-based dataset. As real operating PV systems are affected by atmospheric phenomena, their energy distributions are never perfectly Gaussian, so parametric tests should never be applied. Instead, as ANOVA can be applied for modest violations of its assumptions, the issue consists in evaluating the violation, in order to decide whether it is negligible or not. The proposed methodology, based on hypothesis tests, allows this evaluation. The first two analyses (based on the data of one and three months, respectively) have been carried out by means of the parametric test ANOVA, whereas the third and the fourth analyses have been based on the nonparametric test K-W, because the mandatory ANOVA's assumptions were not satisfied. Moreover, while the first three analyses have not evidenced any anomaly in the PV plant—in fact the energy distributions of the arrays were almost equal—instead the last analysis has shown a not negligible anomaly in the array n. 4. The proposed methodology does not allow identifying the origin of the anomaly, but only to detect and locate it. Finally, the proposed procedure is particularly effective in absence of environmental parameters, i.e., for monitoring PV plants not equipped with a weather station. In this case, the procedure allows extracting the main operating features of the PV plants without adding new hardware; thus, this approach is also cheap. Nevertheless, when a commercial PV plant has to be evaluated, it is mandatory to take into account the environmental parameters; so, if the PV plant is not equipped by a weather station, it is necessary to add this component and to use the monitoring methodologies based on the environmental data, even though this results more expensive.

**Conflicts of Interest:** The author declares no conflict of interest.

## References

1. Xiao, W.; Ozog, N.; Dundorf, W.G. Topology study of photovoltaic interface for maximum power point tracking. *IEEE Trans. Ind. Electron.* **2007**, *54*, 1696–1704. [CrossRef]
2. Mutoh, N.; Inoue, T. A control method to charge series-connected ultraelectric double-layer capacitors suitable for photovoltaic generation systems combining MPPT control method. *IEEE Trans. Ind. Electron.* **2007**, *54*, 374–383. [CrossRef]
3. Grimaccia, F.; Leva, S.; Mussetta, M.; Ogliari, E. ANN Sizing Procedure for the Day-Ahead Output Power Forecast of a PV Plant. *Appl. Sci.* **2017**, *7*, 622. [CrossRef]
4. Dellino, G.; Laudadio, T.; Mari, R.; Mastronardi, N.; Meloni, C.; Vergura, S. Energy Production Forecasting in a PV Plant Using Transfer Function Models. In Proceedings of the 2015 IEEE 15th International Conference on Environment and Electrical Engineering (EEEIC), Roma, Italy, 10–13 June 2015.

5.  Vergura, S.; Pavan, A.M. On the photovoltaic explicit empirical model: Operations along the current-voltage curve. In Proceedings of the 2015 International Conference on Clean Electrical Power (ICCEP), Taormina, Italy, 16–18 June 2015.

6.  Guerriero, P.; di Napoli, F.; Vallone, G.; D'Alessandro, V.; Daliento, S. Monitoring and diagnostics of PV plants by a wireless self-powered sensor for individual panels. *IEEE J. Photovolt.* **2015**, *6*, 286–294. [CrossRef]

7.  Takashima, T.; Yamaguchi, J.; Otani, K.; Kato, K.; Ishida, M. Experimental Studies of Failure Detection Methods in PV module strings. In Proceedings of the 2006 IEEE 4th World Conference on Photovoltaic Energy Conversion, Waikoloa, HI, USA, 7–12 May 2006; Volume 2, pp. 2227–2230.

8.  Breitenstein, O.; Rakotoniaina, J.P.; Al Rifai, M.H. Quantitative evaluation of shunts in solar cells by lock-in thermography. *Prog. Photovolt. Res. Appl.* **2003**, *11*, 515–526. [CrossRef]

9.  Grimaccia, F.; Leva, S.; Dolara, A.; Aghaei, M. Survey on PV Modules' Common Faults after an O&M Flight Extensive Campaign over Different Plants in Italy. *IEEE J. Photovolt.* **2017**, *7*, 810–816.

10. Johnston, S.; Guthrey, H.; Yan, F.; Zaunbrecher, K.; Al-Jassim, M.; Rakotoniaina, P.; Kaes, M. Correlating multicrystalline silicon defect types using photoluminescence, defect-band emission, and lock-in thermography imaging techniques. *IEEE J. Photovolt.* **2014**, *4*, 348–354. [CrossRef]

11. Peloso, M.P.; Meng, L.; Bhatia, C.S. Combined thermography and luminescence imaging to characterize the spatial performance of multicrystalline Si wafer solar cells. *IEEE J. Photovolt.* **2015**, *5*, 102–111. [CrossRef]

12. Vergura, S.; Marino, F. Quantitative and Computer Aided Thermography-based Diagnostics for PV Devices: Part I—Framework. *IEEE J. Photovolt.* **2017**, *7*, 822–827. [CrossRef]

13. Vergura, S.; Colaprico, M.; de Ruvo, M.F.; Marino, F. A Quantitative and Computer Aided Thermography-based Diagnostics for PV Devices: Part II—Platform and Results. *IEEE J. Photovolt.* **2017**, *7*, 237–243. [CrossRef]

14. Mekki, H.; Mellit, A.; Salhi, H. Artificial neural network-based modelling and fault detection of partial shaded photovoltaic modules. *Simul. Model. Pract. Theory* **2016**, *67*, 1–13. [CrossRef]

15. Harrou, F.; Sun, Y.; Kara, K.; Chouder, A.; Silvestre, S.; Garoudja, E. Statistical fault detection in photovoltaic systems. *Sol. Energy* **2017**, *150*, 485–499.

16. Vergura, S.; Carpentieri, M. Statistics to detect low-intensity anomalies in PV systems. *Energies* **2018**, *11*, 30.

17. Ventura, C.; Tina, G.M. Development of models for on line diagnostic and energy assessment analysis of PV power plants: The study case of 1 MW Sicilian PV plant. *Energy Procedia* **2015**, *83*, 248–257. [CrossRef]

18. Silvestre, S.; Kichou, S.; Chouder, A.; Nofuentes, G.; Karatepe, E. Analysis of current and voltage indicators in grid connected PV (photovoltaic) systems working in faulty and partial shading conditions. *Energy* **2015**, *86*, 42–50. [CrossRef]

19. Vergura, S. A Complete and Simplified Datasheet-based Model of PV Cells in Variable Environmental Conditions for Circuit Simulation. *Energies* **2016**, *9*, 326. [CrossRef]

20. Vergura, S. Scalable Model of PV Cell in Variable Environment Condition based on the Manufacturer Datasheet for Circuit Simulation. In Proceedings of the 2015 IEEE 15th International Conference on Environment and Electrical Engineering (EEEIC), Roma, Italy, 10–13 June 2015.

21. Chao, K.H.; Ho, S.H.; Wang, M.H. Modeling and fault diagnosis of a photovoltaic system. *Electr. Power Syst. Res.* **2008**, *78*, 97–105. [CrossRef]

22. Hamdaoui, M.; Rabhi, A.; Hajjaji, A.; Rahmoum, M.; Azizi, M. Monitoring and control of the performances for photovoltaic systems. In Proceedings of the International Renewable Energy Congress, Sousse, Tunisia, 5–7 November 2009.

23. Kim, I.-S. On-line fault detection algorithm of a photovoltaic system using wavelet transform. *Sol. Energy* **2016**, *226*, 137–145. [CrossRef]

24. Rabhia, A.; El Hajjajia, A.; Tinab, M.H.; Alia, G.M. Real time fault detection in photovoltaic systems. *Energy Procedia* **2017**, *11*, 914–923.

25. Plato, R.; Martel, J.; Woodruff, N.; Chau, T.Y. Online fault detection in PV systems. *IEEE Trans. Sustain. Energy* **2015**, *6*, 1200–1207. [CrossRef]

26. Ando, B.; Bagalio, A.; Pistorio, A. Sentinella: Smart monitoring of photovoltaic systems at panel level. *IEEE Trans. Instrum. Meas.* **2015**, *64*, 2188–2199. [CrossRef]

27. Harrou, F.; Sun, Y.; Taghezouit, B.; Saidi, A.; Hamlati, M.E. Reliable fault detection and diagnosis of photovoltaic systems based on statistical monitoring approaches. *Renew. Energy* **2018**, *116*, 22–37. [CrossRef]

28.  International Electrotechnical Commission (IEC). *Photovoltaic System Performance Monitoring—Guidelines for Measurement, Data Exchange and Analysis*, International Standard 61724, 1st ed.; International Electrotechnical Commission (IEC): Geneva, Switzerland, 1998.

29.  Leloux, J.; Narvarte, L.; Trebosc, D. Review of the performance of residential PV systems in Belgium. *Renew. Sustain. Energy Rev.* **2012**, *16*, 178–184. [CrossRef]

30.  Leloux, J.; Narvarte, L.; Trebosc, D. Review of the performance of residential PV systems in France. *Renew. Sustain. Energy Rev.* **2012**, *16*, 1369–1376. [CrossRef]

31.  Available online: http://re.jrc.ec.europa.eu/pvg_tools/en/tools.html (accessed on 20 January 2018).

32.  Rohatgi, V.K.; Szekely, G.J. Sharp inequalities between skewness and kurtosis. *Stat. Probab. Lett.* **1989**, *8*, 297–299. [CrossRef]

33.  Klaassen, C.A.J.; Mokveld, P.J.; van Es, B. Squared skewness minus kurtosis bounded by 186/125 for unimodal distributions. *Stat. Probab. Lett.* **2000**, *50*, 131–135. [CrossRef]

34.  Hogg, R.V.; Ledolter, J. *Engineering Statistics*; MacMillan: Basingstoke, UK, 1987.

35.  Roussas, G. *An Introduction to Probability and Statistical Inference*; Academic Press: Cambridge, MA, USA, 2015.

36.  Hartigan, J.A.; Hartigan, P.M. The dip test of unimodality. *Ann. Stat.* **1985**, *13*, 70–84. [CrossRef]

37.  Gibbons, J.D. *Nonparametric Statistical Inference*, 2nd ed.; M. Dekker: New York, NY, USA, 1985.

38.  Hollander, M.; Wolfe, D.A. *Nonparametric Statistical Methods*; Wiley: Hoboken, NJ, USA, 1973.

*Article*

# Solar Cell Capacitance Determination Based on an RLC Resonant Circuit

**Petru Adrian Cotfas [1,\*], Daniel Tudor Cotfas [1], Paul Nicolae Borza [1], Dezso Sera [2] and Remus Teodorescu [2]**

[1]   Department of Electronics and Computers, Transilvania University of Brasov, Eroilor 29,
      500036 Brasov, Romania; dtcotfas@unitbv.ro (D.T.C.); borzapn@gmail.com (P.N.B.)
[2]   Department of Energy Technology, Aalborg University, Pontoppidanstraede 101,
      DK-9220 Aalborg, Denmark; des@et.aau.dk (D.S.); ret@et.aau.dk (R.T.)
\*    Correspondence: pcotfas@unitbv.ro; Tel.: +40-268-413-000

Received: 14 February 2018; Accepted: 13 March 2018; Published: 16 March 2018

**Abstract:** The capacitance is one of the key dynamic parameters of solar cells, which can provide essential information regarding the quality and health state of the cell. However, the measurement of this parameter is not a trivial task, as it typically requires high accuracy instruments using, e.g., electrical impedance spectroscopy (IS). This paper introduces a simple and effective method to determine the electric capacitance of the solar cells. An RLC (Resistor Inductance Capacitor) circuit is formed by using an inductor as a load for the solar cell. The capacitance of the solar cell is found by measuring the frequency of the damped oscillation that occurs at the moment of connecting the inductor to the solar cell. The study is performed through simulation based on National Instruments (NI) Multisim application as SPICE simulation software and through experimental capacitance measurements of a monocrystalline silicon commercial solar cell and a photovoltaic panel using the proposed method. The results were validated using impedance spectroscopy. The differences between the capacitance values obtained by the two methods are of 1% for the solar cells and of 9.6% for the PV panel. The irradiance level effect upon the solar cell capacitance was studied obtaining an increase in the capacitance in function of the irradiance. By connecting different inductors to the solar cell, the frequency effect upon the solar cell capacitance was studied noticing a very small decrease in the capacitance with the frequency. Additionally, the temperature effect over the solar cell capacitance was studied achieving an increase in capacitance with temperature.

**Keywords:** solar cells; AC parameters; underdamped oscillation; impedance spectroscopy

---

## 1. Introduction

Solar cells represent one of the most important renewable energy sources. There are many studies on increasing the efficiency of the photovoltaic (PV) systems used as renewable energy sources as a solution for the reduction of pollution [1–5]. Due to the nonlinear current-voltage (I-V) characteristic of PV modules, the accurate determination of their electrical parameters is not easy. Extensive studies on the determination of the PV modules' and solar cells' electrical parameters were performed in the last decades. Studies were focused on calculating both the DC and AC [6–16] electrical parameters.

The most widely used method to study the dynamic behavior of the PV modules is the impedance spectroscopy—IS [17,18]. Based on this method, the AC parameters of the solar cell, like capacitance, and dynamic and series resistance can be determined.

The IS method implies the application of a small AC signal (voltage or current) to the device under test (DUT), and the measurement of the resulting AC current or voltage. The DUT impedance can be calculated from the known AC voltage and the current. By varying the AC signal frequency, the DUT impedance spectrum can be found. Based on the DUT equivalent circuit, its components can be determined by using the fitting method (like the series and parallel resistance and the capacitance in the case of solar cells). The IS method has the advantage of offering the information about three important dynamic parameters and can be implemented with dedicated instruments or LCR meters. The disadvantage of this method is the need for an external signal injection into the DUT which can increase the price of the used instruments, especially if the tests are conducted under light conditions. These conditions involve DC power electronics in the instruments in order to bias the PV.

Other methods are based on the time response of the PV modules. These methods are based on the transient effect of the Resistor Capacitor (RC) circuit (discharging effect of the PV capacitance [19,20] or time constant of the PV circuit [21]). The reported time response-based methods require external signal injection into the DUT and are used only in dark conditions.

The dynamic parameters can be used as tools to characterize and diagnose the quality and the degradation status of the PV modules. One of the key dynamic parameters is the electrical capacitance of the PV modules. Oprea et al. [22] found that the PV panel capacitance and parallel resistance are affected by the PV panel degradation state when the potential-induced degradation was studied. Bhat et al. [23] found that the silicon cell capacitance decreases when the solar cell is irradiated with 8 MeV electrons because the carrier concentration decreases when increasing dose. Kim et al. [24] found that the capacitance of the PV panels increases when hot spots appear. Osawa et al. [25] studied the influence of the cracks and interconnected ribbon disconnection defects over the dynamic parameters of the PV modules. Their results showed that the parallel resistance decreases with the cracks number, while the parallel capacitance increases with the cracks number. Increasing the number of the interconnected ribbon disconnections increases the series resistance, slightly decreases the parallel resistance, and does not affect the capacitance of the PV modules.

The solar cell capacitance is a combination of two capacitances: transition capacitance (also known as junction capacitance) and diffusion capacitance connected in parallel. The solar cell capacitance varies function of the cell voltage, level of irradiance, frequency, and temperature [26–33]. The capacitance variation of the silicon solar cell function of the bias voltage presents two regions: in the first region, stretching from reverse to forward bias until the knee voltage, the capacitance presents a small increasing slope (region associated with the transition capacitance), while in the second region, beginning above the knee voltage, the capacitance presents an exponential increase (in the region associated with the diffusion capacitance). The increasing of the irradiance level leads to an increase of the solar cell capacitance [18,30,33]. Increasing the solar cell temperature also increases its capacitance [18,31–33]. Based on the dependence of the solar cell capacitance on the temperature, Anantha Krishna et al. [31] proposed a method for measuring the solar cell blanket in spacecraft.

This paper describes a simple method which allows determining the capacitance of the solar cells. The method is based on the behavior of the RLC (Resistor Inductance Capacitor) circuits. An inductance is used as a variable load for solar cells. The method can be applied under light conditions starting from 1 W/m$^2$. The influences of the irradiance level, frequency, and temperature on the solar cell capacitance are described, as well.

## 2. Materials and Methods

### 2.1. Photovoltaic Cell Modeling

Solar cells characterization can be performed in DC and AC regimes. In the DC case, the solar cell can be described using the equivalent electrical circuit shown in Figure 1. This is based on the one-diode model.

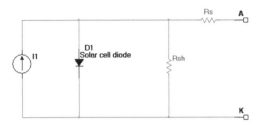

**Figure 1.** The DC equivalent electrical circuit of a solar cell.

For the AC case, the equivalent circuit of the solar cell is shown in Figure 2, where:

- $I_1$ = photogenerated current
- $R_s$ = series resistance
- $R_{sh}$ = shunt resistance
- $R_d(V)$ = dynamic resistance of the diode
- $CD(V, \omega)$ = diffusion capacitance
- $CT(V)$ = transition capacitance
- $\omega$ = signal frequency

The $CT$ capacitance describes the separation of charges in the depletion region and the $CD$ capacitance describes the gradient in the charge density inside the solar cell [18,26].

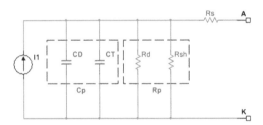

**Figure 2.** The AC equivalent electrical circuit of a solar cell.

The equivalent electrical circuit of the solar cell can be simplified like in Figure 2; the dashed line part, $C_p = CD \,||\, CT$, is the equivalent parallel capacitance, and $R_p = R_{sh} \,||\, R_d$ is the equivalent parallel resistance.

### 2.2. Proposed Method for Determining the Solar Cell Capacitance

The proposed method is based on the RLC circuit behavior. The method supposes having the solar cell under a constant level of irradiance and in the open circuit point. Using a switch, a coil with known inductance $L_1$ and series resistance $RL$ is connected to the solar cell. At the time instance when the switch is closed, a transient process of the coil charging begins. The operating point of the system is moving from the open circuit point towards the short circuit current point. When the coil is completely charged (when $IL \cong I_1$) an oscillation between the solar cell capacitance $C_p$ and external inductor $L_1$ appears, and the coil discharge begins through the circuit resistance and charges the solar cell capacitance. The charging-discharging process is repeated as an underdamped oscillation until a steady state is obtained.

Therefore, when the inductance $L_1$ with series resistance $RL$ is connected to the solar cell's terminals using the $SW1$ switch, the circuit shown in Figure 3 is obtained. The $R_c$ is the resistance of the connection circuit between the solar cell and the coil.

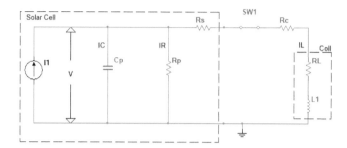

**Figure 3.** The electrical circuit obtained connecting an inductance coil to a solar cell.

The circuit from Figure 3 is an RLC circuit that can be described based on following equation:

$$V(t) = A_0 e^{-\beta t} sin(\omega t + \varphi) \tag{1}$$

or:

$$V(t) = A(t)sin(\omega t + \varphi) \tag{2}$$

where:

$$\omega = \sqrt{\omega_0^2 - \beta^2} \tag{3}$$

is called the resonant frequency, $A_0$ is the initial amplitude of the oscillation, $\varphi$ is the initial phase, $\beta$ is the damping coefficient which is dependent on the capacitance, parallel and series resistances of the solar cell, the coil inductance, and the circuit resistance, $\omega_0 = 1/\sqrt{L_1 C_P}$ represents the resonant frequency of the ideal oscillator, and $A(t) = A_0 e^{-\beta t}$ is the time-dependent oscillation amplitude. If $\beta < \omega_0$ then the voltage across the cell is an underdamped oscillation and the proposed method can be used. If $\beta \geq \omega_0$, then the voltage across the cell presents an overdamped evolution and the method could not be applied.

The logarithmic decrement of the underdamped oscillation is given by the logarithm value of the ratio between two consecutive oscillator amplitudes:

$$\Delta = \ln \frac{A(t)}{A(t+T)} = \ln \frac{A_0 e^{-\beta t}}{A_0 e^{-\beta(t+T)}} = \ln e^{\beta T} = \beta T \tag{4}$$

where $T = \frac{2\pi}{\omega}$ is the period.

The $C_p$ capacitance of the solar cell can be calculated at $f = \omega/2\pi$ frequency, by knowing the coil inductance value, and by determining the frequency and the logarithmic decrement of the resulted oscillation from the measurements:

$$\Delta = \beta T = \frac{2\pi\beta}{\omega} \Rightarrow \beta = \frac{\omega\Delta}{2\pi} \tag{5}$$

$$\omega^2 = \omega_0^2 - \left(\frac{\omega\Delta}{2\pi}\right)^2 = \frac{1}{L_1 C_p} - \left(\frac{\omega\Delta}{2\pi}\right)^2 \tag{6}$$

$$C_p = \frac{1}{L_1\left(\omega^2 + \left(\frac{\omega\Delta}{2\pi}\right)^2\right)} \tag{7}$$

If the coil is changed, a new oscillating circuit is obtained, which has a new resonant frequency.

## 2.3. Simulation and Experimental Setup

The research for this paper is made in two stages. The first stage is based on the simulation of circuits using the National Instruments (NI) Multisim software package as SPICE simulation software.

The circuit used for the simulation is shown in Figure 4. The $SC1$ solar cell is connected through the $R_1$ resistance to a coil with $L_1$ inductance and $RL$ internal resistance known. The solar cell voltage and the current through the circuit are measured with the $XSC1$ oscilloscope. The $R_1$ resistance with known value is used for current measurement. The $R$ resistance and the $S_1$ single pole double throw (SPDT) switch are used for coil discharge.

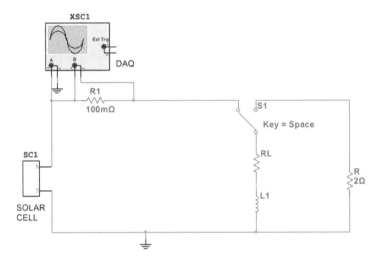

**Figure 4.** The electrical circuit used to study the AC solar cell parameters.

The $SC1$ equivalent electrical circuit used is shown in Figure 1, where the $D_1$ diode is used in order to observe the dynamic behavior of the solar cell.

In order to obtain the solar cell capacitance at different frequencies, several coils with different inductance values are used. The data obtained from the simulation are exported and processed in a software application developed using NI LabVIEW (2014, National Instruments, Austin, TX, USA). The developed application allows selecting the desired part of the underdamped oscillation signal and then, by using the Fourier analysis, the resonant frequency is determined. The application allows calculating the logarithmic decrement based on two consecutive amplitudes of the oscillation. Using Equation (7), the parallel capacitance of the solar cell is calculated.

In the second stage, a commercial encapsulated Si monocrystalline solar cell with an active area of 7.9 cm$^2$, several coils with values between 21 µH and 566 µH, a 12-bit DAQ system (type NI PCI-MIO-16E-1), and a self-built signal conditioning (SBSC) circuit with adjustable amplification are used. The measurement chain is based on the circuit from Figure 4, in which the $XSC1$ oscilloscope was replaced with the DAQ + SBSC system. The measurement and data processing applications were also developed in NI LabVIEW. Consequently, the results of simulation could be compared to the experimental measurements. Additionally, in this stage the IS method [17,18,26] was used to validate the results obtained through the proposed method. The impedance measurements were carried out using a Zahner IM6 system and the associated software Thales (Z2.25, ZAHNER-elektrik, Kronach, Germany). The main characteristics of the Zahner IM6 system are: the frequency range of 10 µHz to 3 MHz, the AC amplitude 1 mV to 1 V, the DC amplitudes of ±14 V and ±2 A, the measurement accuracy ±250 µV and ±0.05% at 2 µA to 100 mA and ±0.5% at <2 µA or >100 mA. The measurements were taken at different levels of irradiance (1, 183, 263, 354, and 642 W/m$^2$), over a frequency range of

1 Hz–100 kHz with 20 points per decade. The excitation signal was a sinusoidal voltage with 25 mV peak and the DC forward bias voltages were the averages values of the data obtained from RLC method (see Section 3.2). The irradiance on the solar cell is obtained using a halogen bulb. The distance between the halogen bulb and the solar cell was modified in order to vary the level of irradiance. The measurements were obtained at the constant temperature of 25 °C. For the Nyquist plot fitting obtained through the IS method, the simplified circuit from Figure 2 was used.

The proposed method was further tested on a Solvis SV36 multi-crystalline Si PV panel. The measurements were done in the Photovoltaic Systems Laboratory of the Department of Energy Technology, Aalborg University.

The inductor connected to the PV panel (see Figure 3) has the following parameters: $L_1 = 1$ mH, $RL = 128$ m$\Omega$. The current measurements are done using a Tektronix DPO4054B digital oscilloscope (Tektronix, Inc., Beaverton, OR, USA) with a TC0030 current probe.

The impedance spectroscopy measurements for the PV panel were performed with a HP 4284A Precision LCR Meter. This LCR meter is capable of a frequency range of 20 Hz–1 MHz and a current or voltage excitation signal range (peak value) of 50 $\mu$A–20 mA and 5 mV–2 V, respectively. The HP 4284A has a measurement accuracy of ±0.05% of full-scale value. The HP 4284A uses the self-balancing bridge method principle to determine the impedances.

The IS measurements are carried out in dark conditions, over a frequency range of 20 Hz–50 kHz (with 1 kHz resolution). The excitation signal is a sinusoidal voltage with 1 V peak. In order to avoid opening the bypass diodes—hence distorting the results—a DC forward bias of 1 V is applied.

## 3. Results and Discussion

### 3.1. Simulation Results

By connecting the $L_1$ inductance to the circuit of the cell by making the $S_1$ switch commutation (Figure 4), the voltage and current waveforms are obtained, as shown in Figure 5. The obtained signals are underdamped oscillating signals. The oscillation is around the values of $V \approx 0$ and $I \approx I_1$ for voltage and for current, respectively. The value of the balance point for the voltage ($V_{pol}$) is dependent on the short circuit current of the solar cell $I_{sc}$ (which is approximately equal to $I_1$) and the series resistance of the circuit ($R_c$ and $R_L$):

$$V_{pol} = I_1(R_c + R_L), \tag{8}$$

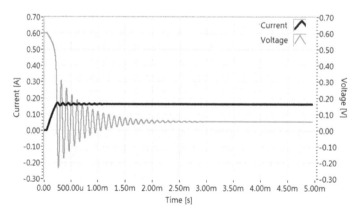

**Figure 5.** The voltage and current waveforms obtained in simulation ($C_p = 0.433$ $\mu$F, $R_{sh} = 777$ $\Omega$, $L_1 = 566$ $\mu$H, $R_s = 143$ m$\Omega$, $R_1 = 100$ m$\Omega$).

The representation of current depending on the voltage (the I-V characteristic) is shown in Figure 6. The red thin curve with square points represents the DC I-V characteristic of the solar cell, while the black thin curve represents the I-V characteristic of the solar cell obtained with the inductance.

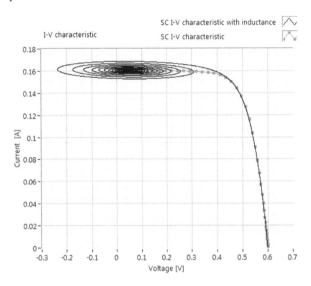

**Figure 6.** The simulated I-V characteristic obtained with the $L_1$ as variable load.

When the $L_1$ coil is connected to the solar cell, it acts as a very large impedance so the I-V characteristic is obtained starting from the $V_{OC}$ point and then continuing to the $I_{SC}$ point where its impedance drops significantly (equal to its internal resistance). From Figure 6 one can observe that $L_1$ acts as a variable load for the solar cell starting from large impedance and varying towards small impedance. The underdamped oscillation of the formed RLC circuit is obtained around the value $V_{pol}$. By determining the frequency of the voltage signal, the logarithmic decrement of the oscillation and by using Equation (7), the capacitance of the solar cell can be determined. The values of the components used in the circuit of the solar cell SC1 were: $R_{sh} = 777$ Ω, $C_p = 433$ nF and $R_s = 143$ mΩ, while $R_1 = 100$ mΩ. Table 1 presents the values resulted after processing the data obtained from simulations. In these simulations the photogenerated current $I_1$ was set at different values.

**Table 1.** Results obtained from simulation using a inductance $L_1 = 21$ μH and $R_L = 223$ mΩ.

| $I_1$ [mA] | $C_p$ [nF] | $V_{pol}$ [mV] |
|---|---|---|
| 46 | 432.3 | 14.7 |
| 66 | 432.5 | 21.2 |
| 89 | 432.4 | 28.6 |
| 161 | 432.2 | 51.8 |

Table 2 shows the results obtained after using different inductances. One can notice that the resonant frequency changes (according to Equation (3)), thus resulting in the possibility of studying the variation of $C_p$ with frequency (the domain being 10 kHz–52 kHz for the chosen coils and for chosen solar cell (SC)).

**Table 2.** Results obtained from simulation using different inductances.

| Inductance [μH] | Frequency [kHz] | $C_p$ [nF] | $V_{pol}$ [mV] |
|---|---|---|---|
| 21 | 52.69 | 432.2 | 51.9 |
| 100 | 24.22 | 432.9 | 51.9 |
| 271 | 14.67 | 432.7 | 51.9 |
| 566 | 10.15 | 432.6 | 51.9 |

The effect that the errors in the inductance and resonant frequency determination have over the solar cell capacitance calculation is studied through simulation. Therefore, if the inductance determination has an error of 10%, the error in the capacitance calculation is 9.1%. In the case of a 10% error in the resonant frequency determination, it introduces a 17.4% error in the capacitance calculation.

*3.2. Experimental Results*

For the solar cell study the measurements were taken at five levels of irradiance. On each level, four different coils were used. The inductances of the used coils are: 21, 100, 271, and 566 μH.

Table 3 shows the results obtained through the analysis of the measurements made on the monocrystalline silicon solar cell.

**Table 3.** The results obtained from measurements.

| Level [W/m²] | Inductance [μH] | Frequency [kHz] | $C_p$ [nF] | $V_{pol}$ [mV] | $I_{sc}$ [mA] |
|---|---|---|---|---|---|
| | 21 | 55.71 | 379.3 | ~0 | ~0 |
| 1.0 | 100 | 25.93 | 381.6 | ~0 | ~0 |
| | 271 | 15.63 | 380.8 | ~0 | ~0 |
| | 566 | 10.99 | 388.7 | ~0 | ~0 |
| | 21 | 55.28 | 394.0 | 14.8 | 46.0 |
| 183 | 100 | 25.45 | 394.2 | 14.5 | 46.0 |
| | 271 | 15.31 | 400.3 | 15.9 | 46.0 |
| | 566 | 10.83 | 400.7 | 16.7 | 46.0 |
| | 21 | 54.78 | 401.0 | 21.4 | 66.0 |
| 263 | 100 | 25.28 | 403.0 | 21.3 | 66.0 |
| | 271 | 15.32 | 400.8 | 22.5 | 66.0 |
| | 566 | 10.70 | 410.2 | 23.6 | 66.0 |
| | 21 | 54.07 | 406.3 | 29.1 | 89.0 |
| 354 | 100 | 25.03 | 410.7 | 28.6 | 89.0 |
| | 271 | 15.13 | 407.2 | 30.5 | 89.0 |
| | 566 | 10.55 | 412.5 | 33.2 | 89.0 |
| | 21 | 52.82 | 428.3 | 54.0 | 161.0 |
| 642 | 100 | 24.45 | 429.9 | 51.9 | 161.0 |
| | 271 | 14.75 | 424.9 | 56.1 | 161.0 |
| | 566 | 10.27 | 429.5 | 60.3 | 161.0 |

The data from Table 3 regarding the solar cell capacitance are plotted as shown in Figure 7. From the curves, one can observe that the capacitance of the solar cell increases with the level of the irradiance, and decreases very slowly with the frequency in the used range, as it was also obtained in [27–30,33].

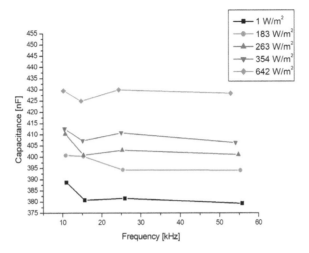

**Figure 7.** The dependence of the SC capacitance function of the frequency.

The comparison between the results obtained, based on the impedance spectroscopy method for solar cell characterization and the results obtained using the proposed method, is shown in Figure 8. For this graph, the average solar cell capacitances over all resonant frequencies for every irradiance level were used; the vertical bars denote the standard deviation for the results obtained through both methods. The differences between the capacitances obtained through the two methods are small. The average value of the errors of the capacitances obtained through the IS and RLC methods is 1.01%. These differences are due to the parasitic capacitance of the external circuit including the coil capacitance, which is connected in parallel with the solar cell capacitance, increasing the determined capacitance, and also due to the measurement error of the coil inductance value and the frequency of the signal.

The proposed method is also verified by determining the photovoltaic panel capacitance and the results obtained again show very good agreement with the reference IS measurement. The PV panel measurements were conducted at a 10 W/m$^2$ irradiance level using a coil with the inductance of 1 mH. The capacitance obtained through the RLC method was 246.4 nF at 10.14 kHz and $V_{pol}$ = 15 mV, while the capacitance obtained through the IS method was 272.5 nF. The results obtained from the two methods are close, with an error of 9.6%. This error can be explained by the differences in the test condition (the level of illumination and the DC forward bias voltage).

The PV panel capacitance has a small value (in comparison with the individual solar cell DUT—see Table 3). This can be explained by the series connection of all solar cells in the panel that implies the series connection of their capacitances. Therefore, even though the area of cells within the Solvis SV36 panel is much higher (~243.4 cm$^2$) compared to the individual cell (7.9 cm$^2$), the resulting overall capacitance of the panel is smaller than that of the cell.

The effect of temperature on the solar cell capacitance was studied at a 183 W/m$^2$ irradiance level using two coils with inductances of 271 μH, and of 21 μH, respectively. Due to the small forward bias voltage (see Table 3) the predominant capacitance is the transition capacitance. The transition capacitance is given by the following equation:

$$CT = \frac{B}{(V_0 - V_d)^n} \tag{9}$$

where $B$ is a constant [32], $V_d$ is cell voltage, $V_0$ is the built-in voltage and n is equal to $1/2$ for abrupt junction and $1/3$ for graded junction. The dependence of $CT$ on temperature can be expressed as a fourth-degree polynomial:

$$CT = K_0 - K_1 T + K_2 T^2 - K_3 T^3 + K_4 T^4 \tag{10}$$

where $K_i$, $i = 1 \ldots 4$, are constants [31].

For the experiment, the temperature was varied in the range of 30–80 °C, in steps of 5 °C, using a PID (Proportional–Integral–Derivative) thermostat. For each step, the temperature was maintained constant for 10 min before measuring. The average solar cell capacitances obtained with the two coils and their standard deviation in function of the temperature are shown in Figure 9. From the graph, one can observe that the capacitance of the studied solar cell increases with the temperature. Similar results are reported in [18,30,31].

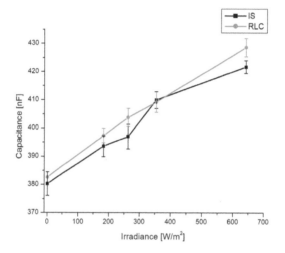

**Figure 8.** SC capacitance obtained through IS and RLC methods.

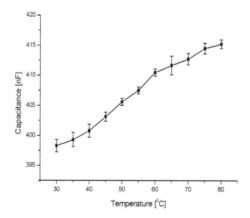

**Figure 9.** The variation of the solar cell capacitance depending on temperature.

## 4. Conclusions

The electric capacitance represents a key dynamic parameter of PV modules and should be studied due to the fact that it can offer information about the health status and quality of the PV modules. In this paper, we proposed a new method, called the RLC method, which allows the determination of the solar cells and PV panels' capacitance. The RLC method is relatively easy to implement and does not require any external signal injection into the DUT, which can significantly decrease the cost of the method. The capacitance values obtained by using this method are in accordance with the results obtained based on the impedance spectroscopy methods. The average error between the two methods is approximately 1% for the capacitance determination of the solar cells and 9.6% for the photovoltaic panels. Using different inductors, the solar cells and PV panels' capacitance could be determined at different frequencies, with only a minor reduction (1.1%) of the estimated capacitance value in the used frequency range (10–56 kHz). Experimental results show that the solar cell capacitance increases from 382 nF to 428 nF as the irradiance increases from 1 W/m$^2$ to 642 W/m$^2$. Additionally, the results show that the capacitance for the studied solar cell increases with the temperature.

The RLC method can be applied under light conditions, starting from irradiance levels below 10 W/m$^2$.

One drawback of the method is that the damping coefficient $\beta$ should always be smaller than the resonant frequency $\omega_0$, in order to obtain underdamped oscillation. Additionally, it is necessary to use coils with different inductance values to get the capacitance over different frequencies.

The accuracy of the solar cell capacitance calculation is strongly affected by the accuracy of the inductance and the resonant frequency determination.

**Author Contributions:** All the authors contributed to the publishing of this paper. Petru Adrian Cotfas contributed to the development of the RLC method, made the simulations and measurements and wrote the paper. Daniel Tudor Cotfas contributed to the development of the RLC method theory and to the writing of the paper and performed the data analysis. Paul Nicolae Borza conducted the IS measurements for the solar cells. Dezso Sera made the IS measurements and data analysis for the PV panel. Remus Teodorescu supervised the research and edited the paper.

**Conflicts of Interest:** The authors declare no conflict of interest.

## References

1. MacGill, I.; Watt, M. Economics of Solar PV Systems with Storage, in Main Grid and Mini-Grid Settings. In *Solar Energy Storage*; Sørensen, B., Ed.; Academic Press: London, UK, 2015; Chapter 10, pp. 225–244, ISBN 978-0-12-409540-3.
2. Gude, V.G.; Nirmalakhandan, N.; Deng, S. Desalination using solar energy: Towards sustainability. *Energy* **2011**, *36*, 78–85. [CrossRef]
3. Das, N.; Wongsodihardjo, H.; Islam, S. Photovoltaic cell modeling for maximum power point tracking using MATLAB/Simulink to improve the conversion efficiency. In Proceedings of the IEEE Power and Energy Society General Meeting (PES 2013), Vancouver, BC, Canada, 21–25 July 2013.
4. Das, N.; Wongsodihardjo, H.; Islam, S. Modeling of multi-junction photovoltaic cell using MATLAB/Simulink to improve the conversion efficiency. *Renew. Energy* **2015**, *74*, 917–924. [CrossRef]
5. Al-Nimr, M.; Al-Ammari, W. A novel hybrid PV-distillation system. *Sol. Energy* **2016**, *135*, 874–883. [CrossRef]
6. Cotfas, D.T.; Cotfas, P.A.; Kaplanis, S. Methods to determine the dc parameters of solar cells: A critical review. *Renew. Sustain. Energy Rev.* **2013**, *28*, 588–596. [CrossRef]
7. Cotfas, D.T.; Cotfas, P.A.; Kaplanis, S. Methods and techniques to determine the dynamic parameters of solar cells: Review. *Renew. Sustain. Energy Rev.* **2016**, *61*, 213–221. [CrossRef]
8. Yadav, P.; Pandey, K.; Bhatt, V.; Kumar, M.; Kim, J. Critical aspects of impedance spectroscopy in silicon solar cell characterization: A review. *Renew. Sustain. Energy Rev.* **2017**, *76*, 1562–1578. [CrossRef]
9. Chan, D.S.H.; Phillips, J.R.; Phang, J.C.H. A comparative study of extraction methods for solar cell model parameters. *Solid-State Electron.* **1986**, *29*, 329–337. [CrossRef]
10. Raj, S.; Kumar, S.A.; Panchal, A.K. Solar cell parameters estimation from illuminated I-V characteristic using linear slope equations and Newton-Raphson technique. *J. Renew. Sustain. Energy* **2013**, *5*, 255–265. [CrossRef]

11. Cubas, J.; Pindado, S.; Manuel, C. Explicit Expressions for Solar Panel Equivalent Circuit Parameters Based on Analytical Formulation and the Lambert W-Function. *Energies* **2014**, *7*, 4098–4115. [CrossRef]

12. Mughal, M.A.; Ma, Q.; Xiao, C. Photovoltaic Cell Parameter Estimation Using Hybrid Particle Swarm Optimization and Simulated Annealing. *Energies* **2017**, *10*, 1213. [CrossRef]

13. Ye, M.; Wang, X.; Xu, Y. Parameter extraction of solar cells using particle swarm optimization. *J. Appl. Phys.* **2009**, *105*, 094502. [CrossRef]

14. Zagrouba, M.; Sellami, A.; Bouaïcha, M.; Ksouri, M. Identification of PV solar cells and modules parameters using the genetic algorithms: Application to maximum power extraction. *Sol. Energy* **2010**, *84*, 860–866. [CrossRef]

15. Hasanien, H.M. Shuffled frog leaping algorithm for photovoltaic model identification. *IEEE Trans. Sustain. Energy* **2015**, *6*, 509–515. [CrossRef]

16. Yu, K.; Chen, X.; Wang, X.; Wang, Z. Parameters identification of photovoltaic models using self-adaptive teaching-learning-based optimization. *Energy Convers. Manag.* **2017**, *145*, 233–246. [CrossRef]

17. Kumar, R.A.; Suresh, M.S.; Nagaraju, J. Measurement and comparison of AC parameters of silicon (BSR and BSFR) and gallium arsenide (GaAs/Ge) solar cells used in space applications. *Sol. Energy Mater. Sol. Cells* **2000**, *60*, 155–165. [CrossRef]

18. Kumar, R.A.; Suresh, M.S.; Nagaraju, J. Silicon (BSFR) solar cell AC parameters at different temperatures. *Sol. Energy Mater. Sol. Cells* **2005**, *85*, 397–406. [CrossRef]

19. Kumar, R.A.; Suresh, M.S.; Nagaraju, J. Time domain technique to measure solar cell capacitance. *Rev. Sci. Instrum.* **2003**, *74*, 3516–3519. [CrossRef]

20. Deshmukh, M.P.; Kumar, R.A.; Nagarajua, J. Measurement of solar cell ac parameters using the time domain technique. *Rev. Sci. Instrum.* **2004**, *75*, 2732–2735. [CrossRef]

21. Chenvidhya, D.; Limsakul, C.; Thongpron, J.; Kirtikara, K.; Jivacate, C. Determination of solar cell dynamic parameters from time domain responses. In Proceedings of the Technical Digest of the 14th International Photovoltaic Science and Engineering Conference (PVSEC14), Bangkok, Thailand, 26–30 January 2004.

22. Oprea, M.I.; Spataru, S.V.; Sera, D.; Poulsen, P.B.; Thorsteinsson, S.; Basu, R.; Andersen, A.R.; Frederiksen, K.H.B. Detection of potential induced degradation in c-Si PV panels using electrical impedance spectroscopy. In Proceedings of the IEEE 43rd Photovoltaic Specialists Conference (PVSC), Portland, OR, USA, 5–10 June 2016; pp. 1575–1579.

23. Bhat, P.S.; Rao, A.; Sanjeev, G.; Usha, G.; Priya, G.K.; Sankaran, M.; Puthanveettil, S.E. Capacitance and conductance studies on silicon solar cells subjected to 8 MeV electron irradiations. *Radiat. Phys. Chem.* **2015**, *111*, 28–35. [CrossRef]

24. Kim, K.A.; Seo, G.S.; Cho, B.H.; Krein, P.T. Photovoltaic Hot-Spot Detection for Solar Panel Substrings Using AC Parameter Characterization. *IEEE Trans. Power Electron.* **2016**, *31*, 1121–1130. [CrossRef]

25. Osawa, S.; Nakano, T.; Matsumoto, S.; Katayama, N.; Saka, Y.; Sato, H. Fault diagnosis of photovoltaic modules using AC impedance spectroscopy. In Proceedings of the IEEE International Conference on Renewable Energy Research and Applications (ICRERA), Birmingham, UK, 20–23 November 2016; pp. 210–215.

26. Kumar, S.; Sareen, V.; Batra, N.; Singh, P.K. Study of C–V characteristics in thin $n^+$-p-$p^+$ silicon solar cell sand induced junction n-p-$p^+$ cell structures. *Sol. Energy Mater. Sol. Cells* **2010**, *94*, 1469–1472. [CrossRef]

27. Kumar, S.; Singh, P.K.; Chilana, G.S. Study of silicon solar cell at different intensities of illumination and wavelengths using impedance spectroscopy. *Sol. Energy Mater. Sol. Cells* **2009**, *93*, 1881–1884. [CrossRef]

28. Burgelman, M.; Nollet, P. Admittance spectroscopy of thin film solar cells. *Solid State Ion.* **2005**, *176*, 2171–2175. [CrossRef]

29. Bayhan, H.; Kavasoğlu, A.S. Admittance and Impedance Spectroscopy on Cu(In,Ga)Se$_2$ Solar Cells. *Turk. J. Phys.* **2003**, *27*, 529–535.

30. Kumar, R.A.; Suresh, M.S.; Nagaraju, J. GaAs/Ge solar cell AC parameters under illumination. *Sol. Energy* **2004**, *76*, 417–421. [CrossRef]

31. Anantha Krishna, H.; Misra, N.K.; Suresh, M.S. Use of solar cells for measuring temperature of solar cell blanket in spacecrafts. *Sol. Energy Mater. Sol. Cells* **2012**, *102*, 184–188. [CrossRef]

32. Mandal, H.; Nagaraju, J. GaAs/Ge and silicon solar cell capacitance measurement using triangular wave method. *Sol. Energy Mater. Sol. Cells* **2007**, *91*, 696–700. [CrossRef]
33. Panigrahi, J.; Singh, R.; Batra, N.; Gope, J.; Sharma, M.; Pathi, P.; Srivastava, S.K.; Rauthan, C.M.S.; Singh, P.K. Impedance spectroscopy of crystalline silicon solar cell: Observation of negative capacitance. *Sol. Energy* **2016**, *136*, 412–420. [CrossRef]

Article

# Shading Ratio Impact on Photovoltaic Modules and Correlation with Shading Patterns

Alonso Gutiérrez Galeano [1,2,*], Michael Bressan [1], Fernando Jiménez Vargas [1,3] and Corinne Alonso [2,3]

[1] Department of Electrical and Electronic Engineering, Universidad de los Andes, Bogotá 111711, Colombia; m.bressan@uniandes.edu.co (M.B.); fjimenez@uniandes.edu.co (F.J.V.)
[2] Doctoral School GEET, Université Toulouse III, F-31400 Toulouse, France; corinne.alonso@laas.fr
[3] LAAS-CNRS, 7 Avenue du Colonel Roche, F-31077 Toulouse, France
[*] Correspondence: a.gutierrez75@uniandes.edu.co; Tel.: +33-561-330-000

Received: 10 February 2018; Accepted: 30 March 2018; Published: 5 April 2018

**Abstract:** This paper presents the study of a simplified approach to model and analyze the performance of partially shaded photovoltaic modules using the shading ratio. This approach integrates the characteristics of shaded area and shadow opacity into the photovoltaic cell model. The studied methodology is intended to improve the description of shaded photovoltaic systems by specifying an experimental procedure to quantify the shadow impact. Furthermore, with the help of image processing, the analysis of the shading ratio provides a set of rules useful for predicting the current–voltage behavior and the maximum power points of shaded photovoltaic modules. This correlation of the shading ratio and shading patterns can contribute to the supervision of actual photovoltaic installations. The experimental results validate the proposed approach in monocrystalline and polycrystalline technologies of solar panels.

**Keywords:** partial shading; photo-generated current; photovoltaic performance; maximum power point; image processing

## 1. Introduction

Nowadays, the integration of photovoltaic (PV) systems into electrical grids is becoming increasingly widespread as a promising alternative distributed-energy resource [1]. Their ease of installation and adaptability have encouraged their integration into urban-area and rural-area energy grids. However, shadows from surrounding structures affect the PV installations, which causes power losses and structural failures [2,3]. Several authors have therefore developed modeling approaches to better understand the impact of shadows on PV systems [4–6]. Despite these important contributions, the observed behavior and harmful conditions suggest the need for improving shadow impact quantification [7,8]. Indeed, innovative modeling and supervision approaches are required to better understand and prevent the production losses in PV systems [9,10]. In addition, innovative approaches can improve the design of power converters and control strategies to reduce the shadow impact [11,12]. As a result, the development of novel methods to quantify and supervise the shadow impact is currently an important issue for improving PV system performance [13,14].

The previously mentioned research area relies on reverse-bias behavior of shaded PV-cells. A widely accepted model was presented by Bishop for describing the shaded PV-cell behavior in reverse-bias [15]. Quaschning et al. extended the model proposed by Bishop to the two-diodes model [16]. Kawamura et al. simulated the previous Bishop model while considering shadow transmittance in order to study the corresponding I–V characteristics [17]. In order to obtain a more dynamic model, Guo et al. investigated the influence of moving shadows on the PV-power characteristics [18]. Afterwards, Olalla et al. simulated large PV systems with high granularity using diffuse irradiance to model the partial shaded effects [19]. In addition, Díaz et al. proposed a generalized and simplified model while considering the shadow geometry [20].

For the study of the shading ratio, Silvestre et al. extend the Bishop model to analyze the performance of PV modules [21]. Jung et al. proposed a mathematical model for the output characteristics of a photovoltaic module including three key factors and the photo-current for a different shading ratios [22]. In Ref. [23], Yong et al. presents a non-disruptive cell-level characterization of a photovoltaic module extracting the shunt resistances and the short-circuit currents of individual cells by using a partial shading technique with two different shading ratios. He et al. study the hot-spot issues in a PV module in different numbers of PV-cells using several shading ratio scenarios [24]. The work presented in Ref. [21] develops a simulation and modeling of PV modules' performance under partial shading for several shadow rates testing single cells in PV modules to analyze the influence of the shadow rate on the most important PV module parameters.

As shown through this brief historical background, the researchers have progressively developed more detailed and extensive approaches to describe the shaded PV system behavior and the influence of the shading ratio. However, research on evolutionary PV installations currently requires accurate but simplified analysis given the variable nature of shadows in real-world applications [25–27].

In this context, our work proposes a more accurate definition of the shading ratio and an innovative experimental set-up to integrate the shadow properties into the shaded PV model. This work includes the analysis of the shading ratio to quantify the shadow impact on PV installations. This shading ratio associates the shadow characteristics of the shaded area and the shading factor. Furthermore, with the help of image processing methods, the proposed approach adds a novel experimental set-up to analyze and supervise the shadow impact using the shading ratio. This analysis provides a set of rules useful for predicting the current–voltage behavior of shaded photovoltaic modules. Additionally, the correlation between the shading ratio and the shadow image patterns allowed for developing a simplified expression to localize the maximum power points (MPPs) in actual shaded conditions. Finally, these correlations were experimentally validated, which provides fundamentals for the applications of image processing methods to quantify and supervise the shadow impact on PV installations. Figure 1 outlines the methodology that uses the shading ratio and image processing.

This paper is organized as follows. Section 2 presents the modeling background. Section 3 describes the proposed approach. In Section 4, simulations of shaded PV modules are analyzed. Section 5 explains in detail the experimental setup for validating and correlating the proposed approach with shadow image patterns. Finally, experimental results are discussed.

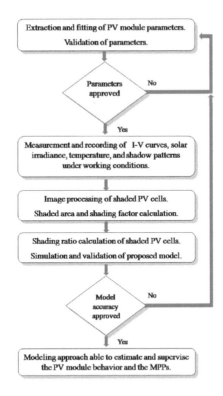

**Figure 1.** Modeling methodology using the shading ratio and image processing.

## 2. Photovoltaic Model for Shaded Conditions

Shaded PV modules have a high risk of structural failures and a high risk of losing power production. Several authors have studied this behavior at level of PV-cells [15,27]. From proposed models, the approach presented by Bishop has allowed for a suitable agreement with controlled tests [15]. However, the complex nature of the shading phenomenon has demonstrated the need for complementing these approaches [28]. This section describes the shaded PV behavior and current modeling methods.

### 2.1. Shaded PV Modules and Modeling Background

A typical partially shaded setup is used for the test in this study, which is shown in Figure 2. This experimental shading test was performed on 14 February 2017 in sunny weather. The ambient temperature was 15 °C and the global solar irradiation in the horizontal plane was 910 W/m² at 1:00 p.m. The experimental results in Figure 2 illustrate the drastic impact on the I–V and P–V curves.The partial shadows can produce multiple maximum power points (MPPs). In addition, studies have shown that these partial shadows can lead to overheating and hot-spot issues [3,29].

Several authors have studied this shaded behavior. Bishop presents a model for the reverse-bias characteristics of shaded solar cells based on previous works regarding the avalanche breakdown theory [15]. The authors propose a numerical simulation [16] and then the authors investigated the I–V characteristic under shadow conditions [17]. The work presented an alternative model for various types of PV-cells [30]. The study in Ref. [21] describes the PV performance in relation with the shadow rate. Thermal stability and hot-spot risks are studied in Ref. [31]. The work in Ref. [18] outlines a study of the shadow movement influence. For shaded PV installations, a discrete I–V model is presented [20].

Other studies have correlated the shaded impact with PV power production [32,33]. In Ref. [34], the authors deal with shaded PV installations in urban environments using 3D modeling. A simplified method is presented in Ref. [9] for simulating the output power of shaded PV systems. However, the complex nature of the shading phenomenon suggests that proposed approaches can be extended to improving the PV module performance [25]. As a first step, the following section describes the model proposed by Bishop at the level of shaded PV-cells.

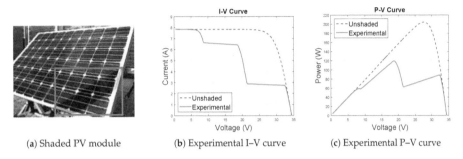

| (a) Shaded PV module | (b) Experimental I–V curve | (c) Experimental P–V curve |

**Figure 2.** Photovoltaic module under partially shaded conditions.

## 2.2. Shaded PV-Cell Model

This section describes the approach proposed by Bishop to model shaded PV-cells in PV modules because of the granularity and the scalability of PV systems [15]. Under shaded conditions, the PV-cells can be forced to carry current in reverse bias. As such, a negative voltage appears at the PV-cell terminals and causes dangerous reverse current to increase [31]. Bishop explains this current multiplication effect through Equation (1) by modeling shaded PV-cells using a nonlinear multiplier factor:

$$I = I_{ph} - I_0 \left[ e^{\left( \frac{V_c + IR_s}{V_t} \right)} - 1 \right] - \frac{V_c + IR_s}{R_p} \left[ 1 + k \left( 1 - \frac{V_c + IR_s}{V_b} \right)^{-n} \right]. \tag{1}$$

Equation (1) represents the relation between the PV-cell current $I$ and the PV-cell voltage $V_c$ [15]. Where $R_s$ is the series resistance associated with conductive losses and $R_p$ is the shunt resistance associated with distributed losses inside of the p-n material. $I_0$ is defined as the inverse saturation current and $V_t$ is the thermal voltage [5]. In the nonlinear multiplier factor, $k$ is the fraction of current involved in avalanche breakdown, $V_b$ the breakdown voltage, and $n$ is the avalanche breakdown exponent. $I_{ph}$ is the photo-generated current given by Equation (2):

$$I_{ph} = [I_{sc\_STC} + (C_{Ti} (T_c - T_{STC}))] \frac{G_i}{G_{STC}}, \tag{2}$$

where $G_i$ is the incident irradiance, $C_{Ti}$ is the thermal current coefficient, and $T_c$ is the cell temperature. $I_{sc\_STC}$, $T_{STC}$, $G_{STC}$ are the short-circuit current, the cell temperature, and the incident irradiance for Standard Test Conditions (25 °C, 1000 W/m$^2$), respectively. Equation (2) becomes the expression for the totally illuminated photo-generated current $I_{ph_{Ti}}$ when $G_i$ is the incident irradiance on the totally unshaded cells.

The model proposed by Bishop is able to describe the PV-cell behavior for completely unshaded and shaded conditions [15]. However, this model disregards the geometric and the optical properties of partial shadows, which can lead to significant loss of accuracy. Indeed, the photo-generated current in Equation (1) depends on a uniform irradiance and partial shading is not discussed by Bishop in Ref. [15]. Some authors have extended the scope of this model to consider shadow properties [20,35]. However, experimental methods to quantify these shadow properties are less widespread in the literature because of shadow complexity [28]. The next section describes the proposed approach for calculating partially shaded PV modules when considering quantifiable shadow characteristics.

## 3. Proposed Approach for Partially Shaded PV Modules

The previous section described a widespread approach to model shaded PV Modules. However, experimental results have shown that this approach can lose accuracy under actual partially shaded conditions [5]. Given the complex nature of the shading phenomenon, the shadow analysis requires the inclusion of the shadow properties without increasing the computational effort due to the scalability of PV systems. These concerns have encouraged the development of the proposed approach through this section.

### 3.1. Partially Shaded PV-Cell Model

Figure 3 shows that, in a PV module, the partially shaded cells have two main shadow features. The first feature is the shadow geometry represented by $a_s + a_i = 1$, where $a_s$ is the fraction of shaded cell area and $a_i$ is the fraction of illuminated cell area. The second shadow feature includes the optical properties of the solar irradiance on the PV module represented by the shadow transmittance $\tau$ and the shading factor $S_f$. The shadow transmittance $\tau$ is defined by the ratio between the scattered irradiance $G_s$ on the shadow and the incident irradiance $G_i$, where $\tau = G_s/G_i$ [18]. $\tau = 0$ means that all the available irradiance is blocked in the interest region. In contrast, $\tau = 1$ means that all the available irradiance shines on the interest region because the scattered irradiance becomes $G_s = G_i$. The shading factor $S_f$ is defined in Equation (3) to describe the shadow opacity [36],

$$S_f = 1 - \frac{G_s}{G_i},\tag{3}$$

where $0 \leq S_f \leq 1$. $S_f = 0$ means that the available irradiance shines on the interest region. In contrast, $S_f = 1$ means that all available irradiance is blocked in the interest region. Then, the relation between $\tau$ and $S_f$ is given by Equation (4):

$$S_f + \tau = 1.\tag{4}$$

Physical meaning of Equation (4) shows that the shadow parameters of shading factor $S_f$ and shadow transmittance $\tau$ are complementary. For instance, a totally shaded PV-cell ($a_s = 1$) with a shading factor $S_f = 0.8$ means that only the 20% of the available irradiance achieves the PV-cell surface, which represents a shadow transmittance $\tau = 0.2$.

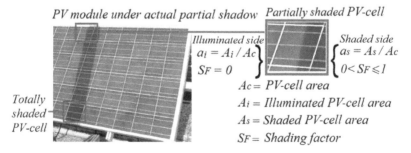

**Figure 3.** Partially shaded PV-cell.

Figure 4a shows a 3D schematic section of a partially shaded PV-cell. In Figure 4, $I_{ph_i}$ and $I_{ph_s}$ represent the photo-generated currents in the illuminated and shaded areas. $I_{ph_T}$ defined as the total photo-generated current. As shown in Figure 4a, electron–hole pairs are generated when photons arrive at the p–n junction in the illuminated area. As a result, a photo-generated current $I_{ph_i}$ is produced in the illuminated area. In contrast, fewer photons can arrive to the p–n junction in the shaded area, which produces lower photo-generated current $I_{ph_s}$ in the shaded area. Therefore, using a simplified approach, the total photo-generated current $I_{ph_T}$ depends on contributions of shaded and unshaded areas, which is defined in Equation (5). Figure 4b shows the equivalent circuit for the photo-generated currents [28].

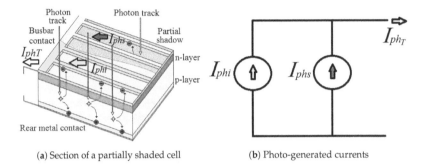

(a) Section of a partially shaded cell        (b) Photo-generated currents

**Figure 4.** Photo-generated currents in a partially shaded PV-cell.

$$I_{ph_T} = I_{ph_i} + I_{ph_s}. \tag{5}$$

Using the current density definition $J = I/A$ for linking the electrical characteristics and the shadow geometric, we obtain Equation (6):

$$I_{ph_T} = J_{ph_i} A_i + J_{ph_s} A_s = J_{ph_i} a_i A_c + J_{ph_s} a_s A_c. \tag{6}$$

Considering the relation between the illuminated and shaded current densities given by the shadow transmittance, $J_{ph_s} = \tau J_{ph_i}$,

$$I_{ph_T} = J_{ph_i} A_c \left( a_i + \tau a_s \right), \tag{7}$$

as described previously $S_f + \tau = 1$ and $a_s + a_i = 1$. Thus,

$$I_{ph_T} = J_{ph_i} A_c \left( 1 - a_s S_f \right). \tag{8}$$

Given that $J_{ph_i}$ represents the photo-generated current produced per unit cell area in the illuminated side and $A_c$ defined as the total PV-cell area, the factor $J_{ph_i} A_c$ can be interpreted as the photo-generated current $I_{ph_{Ti}}$ that should be provided by the PV-cell in totally illuminated conditions. Therefore, Equation (7) can be rewritten as seen below:

$$I_{ph_T} = I_{ph_{Ti}} \left( 1 - a_s S_f \right). \tag{9}$$

The physical meaning of Equation (9) represents that the total photo-generated current $I_{ph_T}$ is proportional to the totally illuminated photo-generated current $I_{ph_{Ti}}$ given a ratio that depends on the shadow properties [28]. Equation (9) shows that the total photo-generated current depends on the shaded area percentage $a_s$ and the shadow opacity $S_f$ but is independent of the shadow shape. Defining this relation by the shading ratio $\delta$, Equation (10) is obtained:

$$\delta = 1 - a_s S_f. \tag{10}$$

Thus, the total photo-generated current $I_{ph_T}$ is given through Equation (11). In the $I_{ph_T}$ expression, the totally illuminated photo-generated current $I_{ph_{Ti}}$ is evaluated using Equation (12) and considering $G_i$ as the incident irradiance in totally unshaded conditions, which was clarified previously in Equation (2):

$$I_{ph_T} = I_{ph_{Ti}} \delta, \tag{11}$$

$$I_{ph_{Ti}} = [I_{sc\_STC} + (C_{Ti}(T_c - T_{STC}))]\frac{G_i}{G_{STC}}. \tag{12}$$

In addition, Equation (13) is defined by considering $I_{sc_{Ti}}$ as the totally illuminated short-circuit current for unshaded cell conditions:

$$I_{ph_{Ti}} \approx I_{sc_{Ti}}. \tag{13}$$

Thus,

$$I_{ph_T} \approx I_{sc_{Ti}} \delta. \tag{14}$$

We propose extending the model presented by Bishop [15] while using $I_{ph_T}$ for reformulating Equation (1) and Equation (15). At this point, it is important to highlight that the shading ratio $\delta$ depends on the quantifiable parameters $a_s$ and $S_f$. Therefore, $\delta$ is also quantifiable. Figure 5a outlines the current–voltage behavior of a shaded PV-cell according to Equation (15). The equivalent PV-cell circuit is shown in Figure 5b. This simplified $\delta$ factor improves the description scope of shaded PV systems including measurable shadow features without needing to increase the computational effort.

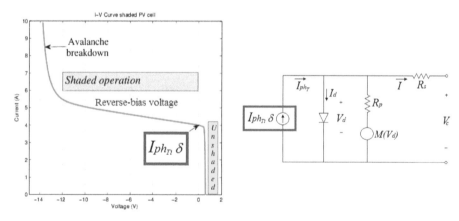

(a) Current-voltage behavior of partially shaded cell

(b) Modified equivalent PV-cell circuit using the shading ratio

**Figure 5.** Equivalent circuit and I–V curve of the partially shaded PV-cell.

$$I = \underbrace{I_{ph_{Ti}}\delta}_{I_{ph}} - I_0\left[e^{\left(\frac{V_c + IR_s}{V_t}\right)} - 1\right] - \frac{V_c + IR_s}{R_p}\left[1 + k\left(1 - \frac{V_c + IR_s}{V_b}\right)^{-n}\right]. \tag{15}$$

Equation (15) is a nonlinear equation that can be solved using numerical methods. The numerical method usually employed to solve these types of equations is the Newton–Raphson method [4]. The method starts with a function $f(V_c)$ defined as $f(V_c) = 0$ as rewritten below in Equation (16):

$$f(V_c) = I_{ph_{Ti}}\delta - I - I_0\left[e^{\left(\frac{V_c + IR_s}{V_t}\right)} - 1\right] - \frac{V_c + IR_s}{R_p}\left[1 + k\left(1 - \frac{V_c + IR_s}{V_b}\right)^{-n}\right]. \tag{16}$$

Given that the function satisfies the condition $f'(V_c) \neq 0$, the following iterative process is repeated until a sufficiently accurate value is reached:

$$V_{cn+1} = V_{cn} - \frac{f(V_{cn})}{f'(V_{cn})}. \tag{17}$$

The solution of the iterative process in Equation (17) describes the PV-cell voltage given the influence of the shading ratio $\delta$ and a known $I$ current. The solution of this iterative process is performed for a range of $I$ currents from 0 to $I_{sc}$ and for the respective shading ratios $\delta$ of shaded PV cells. This method allows for calculating the voltage at the PV-cell level under several working conditions. Nevertheless, a series of connections of PV-cells form PV modules and it is required to go in depth about this aspect. The following section presents a systematic perspective to analyze the influence of the shading ratio on PV modules.

### 3.2. Influence of the Shading Ratio $\delta$ on the PV Module Behavior

This section relates the previous proposed approach with shadow patterns to extend the shadow impact analysis at the level of PV modules. Series connections of PV-cells form PV modules, and PV module manufacturing usually connects by-pass diodes to groups of PV-cells for decreasing the damage risk [29]. Thus, the voltage in a PV module $V_p$ with $m$ groups of $q$ PV-cells and by-pass diode voltage $V_{BD}$ is given by Equation (18):

$$V_p = \sum_{j=1}^{m} V_{G_j} \text{ where } V_{G_j} = \begin{cases} \sum_{i=1}^{q} V_{c_i} \text{ if } \sum_{i=1}^{q} V_{c_i} \geq 0, \\ V_{BD} \text{ if } \sum_{i=1}^{q} V_{c_i} < 0. \end{cases} \tag{18}$$

The PV-cell voltages $V_{c_i}$ come from the solution of the nonlinear Equation (16) by applying the numerical Newton–Raphson method of Equation (17). In addition, the parameters $I_0$, $R_s$, $V_t$, and $R_p$ of Equation (16) have been extracted according to the iterative methods presented in Ref. [5]. The parameters $k$ and $n$ of the multiplier factor proposed by Bishop in Ref. [15] have been extracted using nonlinear curve fitting methods from experimental results in shaded conditions with unconnected by-pass diodes. The parameter $V_b$ depends on the PV module technology and it has been fitted according to operation regions proposed in Ref. [29].

Solutions of Equations (15) and (18) for a group of twenty cells with a single shaded cell provides the results in Figure 6. As shown in Figure 6, a partial shadow in a single cell can change the normal behavior of the group drastically. Denoting $I_{dv}$ as the divergence current where the I–V curve diverges of normal operation in shaded conditions given by $\delta < 1$, and the comparison of results in Figure 6a,b allows for deducing the behavior of $I_{dv}$ described by Equation (19):

$$I_{dv} \approx I_{ph_{Ti}} \delta \qquad for \ \delta < 1. \tag{19}$$

In addition, the totally illuminated short-circuit current was considered in Equation (13) as $I_{sc_{Ti}} \approx I_{ph_{Ti}}$. Then,

$$I_{dv} \approx I_{sc_{Ti}} \delta \qquad for \ \delta < 1. \tag{20}$$

Equation (19) is deduced because the voltage in the shaded PV-cell begins to be negative when the PV-cell current is higher than $I_{ph_{Ti}} \delta$, which leads to a prominent change of the I–V curve. If the PV-cell current follows increasing, the PV-cell voltage is each time more negative until achieving the activation of the by-pass voltage. In this operation condition, the shaded PV-cell dissipates power due to the negative voltage and risk of damage can arise. Figure 6 shows that the situation can get worse if the shading ratio is higher because the dissipate power increases. This situation can induce hot-spots if the partial shadows are small and permanent. Failures of this type have been reported in literature and require preventive actions to avoid the deterioration of the PV system performance [37].

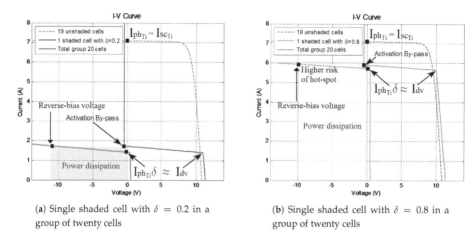

(a) Single shaded cell with $\delta = 0.2$ in a group of twenty cells

(b) Single shaded cell with $\delta = 0.8$ in a group of twenty cells

**Figure 6.** Influence of $\delta$ in a group of twenty cells with a single shaded cell.

Figure 6 also allows for deducing the relation between the shaded PV-cells in a group with the lowest shading ratio. Assuming a case in which the shaded PV-cells of Figure 6a,b are in the same group, the lowest shading ratio of Figure 6a would lead the group toward the by-pass activation condition. Therefore, the shading ratio of the Figure 6b would have a minimum impact in the divergence current because the by-pass diode is already active. This operation principle can be extended to several shaded cells with different shading ratios because the lowest shading ratio is the first to activate the by-pass diode.

Figure 6 shows that the reverse-bias voltage is critical for the structural healthy of the PV-module. For that reason, in Figure 7, the I–V behavior is depicted in reverse-bias condition for several shaded cells. In this case, one PV-cell has a higher slope because the proximity of the breakdown voltage. In contrast, the illustrative example of Figure 7 shows that increasing the $N$ shaded PV-cell multiplies the negative voltage $N$ times because the PV-cell are connected in series. Therefore, the slope in the negative region decrease and for a given current interval $Slope = \triangle I / (N * \triangle V)$.

Figure 8 extends the analysis to several shaded PV cells in a PV module. These figures show the interrelation between the divergence currents and the maximum power points (MPPs). As shown in Figure 8, $z$ represents the index for the lowest shading ratios $\delta_z$ in each group where $z = \{0, 1, 2, .., g - 1\}$ and $g$ is the total number of groups connected in series. $V_{mz}$ and $I_{mz}$ are the voltages and currents at the MPPs. The relation between the MPPs and the lowest shading ratios $\delta_z$ is given by the behavior of the divergence currents $I_{dvz}$ and the local MPPs. Figure 8 allows for deducing that $I_{dvz} \approx I_{mz}$ because the MPPs arise around the current divergence. Nevertheless, an exception to this pattern is presented in unshaded groups where $I_{mz} \approx I_{mp}$.

$\triangle V_z$ is defined in Equation (21) as a proportional relation between the voltage difference $V_{oc} - V_{mp}$ and the corresponding shading ratio $\delta_z$ for the shading ratios arranged from the lower to the higher $\delta_z < \delta_{z+1}$. The physical meaning of Equation (21) represents that the voltage displacement of $V_{mz}$ in relation to the local MPPs in an unshaded condition is associated with the shading ratio $\delta_z$.

$$\triangle V_z \approx (V_{oc} - V_{mp}) \, \delta_z \quad for \; z = \{0, 1, ..., g - 1\} \, , g = number \; of \; groups, and \; \delta_z < \delta_{z+1}. \tag{21}$$

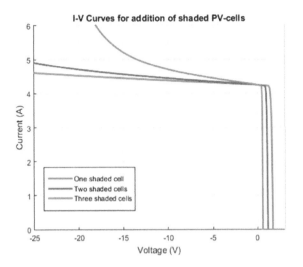

**Figure 7.** Simulation of I–V Curves for addition of shaded PV-cells.

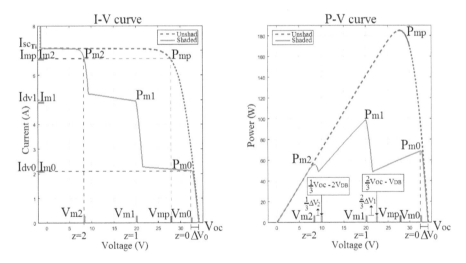

**Figure 8.** I–V and P–V curves for a partially shaded PV module.

All groups are connected in series and each group proportionally contributes to the open circuit voltage $V_{oc}$. For this reason, the voltage $V_{mz}$ at the local MPPs is expressed as a fraction of $V_{oc}$ and $\triangle V_z$. For the illustrative example of the Figure 8, the expressions for the voltages $V_{mz}$ at the local MPPs are given from Equation (22) to Equation (24), where $V_{BD}$ is the forward by-pass diode voltage which displaces the proportion of $V_{oc}$:

$$V_{m0} = V_{oc} - \triangle V_0, \tag{22}$$

$$V_{m1} = \left(\frac{2}{3}V_{oc}\right) - \left(\frac{2}{3}\triangle V_1\right) - V_{BD}, \tag{23}$$

$$V_{m2} = \left(\frac{1}{3}V_{oc}\right) - \left(\frac{1}{3}\triangle V_2\right) - 2V_{BD}. \tag{24}$$

A general expression of $V_{mz}$ is deduced in Equation (25) for $g$ groups and $z = \{0, 1, ..., g-1\}$,

$$V_{mz} = \left(\frac{g-z}{g}\right)V_{oc} - \left(\frac{g-z}{g}\right)\triangle V_z - (zV_{BD}). \tag{25}$$

For $\delta_z < 1$ and $\delta_z < \delta_{z+1}$,

$$P_{mz} = V_{mz}I_{mz} \approx V_{mz}I_{dvz} \approx V_{mz}I_{sc_{Ti}}\delta_z, \tag{26}$$

$$P_{mz} \approx \left[\left(\left(\frac{g-z}{g}\right)(V_{oc} - \triangle V_z)\right) - (zV_{BD})\right]I_{sc_{Ti}}\delta_z. \tag{27}$$

For unshaded groups $\delta_z = 1$ and $P_{mz}$ is given by Equation (28),

$$P_{mz} = V_{mz}I_{mp}, \tag{28}$$

$$P_{mz} \approx \left[\left(\left(\frac{g-z}{g}\right)V_{mp}\right) - (zV_{BD})\right]I_{mp}. \tag{29}$$

Equations (27) and (29) allow a fast approximation to the MPPs for known shadow patterns and unshaded operation parameters. The procedure to evaluate the MPPs is described as follows:

**Step 1:**  Determination of the lowest shading ratios $\delta_z$ in each group, arrangement of shading ratios from the lower to the higher $\delta_z < \delta_{z+1}$.

**Step 2:**  Evaluation of $V_{mp}$, $I_{mp}$, $I_{sc_{Ti}}$, and $V_{oc}$ from unshaded condition, considering $V_{BD} \approx 0.7V$.

**Step 3:**  Calculation of $P_{mz}$ for $z = \{0, 1, ..., g-1\}$ using Equation (27) if $\delta_z < 1$ or Equation (29) if $\delta_z = 1$.

**Step 4:**  In the special case of $\delta_z = \delta_{z+1}$, the sequence of values for $P_{mz}$ and $P_{mz+1}$ are evaluated normally. However, only the highest value of power defines the region for the local MPP.

Given the proposed modeling approach through this section, the next stage will analyze the simulation of shaded PV modules.

## 4. Simulation Analysis of Shaded PV Modules

The cases of shadow patterns in this section have been selected to illustrate the potential features of proposed approaches in simulation. First, two cases describe the impact of single shaded cells scattered in several groups and the impact of shaded cells grouped in a single group. Then, two cases are intended to show the shadow movement impact. The final simulation targets a shaded PV string.

The simulations have been performed in a conventional computational platform by solving Equations (16)–(18) according to the lineaments presented in Section 3. In addition, the simulated shading ratios $\delta$ are set for analysis and further correlation with experimental patterns. The shaded PV-module images in this section have only a character illustrative and do not represent any software in particular.

### 4.1. Simulation of Partially Shaded PV Modules

The nominal parameters of the simulated PV modules are $I_{sc} = 8.3A$ and $V_{oc} = 37.3$ V with simulation conditions of incident irradiance $G_i = 850 \, W/m^2$ and cell temperature $T_c = 45\,°C$. The cases consider a uniform shading factor $S_f = 0.8$. We also consider a conventional PV module with sixty cells distributed in groups of twenty cells connected to by-pass diodes [3].The analysis uses a matrix notation $ij$ where $\delta_{ij}$ represents the shading ratio of a PV-cell in the relative position $ij$ in a PV module.

The first case depicted in Figure 9 shows all groups with a single shaded cell. This simulation is intended to study the impact of single shaded cells in the normal current–voltage behavior. In Figure 9, the PV module current $I_{PV}$ is normalized in ratio to $I_{scT_i} = 7.1$ A. Therefore, on the y-axis, the $I_{Norm} = I_{PV}/I_{scT_i}$. This simulation case shows that the lowest divergence current $I_{dv}$ is proportional to the shaded cells with the lowest value of $\delta$. For instance, the first divergence current $I_{dv_0}$ in Figure 9b is caused by the PV-cell with $\delta_{2.10} = 0.20$ of group one. Figure 9 confirms that the divergence current $I_{dvz}$ due to each group is close to $I_{dvz} = \delta_z I_{scT_i}$, where $\delta_z$ depends on the shaded cell $C_{ij}$ with the lowest value of $\delta$ in the group.

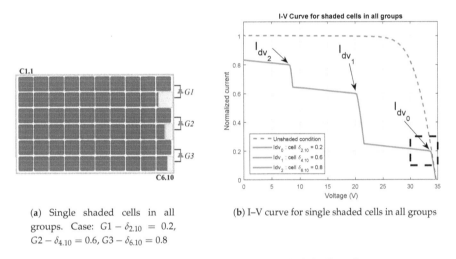

(a) Single shaded cells in all groups. Case: $G1 - \delta_{2.10} = 0.2$, $G2 - \delta_{4.10} = 0.6$, $G3 - \delta_{6.10} = 0.8$

(b) I–V curve for single shaded cells in all groups

**Figure 9.** Study case of shadow pattern for single shaded cells in all groups.

Figure 10 depicts three sub-cases of distributed shaded cells in a single group. As shown in Figure 10a, the first sub-case has one shaded cell with $\delta = 0.2$. The second sub-case has two shaded cells with $\delta = 0.2$ and the other two cells with $\delta = 0.4$. The third sub-case has five cells with $\delta = 0.2$ and the other two cells with $\delta = 0.4$. The simulation results show that the lowest value of $\delta$ in a group with several shaded cells causes the divergence current $I_{dv}$. In addition, Figure 10b illustrates that a greater number of shaded cells in a group causes a decrease in the I–V curve slope. This phenomenon is due to the behavior of the shaded cells in the reverse-bias as described previously. As a result, a single shaded cell in a group has a higher I–V curve slope and more risk of hot-spots than a group with several shaded cells because the reverse-bias voltage and power dissipation distribution [2,3].

Figures 9 and 10 allowed for the analysis of single shaded PV cells and single shaded groups. However, the shadow displacement in daily conditions can generate several irregular shadow patterns. To describe this more realistic aspect, Figures 11 and 12 illustrate two irregular shadow patterns from a hypothetical pole, antenna, or chimney.

Figure 11a shows a diagonal shadow pattern and the associated shading ratios. In this case, Figure 11b shows that the group three with $\delta_{6.9} = \delta_{5.10} = 0.28$ produces the lowest divergence current $I_{dv_0} = 0.28 I_{scT_i} \approx 2A$, and the group two with $\delta_{4.10} = 0.7$ produces the divergence current $I_{dv_1} = 0.7 I_{scT_i} \approx 5A$. After finding $I_{dv_0}$ and $I_{dv_1}$, the maximum power points (MPPs) are calculated using Equations (27) and (29) as described in Section 3.2. Table 1 lists these approximate and simulated MPPs. The results in Table 1 show a suitable agreement between actual and estimated MPPs. This simplified method allows for quickly identifying the global MPP and its source.

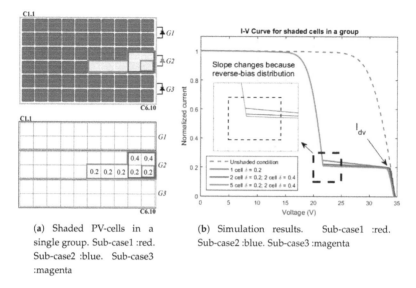

(**a**) Shaded PV-cells in a single group. Sub-case1 :red. Sub-case2 :blue. Sub-case3 :magenta

(**b**) Simulation results. Sub-case1 :red. Sub-case2 :blue. Sub-case3 :magenta

**Figure 10.** Representation of three sub-study cases with shadow patterns distributed in a single group.

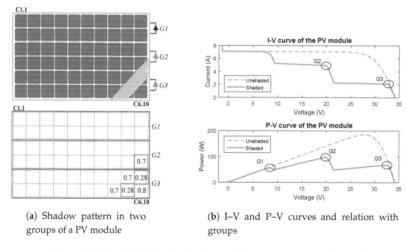

(**a**) Shadow pattern in two groups of a PV module

(**b**) I–V and P–V curves and relation with groups

**Figure 11.** I–V and P–V behavior for diagonal shadow pattern in a PV module.

**Table 1.** Maximum power points—Case: diagonal shadow.

| Parameter | Values | | |
|---|---|---|---|
| $z$ | 0 | 1 | 2 |
| $\delta_z$ | 0.28 | 0.7 | 1.0 |
| Group | G3 | G2 | G1 |
| $P_{mz}$ approx.(W) | 64.09 | 95.02 | 53.65 |
| $P_{mz}$ simul.(W) | 68.06 | 98.39 | 55.27 |
| Rel. error | 0.06 | 0.03 | 0.03 |

Figure 12a describes a pattern in all vertical groups. This figure shows that the divergence currents depend on the lowest $\delta$ in each group. In addition, the other shaded cells impact the slope of the

I–V curve without relevant contribution to the $I_{dv}$. Table 2 shows the simplified calculation of the approximate MPPs using Equations (27) and (29) of Section 3.2. These results show that the group G1 provides the global MPPs, which agrees with the simulation results. Therefore, Tables 1 and 2 confirm that the MPPs can be localized from the lowest $\delta$ in each group and the parameters of unshaded operation without an exhaustive calculation from all of the shaded PV cells.

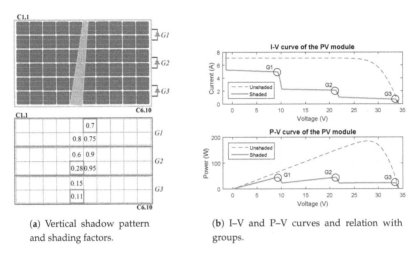

(a) Vertical shadow pattern and shading factors.

(b) I–V and P–V curves and relation with groups.

**Figure 12.** I–V and P–V behavior of the vertical shadow pattern in a PV module.

**Table 2.** Maximum power points—Case: vertical shadow.

| Parameter | Values | | |
|---|---|---|---|
| $z$ | 0 | 1 | 2 |
| $\delta_z$ | 0.11 | 0.28 | 0.7 |
| Group | G3 | G2 | G1 |
| $P_{mz}$ *approx.*(W) | 26.02 | 41.54 | 43.05 |
| $P_{mz}$ *simul.*(W) | 24.07 | 44.53 | 45.72 |
| *Rel. error* | 0.081 | 0.067 | 0.058 |

### 4.2. Simulation of Partially Shaded PV String

Figure 13 depicts the final studied case at level of PV string. To facilitate understanding, this figure highlights the most significant shaded PV cells in an irregular shadow pattern. The simulations results allow identifying four regions. The region $R_1$ depends on G1.2 and G2.2. In this region, PV1 $\delta_{3.1}$ and PV2 $\delta_{4.1}$ cause the lowest $I_{dv0}$ in the PV string. $R_1$ is extended by around 20V because the bypass activation of two groups. G1.1 and G1.3 produce region $R_2$. The divergence current $I_{dv1}$ in $R_2$ is proportional to the 40% of $I_{scT_i}$, which is caused by PV1 $\delta_{1.1} = \delta_{2.1} = \delta_{5.1} = \delta_{6.1} = 0.4$. Region $R_3$ is produced by G3.2 with the single PV-cell PV3 $\delta_{3.1} = 0.6$. The bypass activation point and the I–V curve slope are higher in region $R_3$; therefore, this single cell is more vulnerable to dissipating power and generating hot-spots (see Figure 6b). Finally, $R_4$ depends on the unshaded PV groups and provides the highest MPP of all regions.

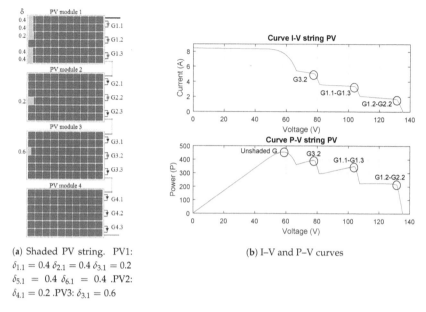

(a) Shaded PV string. PV1:
$\delta_{1.1} = 0.4\ \delta_{2.1} = 0.4\ \delta_{3.1} = 0.2$
$\delta_{5.1} = 0.4\ \delta_{6.1} = 0.4$ .PV2:
$\delta_{4.1} = 0.2$ .PV3: $\delta_{3.1} = 0.6$

(b) I–V and P–V curves

**Figure 13.** I–V and P–V curves of a shaded PV string.

Table 3 lists the MPPs for the studied PV string. These results illustrate a special case of Equations (27) and (29) to evaluate the approximate MPPs where equal $\delta_z$ appear in different groups. For this case, the sequence of values in Equations (27) and (29) are evaluated normally; however, only the highest MPP of equal $\delta_z$ is taken into account to define the MPP region and the global MPP. Finally, the results in Figure 13 and Table 3 confirm that this simplified methodology provides a suitable approximation to the MPPs at the level of PV strings. The next section summarizes the main identified findings.

**Table 3.** Maximum power points—Case: shaded string

| Parameter | Values | | | | | |
|---|---|---|---|---|---|---|
| $z$ | 0 | 1 | 2 | 3 | 4 | 5 |
| $\delta_z$ | 0.2 | 0.2 | 0.4 | 0.4 | 0.6 | 1.0 |
| Group | G1.2 | G2.2 | G1.1 | G1.3 | G3.2 | Unsh.G. |
| $P_{mz}$ approx.(W) | 216.1 | 197.1 | 340.5 | 304.2 | 382.5 | 464.2 |
| $P_{mz}$ simul.(W) | 220.9 | – | 348.8 | – | 391.30 | 453.2 |
| Rel. error | 0.022 | – | 0.024 | – | 0.023 | 0.024 |

*4.3. Identified Patterns between the Shading Ratio and the PV Module Behavior*

The following findings highlight the patterns identified from the interaction between the shading ratio and the partial shadows.

- The divergence currents $I_{dvz}$ are proportional to the lowest shading ratio $\delta_z$ in each shaded PV group. Thus, $I_{dvz} \approx \delta_z I_{scT_i}$ for $\delta_z < 1$.
- Shaded cells have a minimal impact on the I–V curve if their shading ratio is greater than the lowest shading ratio in the same group.
- In a group, shaded cells with shading ratios close to the lowest shading ratio have a lower overheating risk because the reverse bias voltage is distributed between them.

- A single shaded cell in a group with higher shading ratio has a greater probability of being a hot-spot because of the power dissipation despite the by-pass diodes.
- The MPPs can be quickly identified from the lowest shading ratio in each group and the parameters for unshaded conditions.
- The above-mentioned patterns can be extended at the level of PV strings.

The next section presents the experimental tests to validate the proposed approach correlating the current voltage-behavior with shadow image patterns.

## 5. Experimental Validation and Discussion

This section describes the experimental setup for validating the analysis proposed in Section 3. In addition, this section outlines an experimental procedure to quantify the shading ratio using image processing methods. Experimental results are discussed.

### 5.1. Test for Partially Shaded PV Modules

The developed experiments consider two shadow cases as depicted in Figures 14 and 15. Furthermore, the tests use monocrystalline and polycrystalline PV modules to compare these common commercial technologies. The PV module characteristics are listed in Table 4.

**Table 4.** Photovoltaic modules under testing.

| Type<br>Ref. | Monocrystalline<br>Tenesol TE 2200 | Polycrystalline<br>Yingli solar YL290p-35b |
|---|---|---|
| *Electrical parameters at STC* | | |
| Maximum Power ($P_{mp}$) | 250 Wp | 290 Wp |
| Voltage at $P_{mp}$ ($V_{mp}$) | 30.3 V | 35.8 V |
| Current at $P_{mp}$ ($I_{mp}$) | 8.3 A | 8.1 A |
| Open circuit voltage ($V_{oc}$) | 37.3 V | 45.3 V |
| Short-circuit current ($I_{sc}$) | 8.6 A | 8.62 A |

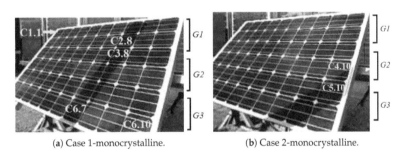

(a) Case 1-monocrystalline.  (b) Case 2-monocrystalline.

**Figure 14.** Experimental tests for monocrystalline PV module.

Figure 16a depicts the experimental setup performed in the platform ADREAM of the Laboratory for Analysis and Architecture of Systems (LAAS-CNRS) in Toulouse, France (43°33′44.3″N 1°28′38.3″E). In this setup, an I–V curve tracer (model MP-160, EKO Instruments, Tokyo, Japan) is used to detect the current–voltage signals. Furthermore, a pyrometer (model SP-Lite, Kipp&Zonen, Delft, the Netherlands) monitors the solar irradiance and a thermographic camera periodically measures the PV module temperature.

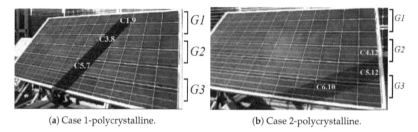

(a) Case 1-polycrystalline.                    (b) Case 2-polycrystalline.

**Figure 15.** Experimental tests for polycrystalline PV module.

Simultaneously, a digital camera records the shadow pattern, which is shown in Figure 16b. The analysis of the shaded PV-cell areas is performed using image processing methods after contour selection. The selected image is converted from gray-scale image to a binary image through digital processing based on the histogram and Otsu's method [38]. Finally, the shaded area is calculated using Equation (30) where $p_b$ is the total number of black pixels and $p_w$ is the total number of white pixels:

$$a_s = \frac{p_b}{p_b + p_w}.\tag{30}$$

The experimental test is described as follows:

**Step 1:** Simultaneous measurements and recording of I–V curves, solar irradiance, PV module temperature, and shadow patterns.

**Step 2:** Selection of synchronized I–V curves and image shadow patterns for analysis.

**Step 3:** Image processing for measurement of shaded PV-cell areas in selected shadow patterns.

**Step 4:** Shading ratio calculation for the PV-cell with the largest shaded area $a_{sL}$ using Equation (31). $I_{ph_{Ti}}$ is calculated using Equation (12). $I_{dv_L}$ is the first divergence-current point in the experimental I–V curve.

**Step 5:** Calculate the shading factor $S_f$ for the PV-cell with the largest shaded area $\delta_L$ using Equation (32). In this experimental setup, the shading factor is considered uniform on the shaded cell because the I–V curve measurements and the shaded PV module image recording are synchronized.

**Step 6:** Evaluate the shading ratio $\delta_{ij}$ for each shaded PV-cell.

**Step 7:** The calculated shading ratios are used to evaluate the I–V and P–V characteristics of the PV modules:

$$\delta_L = \frac{I_{dv_L}}{I_{ph_{Ti}}},\tag{31}$$

$$S_f = \frac{1 - \delta_L}{a_{sL}}.\tag{32}$$

Table 5 summarizes the parameters and values for calculating the shading factor. The shading ratios $\delta_{ij}$ are calculated for each shaded PV-cell while considering their shaded area $a_{ij}$ and the same shading factor $S_f$. Tables 6 and 7 list the shaded cell areas $a_{ij}$ obtained after image processing and the calculated shading ratios $\delta_{ij}$. These shading ratios are used to evaluate the I–V and P–V characteristics for the PV modules. Tables 8 and 9 summarize the MPPs. Lastly, the experimental and calculated I–V curves are depicted in Figures 17 and 18.

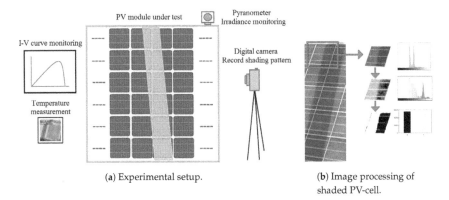

(a) Experimental setup.

(b) Image processing of shaded PV-cell.

**Figure 16.** Experimental setup for model validation and image processing.

**Table 5.** Shading factor results for the PV modules under testing.

| Type | Monocrystalline | | Polycristalline | |
|---|---|---|---|---|
| | Case 1 | Case 2 | Case 1 | Case 2 |
| $G_i$ | 820 W/m$^2$ | 910 W/m$^2$ | 710 W/m$^2$ | 540 W/m$^2$ |
| $T_c$ | 31 °C | 31 °C | 31 °C | 30 °C |
| $I_{ph_{Ti}}$ | 7.07 A | 7.85 A | 6.14 A | 4.67 A |
| $I_{dv_L}$ | 2.16 A | 2.7 A | 1.33 A | 1.66 A |
| $a_{sL}$ | 0.98 | 0.91 | 0.97 | 0.90 |
| $\delta_L$ | 0.31 | 0.34 | 0.22 | 0.36 |
| $S_f$ | 0.70 | 0.72 | 0.8 | 0.71 |

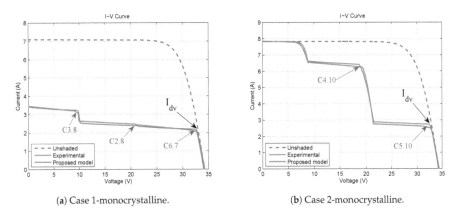

(a) Case 1-monocrystalline.

(b) Case 2-monocrystalline.

**Figure 17.** I–V curves for test with monocrystalline PV module.

**Table 6.** Shaded area and shading ratio for monocrystalline PV module Tenesol TE-2200.

| Group | Case 1 | | | Case 2 | | |
|---|---|---|---|---|---|---|
| | $C_{ij}$ | $a_{ij}$ | $\delta_{ij}$ | $C_{ij}$ | $a_{ij}$ | $\delta_{ij}$ |
| 1 | C1.8 | 0.94 | 0.34 | | | |
| | C2.8 | 0.96 | 0.33 | | | |
| 2 | C3.7 | 0.17 | 0.88 | C4.10 | 0.25 | 0.82 |
| | C3.8 | 0.80 | 0.44 | | | |
| | C4.7 | 0.40 | 0.72 | | | |
| | C4.8 | 0.50 | 0.65 | | | |
| 3 | C5.7 | 0.80 | 0.44 | C5.9 | 0.35 | 0.75 |
| | C5.8 | 0.20 | 0.86 | C5.10 | 0.91 | 0.34 |
| | C6.7 | 0.98 | 0.31 | C6.8 | 0.40 | 0.72 |
| | | | | C6.9 | 0.89 | 0.37 |
| | | | | C6.10 | 0.15 | 0.89 |

**Table 7.** Shaded area and shading ratio for polycrystalline PV module YL290p-35b.

| Group | Case 1 | | | Case 2 | | |
|---|---|---|---|---|---|---|
| | $C_{ij}$ | $a_{ij}$ | $\delta_{ij}$ | $C_{ij}$ | $a_{ij}$ | $\delta_{ij}$ |
| 1 | C1.9 | 0.68 | 0.46 | | | |
| | C2.8 | 0.64 | 0.49 | | | |
| | C2.9 | 0.66 | 0.47 | | | |
| 2 | C3.8 | 0.96 | 0.23 | C4.12 | 0.19 | 0.86 |
| | C3.9 | 0.11 | 0.91 | | | |
| | C4.7 | 0.65 | 0.48 | | | |
| | C4.8 | 0.65 | 0.48 | | | |
| 3 | C5.6 | 0.10 | 0.92 | C5.10 | 0.21 | 0.85 |
| | C5.7 | 0.97 | 0.22 | C5.11 | 0.80 | 0.43 |
| | C5.8 | 0.11 | 0.91 | C5.12 | 0.90 | 0.36 |
| | C6.6 | 0.65 | 0.48 | C6.8 | 0.22 | 0.84 |
| | C6.7 | 0.65 | 0.48 | C6.9 | 0.78 | 0.45 |
| | | | | C6.10 | 0.90 | 0.36 |
| | | | | C6.11 | 0.37 | 0.74 |

**Table 8.** Maximum power points—Monocrystalline.

| Parameter | Case 1 | | | Case 2 | | |
|---|---|---|---|---|---|---|
| $z$ | 0 | 1 | 2 | 0 | 1 | 2 |
| $\delta_z$ | 0.31 | 0.33 | 0.44 | 0.34 | 0.82 | 1.0 |
| Group | G3 | G1 | G2 | G3 | G2 | G1 |
| $P_{mz}$ *approx.*(W) | 70.3 | 48.3 | 28.7 | 85.5 | 120.3 | 59.2 |
| $P_{mz}$ *exper.*(W) | 70.2 | 51.6 | 31.1 | 88.2 | 119.4 | 60.2 |
| *Rel. error* | 0.001 | 0.064 | 0.078 | 0.031 | 0.008 | 0.017 |

**Table 9.** Maximum power points—Polycrystalline.

| Parameter | Case 1 | | | Case 2 | | |
|---|---|---|---|---|---|---|
| $z$ | 0 | 1 | 2 | 0 | 1 | 2 |
| $\delta_z$ | 0.22 | 0.23 | 0.46 | 0.36 | 0.86 | 1.0 |
| Group | G3 | G2 | G1 | G3 | G2 | G1 |
| $P_{mz}$ *approx.*(W) | 55.5 | 37.8 | 33.4 | 68.4 | 93.8 | 44.9 |
| $P_{mz}$ *exper.*(W) | 55.2 | 42.1 | 36.9 | 67.8 | 94.9 | 45.7 |
| *Rel. error* | 0.006 | 0.103 | 0.094 | 0.01 | 0.01 | 0.02 |

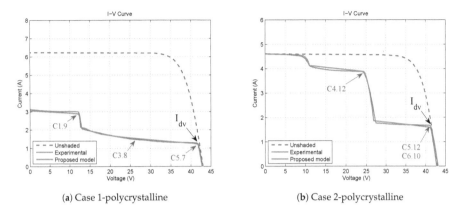

(a) Case 1-polycrystalline

(b) Case 2-polycrystalline

**Figure 18.** I–V curves for tests with polycrystalline PV module.

## 5.2. Discussion of Results

The experimental results confirm the correlation between the shading ratio $\delta$ and the I–V curve. For instance, Table 6 and Figure 17 experimentally show that the first divergence points in the I–V curves are caused by the lowest shading ratios of PV-cells C6.7 and C5.10. Additionally, Table 7 and Figure 18 allow for validating this interpretation.

Considering a uniform $S_f$, the results also demonstrate that the smaller shaded cell areas in comparison with the larger shaded cell areas in the same group provide a minimal contribution to the I–V curve. For instance, the PV-cells C3.7 and C5.8 of case 1-monocrystalline (Table 6) have a minimal impact on the I–V characteristics of Figure 17a. In contrast, shaded cells with small shaded areas are able to modify the I–V curve if they have the lowest shaded area in the group. For example, the PV-cell C4.12 in Case 2-polycrystalline is able to cause changes in the I–V curve of Figure 18b.

Figure 18a,b show that shaded PV-cells in a group, with shading ratios near to the lowest shading coefficient in the group, produce I–V curves with lower slopes because the behavior of the reverse-bias voltage. For instance, PV-cells C5.12 and C6.10 of Case 2-polycrystalline cause a lower slope than caused by the PV-cell C4.12 in Case 2-polycrystalline. Therefore, PV-cell C4.12 has more risk of dissipating power. Table 10 lists the slopes for the case 2-monocrystalline and the case 2-polycrystalline, which have a single shaded cell in a group. Results in Table 10 show a slight difference in the case of monocrystalline but a more significant difference in the case of polycrystalline. Authors also have addressed a detailed experimental study about the partial shading and the slope identification, which has been reported in Ref. [3]:

$$MSE = \frac{1}{n} \sum_{i=1}^{n} (e_i)^2. \tag{33}$$

**Table 10.** Slopes of I–V curves.

| Region | Monocryst. Case 2 | Region | Polycryst. Case 2 |
|---|---|---|---|
| 8 V–18 V | 37 mA/V | 12 V–24 V | 20.37 mA/V |
| 22 V–32 V | 33.8 mA/V | 28 V–42 V | 11.11 mA/V |

We use the mean square error (MSE) to assess the modeling accuracy based on the shading ratio. The MSE values listed in Table 11 illustrate the model accuracy for the experimental and simulated

I–V curves depicted in Figures 17 and 18. Table 11 shows that the proposed approach is suitable for describing the current–voltage behavior of partially shaded PV modules in both monocrystalline and polycrystalline technologies. However, the behavior of polycrystalline modules slightly varies from monocrystalline modules because of the lower breakdown voltage in polycrystalline technology [29]. This phenomenon is more appreciable in the region from 13 V to 40 V of Figure 18a, which can lead to higher risk of hot-spots [29].

**Table 11.** Mean squared error (MSE) from simulated and experimental I–V curves for model validation.

|  | Monocrystalline | | Polycrystalline | |
|---|---|---|---|---|
| Error calculation | Case 1 | Case 2 | Case 1 | Case 2 |
| MSE | 0.92 | 1.07 | 0.89 | 1.01 |

The MSE of model validation in Table 11 also shows a slightly difference for case 1 and case 2 in both technologies. This difference can be produced by several factors such as changes in the internal parameters, current path, or leakage currents. Indeed, some authors have shown that series and shunt resistances are affected by the irradiance conditions [21]. Soto et al. suggest that the series resistance depends on the irradiance level because its value decreases for lower irradiance and even can suffer negative values [39]. Earlier works also indicate negative values for the series resistance under low irradiance [40]. In Ref. [21], the series resistance increases at the same time as the shadow rate, which increases the amount of power dissipated by the series resistance. Nevertheless, most authors consider these variations less relevant by treating the series resistance independent of the incident irradiance and temperature and obtaining sufficient accuracy [41,42]. In contrast, the study of the low irradiance conditions on the shunt resistances have been more widespread in literature because of the strong impact of the reverse-bias conditions [43,44].

For the local maximum power points (MPPs), the results from Tables 6 to 9 show the integration of image processing methods with the proposed modeling for fast localization of the global MPP. The approximate MPPs for monocrystalline and polycrystalline cases are calculated using Equations (27) and (29), and results are registered in Tables 8 and 9. Indeed, these results highlight the correlation between the lowest shading ratios and the MPP calculation. These characteristics of simplified and fast localization of MPPs are potentially applicable to current supervision methods of power production based on image recognition [13,14].

Finally, the proposed methodology through this section can contribute to the supervision strategies based on image processing by considering the following findings in terms of shaded areas:

- Shaded cells with the highest shaded area in each group cause the divergence currents.
- Several shaded cells in a single PV group negligibly modify the operation point imposed by the PV-cell with the highest shaded area.
- Localized shadows on single shaded cells in a group are more harmful because overheating can arise.
- Uniform shadows on several cells of the same group cause less structural risks.
- The MPPs can be quickly localized considering the shaded PV-cells with the highest shaded areas in each group.

*5.3. Comparison with Other Approaches*

In this section, the contributions presented through this paper are compared with the existing schemes in literature. Methodologies in Table 12 address the PV modeling concerns using different perspectives. Ref. [15] describes the reverse-bias behavior using a nonlinear multiplication factor associated with the shunt resistance current. However, the impact of partial shadows is not discussed. The second approach proposes a discrete method to ensure convergence [20]. This paper presents a generalized method mainly based on the Bishop modeling to simulate the electrical behavior of PV installations by discretizing currents and voltages in PV systems. In contrast, quantification methods of shadow parameters are out of this paper's scope [20]. The authors of Ref. [35] integrate tools to forecast PV energy production. The PV installation is described at a high-granularity single-cell level and the non-intuitive influence of small-area shadows is predicted. The authors highlight the high impact of small shadows in power production. However, the structural healthy is not covered [35].

The approach in [45] develops a fast computing method to emulate shaded PV modules. In this paper, the PV module performance is analyzed for parallel and series connections of PV-cells exposed to equivalent external conditions by using the Brune method. However, this approach overlooks the influence of the reverse-bias behavior. The authors of Ref. [46] describe the shaded PV behavior using the two-diodes model. The accurateness of the modeling technique is validated by real-time simulator data and compared with the neural network approach and the single-diode model. However, this approach disregards the impact of partially shaded PV cells. The methodology in Ref. [33] presents an accurate and simplified expression for MPPs at a multi-string level. The PV array is simulated by employing an enhanced version of the single-diode model and reformulated in an explicit manner with the Lambert W function. However, the irradiance on shaded PV groups is considered uniform.

In comparison with these approaches, the distinctive aspect of our work is to develop and study a methodology for quantifying a ratio able to describe the shaded behavior without increasing the computational complexity. Additionally, the proposed methodology provides a useful expression to fast determination of MPPs using image processing methods and unshaded parameters. Nevertheless, the proposed approach can be improved by studying other PV-cell parameters and applying image recognition methods for estimating non-uniform shading factors.

**Table 12.** Comparison with existing schemes in literature.

| Ref. | Accur. | Characteristic | Advantage | Comment |
|------|--------|----------------|-----------|---------|
| [15] | Med. | Non-linear factor | Reverse-bias behavior model | Not Partial shading |
| [20] | High | Discrete method | Convergence and processing time | Partial shading [1] |
| [35] | High | Integration tools | Energy prod. with shadow model | Impact structural healthy |
| [45] | Med. | Matrix equations | Fast computing—array emulation | Not reverse-bias [1] |
| [46] | Med. | Two-diode model | Fast computing | Not reverse-bias [1] |
| [33] | High | MPPs Multistring | Simplified MPPs expression | Uniform irrad. in groups [1] |
| Prop. | High | shading ratio | Simplified. Correlation I–V. MPPs | Other PV parameters |
|  |  |  | Quantification shadow parameters | Non-uniform shading factor |

[1] No method quantifying shadow.

## 6. Conclusions

This paper presented a complementary approach to describe the behavior of partially shaded PV modules. The proposed approach presented a more accurate definition of the shading ratio $\delta$ that is suitable for describing the relation between the shaded area and the shading factor with the partial shading behavior. The studied approach specified a methodology able to quantify experimentally the shadow characteristics and the shading ratio $\delta$. Furthermore, the analysis of the results allowed us to establish the interrelation between the shadow patterns and changes in I–V and P–V characteristics. A simplified expression was developed to quickly calculate MPPs using the lowest shading ratio in each group and the normal operation parameters. The experimental results validated the proposed approach in monocrystalline and polycrystalline technologies. Further analysis should consider non-uniform shading factors and other PV-cell parameters such as the series and the shunt resistances. In future work, a supervision method should be developed by integrating image-processing methods to the output power monitoring in PV installations.

**Acknowledgments:** The authors acknowledge the support provided by the Andes University through the *Crédito Condonable para Doctorado*, the Laboratory for Analysis and Architecture of Systems (LAAS-CNRS), and the Paul Sabatier University in the framework of an international joint supervision PhD. thesis.

**Author Contributions:** A.G.G. has conceived the shading ratio approach and developed the experiments. C.A., M.B., and F.J.V. have contributed to the revision of the manuscript.

**Conflicts of Interest:** The authors declare no conflict of interest.

## Nomenclature

| | |
|---|---|
| $\delta$ | Shading ratio |
| $\tau$ | Shadow transmittance |
| $a_i$ | Percentage of illuminated area |
| $a_s$ | Percentage of shaded area |
| $A_c$ | Cell area |
| $A_i$ | Illuminated area |
| $A_s$ | Shaded area |
| $C_{Ti}$ | Thermal current coefficient |
| $C_{Tv}$ | Thermal voltage coefficient |
| $G_i$ | Incident irradiance |
| $G_s$ | Irradiance in shaded area |
| $G_{STC}$ | Irradiance for $STC$ |
| $I$ | Cell current |
| $I_{dv}$ | Divergence current |
| $I_{mp}$ | Current at MPP |
| $I_{ph}$ | Photo-generated current |
| $I_{ph_T}$ | Total $I_{ph}$ |
| $I_{ph_{Ti}}$ | Completely illuminated $I_{ph}$ |
| $I_{ph_{Ts}}$ | Completely shaded $I_{ph}$ |
| $I_o$ | Inverse saturation current |
| $I_{sc\_STC}$ | Short-circuit current for $STC$ |
| $J$ | Current density |
| $J_{ph}$ | Photo-current density |
| $k$ | Fraction of current in avalanche |
| $n$ | Avalanche breakdown exponent |
| $MPP$ | Maximum Power Point |
| $P_{mp}$ | Power at MPP |
| $PV$ | Photovoltaic |
| $R_p$ | Shunt resistance |
| $R_s$ | Series resistance |
| $S_f$ | Shading factor |
| $STC$ | Standard Test Condition |
| $T_c$ | Cell temperature |
| $T_{STC}$ | Temperature for $STC$ |
| $V_b$ | Breakdown voltage |
| $V_{BD}$ | By-pass diode voltage |
| $V_c$ | Cell voltage |
| $V_G$ | Group voltage |
| $V_{mp}$ | Voltage at MPP |
| $V_p$ | Module voltage |
| $V_t$ | Thermal voltage |

# References

1. Toledo, O.M.; Filho, D.O.; Diniz, A.S.A.C.; Martins, J.H.; Vale, M.H.M. Methodology for evaluation of grid-tie connection of distributed energy resources - Case study with photovoltaic and energy storage. *IEEE Trans. Power Syst.* **2013**, *28*, 1132–1139.

2. Brooks, A.E.; Cormode, D.; Cronin, A.D.; Kam-Lum, E. PV system power loss and module damage due to partial shade and bypass diode failure depend on cell behavior in reverse bias. In Proceedings of the 2015 IEEE 42nd Photovoltaic Specialist Conference (PVSC), New Orleans, LA, USA, 14–19 June 2015; pp. 1–6.

3. Bressan, M.; Basri, Y.E.; Galeano, A.; Alonso, C. A shadow fault detection method based on the standard error analysis of I–V curves. *Renew. Energy* **2016**, *99*, 1181–1190.

4. Batzelis, E.; Georgilakis, P.; Papathanassiou, S. Energy models for photovoltaic systems under partial shading conditions: A comprehensive review. *IET Renew. Power Gener.* **2015**, *9*, 340–349.

5. Jena, D.; Ramana, V.V. Modeling of photovoltaic system for uniform and non-uniform irradiance: A critical review. *Renew. Sustain. Energy Rev.* **2015**, *52*, 400–417.

6. MacAlpine, S.; Deline, C.; Erickson, R.; Brandemuehl, M. Module mismatch loss and recoverable power in unshaded PV installations. In Proceedings of the 2012 38th IEEE Photovoltaic Specialists Conference, Austin, TX, USA, 3–8 June 2012; pp. 1388–1392.

7. Hidalgo-Gonzalez, P.L.; Brooks, A.E.; Kopp, E.S.; Lonij, V.P.; Cronin, A.D. String-Level (kW-scale) IV curves from different module types under partial shade. In Proceedings of the 2012 38th IEEE Photovoltaic Specialists Conference, Austin, TX, USA, 3–8 June 2012; pp. 1442–1447.

8. Daliento, S.; Chouder, A.; Guerriero, P.; Pavan, A.M.; Mellit, A.; Moeini, R.; Tricoli, P. Monitoring, Diagnosis, and Power Forecasting for Photovoltaic Fields: A Review. *Int. J. Photoenergy* **2017**, *2017*, 13.

9. Bai, J.; Cao, Y.; Hao, Y.; Zhang, Z.; Liu, S.; Cao, F. Characteristic output of PV systems under partial shading or mismatch conditions. *Sol. Energy* **2015**, *112*, 41–54.

10. Zhao, Q.; Shao, S.; Lu, L.; Liu, X.; Zhu, H. A New PV Array Fault Diagnosis Method Using Fuzzy C-Mean Clustering and Fuzzy Membership Algorithm. *Energies* **2018**, *11*, 238.

11. Lahouar, F.E.; Hamouda, M.; Slama, J.B.H. Design and control of a grid-tied three-phase three-level diode clamped single-stage photovoltaic converter. In Proceedings of the 2015 Tenth International Conference on Ecological Vehicles and Renewable Energies (EVER), Monte Carlo, Monaco, 31 March–2 April 2015; pp. 1–7.

12. Golroodbari, S.Z.M.; de Waal, A.C.; van Sark, W.G.J.H.M. Improvement of Shade Resilience in Photovoltaic Modules Using Buck Converters in a Smart Module Architecture. *Energies* **2018**, *11*, 250.

13. Triki-Lahiani, A.; Abdelghani, A.B.B.; Slama-Belkhodja, I. Fault detection and monitoring systems for photovoltaic installations: A review. *Renew. Sustain. Energy Rev.* **2017**, doi:10.1016/j.rser.2017.09.101.

14. Madeti, S.R.; Singh, S. A comprehensive study on different types of faults and detection techniques for solar photovoltaic system. *Sol. Energy* **2017**, *158*, 161–185.

15. Bishop, J. Computer simulation of the effects of electrical mismatches in photovoltaic cell interconnection circuits. *Sol. Cells* **1988**, *25*, 73–89.

16. Quaschning, V.; Hanitsch, R. Numerical simulation of current–voltage characteristics of photovoltaic systems with shaded solar cells. *Sol. Energy* **1996**, *56*, 513–520.

17. Kawamura, H.; Naka, K.; Yonekura, N.; Yamanaka, S.; Kawamura, H.; Ohno, H.; Naito, K. Simulation of I–V characteristics of a PV module with shaded PV cells. *Sol. Energy Mater. Sol. Cells* **2003**, *75*, 613–621.

18. Guo, S.; Walsh, T.M.; Aberle, A.G.; Peters, M. Analysing partial shading of PV modules by circuit modelling. In Proceedings of the 2012 38th IEEE Photovoltaic Specialists Conference, Austin, TX, USA, 3–8 June 2012; pp. 2957–2960.

19. Olalla, C.; Clement, D.; Choi, B.S.; Maksimovic, D. A branch and bound algorithm for high-granularity PV simulations with power limited SubMICs. In Proceedings of the 2013 IEEE 14th Workshop on Control and Modeling for Power Electronics (COMPEL), Salt Lake City, UT, USA, 23–26 June 2013; pp. 1–6.

20. Díaz-Dorado, E.; Cidrás, J.; Carrillo, C. Discrete I–V model for partially shaded PV-arrays. *Sol. Energy* **2014**, *103*, 96–107.

21. Silvestre, S.; Chouder, A. Effects of shadowing on photovoltaic module performance. *Prog. Photovolt. Res. Appl.* **2008**, *16*, 141–149.

22. Jung, T.H.; Ko, J.W.; Kang, G.H.; Ahn, H.K. Output characteristics of PV module considering partially reverse biased conditions. *Sol. Energy* **2013**, *92*, 214–220.
23. Kim, Y.S.; Kang, S.M.; Johnston, B.; Winston, R. A novel method to extract the series resistances of individual cells in a photovoltaic module. *Sol. Energy Mater. Sol. Cells* **2013**, *115*, 21–28.
24. He, W.; Liu, F.; Ji, J.; Zhang, S.; Chen, H. Safety Analysis of Solar Module under Partial Shading. *Int. J. Photoenergy* **2015**, *2015*, 8.
25. MacAlpine, S.; Deline, C.; Dobos, A. Measured and estimated performance of a fleet of shaded photovoltaic systems with string and module-level inverters. *Prog. Photovolt. Res. Appl.* **2017**, *25*, 714–726.
26. Jung, J.H.; Ahmed, S. Real-time simulation model development of single crystalline photovoltaic panels using fast computation methods. *Sol. Energy* **2012**, *86*, 1826–1837.
27. Deline, C.; Dobos, A.; Janzou, S.; Meydbray, J.; Donovan, M. A simplified model of uniform shading in large photovoltaic arrays. *Sol. Energy* **2013**, *96*, 274–282.
28. Gutierrez, A. Study of Photovoltaic System Integration in Microgrids through Real-Time Modeling and Emulation of its Components Using HiLeS. Ph.D. Thesis, Université de Toulouse III, Toulouse, France, Universidad de los Andes, Bogotá, Colombia, 2017.
29. Herrmann, W.; Wiesner, W.; Vaassen, W. Hot spot investigations on PV modules-new concepts for a test standard and consequences for module design with respect to bypass diodes. In Proceedings of the Conference Record of the Twenty Sixth IEEE Photovoltaic Specialists Conference, Anaheim, CA, USA, 29 September–3 October 1997; pp. 1129–1132.
30. Alonso-Garcia, M.; Ruiz, J. Analysis and modelling the reverse characteristic of photovoltaic cells. *Sol. Energy Mater. Sol. Cells* **2006**, *90*, 1105–1120.
31. Wendlandt, S.; Drobisch, A.; Tornow, D.; Friedrichs, M.; Krauter, S.; Grunow, P. Operating principle of shadowed C-SI solar cell in PV-modules. In Proceedings of the Solar World Congress (SWC) 2011, Kassel, Germany, 28 August–2 September 2011; pp. 1–10.
32. Deline, C. Partially shaded operation of a grid-tied PV system. In Proceedings of the 2009 34th IEEE Photovoltaic Specialists Conference (PVSC), Philadelphia, PA, USA, 7–12 June 2009; pp. 1268–1273.
33. Psarros, G.N.; Batzelis, E.I.; Papathanassiou, S.A. Partial Shading Analysis of Multistring PV Arrays and Derivation of Simplified MPP Expressions. *IEEE Trans. Sustain. Energy* **2015**, *6*, 499–508.
34. Poshtkouhi, S.; Palaniappan, V.; Fard, M.; Trescases, O. A General Approach for Quantifying the Benefit of Distributed Power Electronics for Fine Grained MPPT in Photovoltaic Applications Using 3-D Modeling. *IEEE Trans. Power Electron.* **2012**, *27*, 4656–4666.
35. D'Alessandro, V.; Napoli, F.D.; Guerriero, P.; Daliento, S. An automated high–granularity tool for a fast evaluation of the yield of PV plants accounting for shading effects. *Renew. Energy* **2015**, *83*, 294–304.
36. Villa, L.F.L.; Ho, T.P.; Crebier, J.C.; Raison, B. A Power Electronics Equalizer Application for Partially Shaded Photovoltaic Modules. *IEEE Trans. Ind. Electron.* **2013**, *60*, 1179–1190.
37. Kim, K.A.; Krein, P.T. Reexamination of Photovoltaic Hot Spotting to Show Inadequacy of the Bypass Diode. *IEEE J. Photovolt.* **2015**, *5*, 1435–1441.
38. Jahne, B. In *Digital Image Processing*, 5th ed.; Springer: Berlin, Germany, 2002.
39. Soto, W.D. Improvement and validation of a model for photovoltaic array performance. Master's Thesis, Mechanical Engineering, University of Wisconsin-Madison, Madison, WI, USA, 2004.
40. Chan, D.; Phillips, J.; Phang, J. A comparative study of extraction methods for solar cell model parameters. *Solid-State Electron.* **1986**, *29*, 329–337.
41. Soto, W.D.; Klein, S.; Beckman, W. Improvement and validation of a model for photovoltaic array performance. *Sol. Energy* **2006**, *80*, 78–88.
42. Ruschel, C.S.; Gasparin, F.P.; Costa, E.R.; Krenzinger, A. Assessment of PV modules shunt resistance dependence on solar irradiance. *Sol. Energy* **2016**, *133*, 35–43.
43. De Blas, M.; Torres, J.; Prieto, E.; Garcia, A. Selecting a suitable model for characterizing photovoltaic devices. *Renew. Energy* **2002**, *25*, 371–380.
44. Daliento, S.; Napoli, F.D.; Guerriero, P.; d'Alessandro, V. A modified bypass circuit for improved hot spot reliability of solar panels subject to partial shading. *Sol. Energy* **2016**, *134*, 211–218.

*Energies* **2018**, *11*, 852

45. Kadri, R.; Andrei, H.; Gaubert, J.P.; Ivanovici, T.; Champenois, G.; Andrei, P. Modeling of the photovoltaic cell circuit parameters for optimum connection model and real-time emulator with partial shadow conditions. *Energy* **2012**, *42*, 57–67.

46. Ishaque, K.; Salam, Z.; Taheri, H.; Syafaruddin. Modeling and simulation of photovoltaic (PV) system during partial shading based on a two-diode model. *Simul. Model. Pract. Theory* **2011**, *19*, 1613–1626.

*Article*

# PV System Performance Evaluation by Clustering Production Data to Normal and Non-Normal Operation

Odysseas Tsafarakis [1,*], Kostas Sinapis [2] and Wilfried G. J. H. M. van Sark [1]

[1] Copernicus Institute, Utrecht University, Heidelberglaan 2, 3584 CS Utrecht, The Netherlands; w.g.j.h.m.vansark@uu.nl
[2] Solar Energy Application Centre, High Tech Campus 21, 5656AE Eindhoven, The Netherlands; sinapis@seac.cc
* Correspondence: O.Tsafarakis@uu.nl; Tel.: +31-30-253-55385

Received: 30 January 2018; Accepted: 8 April 2018; Published: 18 April 2018

**Abstract:** The most common method for assessment of a photovoltaic (PV) system performance is by comparing its energy production to reference data (irradiance or neighboring PV system). Ideally, at normal operation, the compared sets of data tend to show a linear relationship. Deviations from this linearity are mainly due to malfunctions occurring in the PV system or data input anomalies: a significant number of measurements (named as outliers) may not fulfill this, and complicate a proper performance evaluation. In this paper a new data analysis method is introduced which allows to automatically distinguish the measurements that fit to a near-linear relationship from those which do not (outliers). Although it can be applied to any scatter-plot, where the sets of data tend to be linear, it is specifically used here for two different purposes in PV system monitoring: (1) to detect and exclude any data input anomalies; and (2) to detect and separate measurements where the PV system is functioning properly from the measurements characteristic for malfunctioning. Finally, the data analysis method is applied in four different cases, either with precise reference data (pyranometer and neighboring PV system) or with scattered reference data (in plane irradiance obtained from application of solar models on satellite observations).

**Keywords:** photovoltaic (PV) systems monitoring; malfunction detection; data analysis; PV systems; cluster analysis

## 1. Introduction

With the continued increase in photovoltaic (PV) installations throughout the world, their proper functioning is becoming more and more important. Clearly, at high solar irradiation the generated amount of energy is high, while this depends on the actual condition of the PV system including proper system design and operational issues leading to energy loss. Any operational problems must be detected fast to limit the associated energy loss. Consequently, monitoring of PV systems is an important topic, both for scientists and owners/investors of residential and medium to large-scale size systems since it can give insights in the operation of systems and their performance, while it also allows detecting any malfunctions that may occur. The most common performance assessment of a studied PV system is the comparison of its energy production with a reference source. In this paper two reference sources are studied: Global Tilted Irradiance (GTI) and energy production of a neighboring PV system.

## 1.1. Importance of Reference Data in PV Monitoring

In large PV systems monitoring is performed by using hardware such as pyranometers, small scale meteorological stations, data loggers and other intelligent monitoring devices. As a result, a variety of data-sets (for instance tilted irradiance, air/module temperature and wind speed) are available apart from the operational data of the PV system, i.e., power, voltage, current etc. Unfortunately, these systems are quite expensive and this type of monitoring is affordable only for large scale installations. On the other hand, small PV systems on rooftops, which constitute approximately 70% of the PV systems in the Netherlands [1], are monitored either through the inverter or a simple power measuring data-logger. The only available data are the ones extracted from the PV system itself (always power and depending on the data-logger, current, voltage etc.).

The most common performance assessment of a studied PV system is the comparison of its energy production with a reference source, such as the tilted irradiance, or plane-of-array irradiance. As mentioned above, in case of small-scale installations pyranometers are rare, hence in this paper next to Global Tilted Irradiance (GTI), the energy production of a neighboring PV system is studied as potential reference data.

The GTI can be obtained from different sources, usually using a pyranometer mounted on the PV system. Since it is the most accurate (standard pyranometers have the highest accuracy of ~2.5% [2]) it has been used in a variety of studies for PV systems performance characterization and/or fault detection [3–7].

Another method to obtain the GTI is the use of data from local meteorological stations or satellite data. These data are usually the global horizontal irradiance (GHI) and sometimes the diffuse horizontal irradiance (DHI). There is a large variety of models, like the Perez [8], the HDKR (Hay, Davies, Klucher, Reindl) [9,10] or the Olmo model [11], which can calculate the GTI of a PV system from its GHI and DHI. This method is less accurate compared to the pyranometer, and includes plenty of data input anomalies, due to geospatial reasons and inaccuracies of the solar models. However, despite the fact of low data accuracy, this method is the only GTI data source for residential PV systems and has been used in various studies [12–16].

The data from a neighboring PV system could be obtained from a single panel in case of a system with micro-inverters or power optimizers, from the PV system of a neighboring rooftop, in case of a residential PV system and from another inverter, in case of a large solar park. It is not such a popular method for performance evaluation, however it has been used in a few older studies about monitoring of residential PV systems [17] in large scale. Recently, a new method was introduced for automatic fault detection by monitoring identical sets (sister arrays) connected to the same inverter of PV system [18].

The relationship of both reference sources with the power of a studied PV system tends to be linear, albeit using different equations. However, in many cases deviations are observed from the expected linear relationships, thus leading to erroneous performance evaluations. In this paper, a method is introduced that allows to automatically detect those measurements that fit in the correct relationship between the power output of a PV system and the reference source data. As an outcome, the proposed method will cluster the measurements in two groups:

1. The inliers, the measurements that fit the linear regression model and which will be used for the real performance evaluation of a PV system.
2. The outliers, the measurements that do not fit the linear regression model and after further study could be used for the detection of any occurred malfunctions.

## 1.2. Performance Evaluation

### 1.2.1. Performance Ratio

The Performance Ratio (PR) [19] is a broadly used indicator for the performance characterization of PV system. It has been used in studies regarding the performance analysis of PV systems [12–14,20], comparison of different type of PV systems [12] but also in studies regarding malfunction detection [6,21]. It is a dimensionless indicator, the ratio of the system's yield ($Y_f$) to the system's reference Yield ($Y_R$).

$$PR = \frac{Y_f}{Y_R},\tag{1}$$

where

$$Y_f = \frac{E_{AC}}{P_{peak}}\left[\frac{Whr}{W}\right],\tag{2}$$

$$Y_R = \frac{GTI}{1000}\left[\frac{Whr/m^2}{W/m^2}\right],\tag{3}$$

in which $E_{AC}$ is the generated amount of energy and $P_{peak}$ the installed capacity. A system with $PR$ higher than 70% is considered to performing well and above 80% as excellent [2,12].

In the scatterplot $Y_f$ vs. $Y_R$, the indicators tend to follow a linear relationship (LR) and the slope of this linear regression is the Performance Ratio. In previous studies, changes in the slope of the LR is an indicator for the existence of malfunction [7]. Then again, the relationship between $Y_f$ and $Y_R$ is not strictly linear due to the fact that the efficiency of solar panels decreases as its temperature increases which affects the linear relationship. With this in mind, in this paper the typical linear regression function ($Y = a + bX$) is not calculated and in fact never used. Having said that, the term linear relationship is used for better understanding, in order to describe the near-linear relationship of $Y_f$ and $Y_R$.

### 1.2.2. Comparison of Neighboring PV Systems

In the case of data from a neighboring PV system, the process is more straightforward. If the systems have the same capacity (for instance, same parts of solar parks, commercial PV systems on same rooftops or PV system with micro-inverters/power optimizer) the power outputs of the systems can be directly compared. If the compared systems have different capacities, their system yields ($Y_f$) are compared, since $Y_f$ is actually the normalized production. In both cases (same or different capacity) their relationship is expected to be linear with a linear regression line with a constant slope which defines the characteristic relationship of the compared PV systems. If the slope is almost equal to 1 then they have the same performance. If it is different than 1 then one of the systems is performing better. Any unexpected change in the slope denotes that the performance of the one of the PV systems is reducing.

### 1.3. Research Purpose and Paper Organization

The purpose of this paper is to introduce an algorithm, which will study a malfunctioning PV system (referred as studied PV from now on) by comparing its production with reference data (referred as reference data), from the sources mentioned above and calculate the expected energy loss due to these measurements.

This paper introduces a cluster analysis algorithm, applicable to any scatter plot where the data to be analyzed show a near-linear relation. The aim of the algorithm is to automatically detect and distinguish measurements that are following the linear relationship from the ones which are not. As an outcome, the proposed algorithm will cluster the measurements in two groups: inliers and outliers, as mentioned above. The introduced algorithm is applied on PV system power data, in particular to compare "System Power with Reference Power" and it is used for two different purposes:

1.  To detect and exclude any data input anomalies during the monitoring process, especially in case of residential PV systems where GTI is obtained by satellite observations and solar models.
2.  To detect and separate measurements where the PV system is functioning properly from the measurements that show that the PV system is malfunctioning or shaded. Measurements showing

proper functioning can then be used for the performance analysis while the rest can be further studied for malfunction characterization.

The aim of this algorithm is to be used in larger researches regarding monitoring of large numbers of residential PV systems. Hence only power output is used, which is the most common data provided by residential PV systems. Other PV system measurements (voltage or current) can be used for malfunction characterization, especially on outliers of purpose 2.

This paper is organized as follows. Section 2 discusses the reasons why measurements could not follow the linear relationship, for every case of reference data, and distinct them to data input anomalies and PV system failures. Section 3 provides the description of the proposed algorithm. In Section 4, the algorithm is applied in four different cases and the results are discussed. Finally, Section 5 summarizes the main findings of the paper.

## 2. Data Outliers in Performance Evaluation

For a variety of reasons measurements of system and reference yield do not follow a linear relationship and as a consequence the calculation of the performance evaluation may be erroneous. In this section the reasons of the existence of outliers for each type of reference data are described; they are presented in Table 1.

**Table 1.** The reasons of outliers per reference data (ranked by the most possible). The second and the third reference data are used for the monitoring of residential PV systems on rooftops, in case of the Netherlands, where different objects might create shadow. To that end, for these reference data, faults due to surroundings are assumed to be the most common reason of outliers rather than malfunctions of the system. In contrast, pyranometers are used at large scale installations where surrounding objects are quite rare.

| Reference Data | | Reason of Outliers (Shorted by the Most Frequent) |
|---|---|---|
| Direct measured GTI (Pyranometer, ref. cell) | 1. | Malfunction on the system (string fault, MPPT error, etc.) |
| | 2. | Faults due to surrounding area (shadow, snow, reflections) |
| Indirect measured GTI (Satellite/local weather stations + solar models) | 1. | Geospatial reasons |
| | 2. | Faults due to surrounding area (shadow, snow, reflections) |
| | 3. | Malfunction on the system |
| Neighbouring PV systems | 1. | Faults due to surrounding area (shadow, snow, reflections) |
| | 2. | Malfunction on the system (string fault, MPPT error etc.) |

### 2.1. Irradiance Data from Pyranometers

In this case, the GTI irradiance measurements are accurate within 2.5% [2], as the pyranometers are installed at the same tilt angle as the module. The majority of erroneous performance evaluations then is due to an existence of a malfunction in the system. Provided that the malfunction causes a constant energy loss in the system (e.g., detached string, broken panel/inverter) the production is reduced, usually significantly, which is clear from the determined *PR* value.

Then again, in the event that a malfunction causes a changing energy loss, the PV system could operate either normally or not, depending usually on the level of the irradiance. Such malfunctions are partial shading [7,21,22] (which could lead to the creation of hot-spots [23]), losses due to maximum power point tracking (MPPT) errors in the inverter [21], grounding fault [24] and overheated modules [6]. In these cases, the majority of the measurements will follow a linear regression, while some measurements will not. However, the reduction of the *PR* value will not be significant and malfunctions can even remain undetected if the "$Y_f$ vs. $Y_R$" plot of the systems is not studied at high time resolution (minutes).

## 2.2. Irradiance Data from Other Sources

As mentioned above, other sources of GHI data could be from local meteorological stations or satellites and subsequent processing to GTI data using solar models. In this case, the data would have more "noise", mainly due to the distance between the studied PV system and the measuring device as well as due to the uncertainty of the used solar models.

An increased uncertainty in GHI can be due to the fact that most of the meteorological stations are located outside of the cities, and could be easily some km away from the studied PV system. Thus the larger the distance between the station and the PV system, the higher the possibility that a cloud that effects the PV system can remain undetected by the station and vice versa. In case of satellite data, satellites such as MeteoSat 10 [25], are providing solar irradiance data of spatial grids (3 × 3 km in case of MeteoSat) that are to coarse to capture the effects of small clouds that reduce irradiance locally. Moreover, the irradiance of each grid is not measured constantly but once every 15 min. Thus any major changes within these 15 min, for instance one moving cloud on a sunny day, can remain undetected thus yielding incorrect reference data.

An increased uncertainty in GTI can be caused through its calculation procedures in the used solar model(s). For GTI calculations, the measured GHI firstly is separated to its components, DHI and to DNI (direct normal irradiance). Secondly, the impact of each component on the tilted surface is calculated and the sum of the impacts is the GTI. Comparisons of solar models at different locations prove the accuracy of these models, however they note the possibility of faults, while it is also clear that the calculation accuracy of GTI is strongly based on the accuracy of the inputs, DHI and DNI [26–28]. However, as DHI measurements are not common other models have to be used for the separation of GHI, the GHI separation models. The most used separation models are DIRINT [29], DISC [30] and Erbs [31]. Recently a more modern approach was introduced [32], where it was also stated that such models are empirical and local and yield an extra possibility of error in the calculation of GTI, and thus in erroneous assessment of performances. In Figure 1 the system yield of a commercial rooftop PV system is compared with the reference yield, obtained by satellite observations and the use of the HDKR Model [9,10]. It is obvious that the strong majority of the measurements are following a linear relationship. However, a large number of measurements are clearly outside the linear trend giving the impression to an observer that the system produces high values of electricity under very low radiation and vice versa.

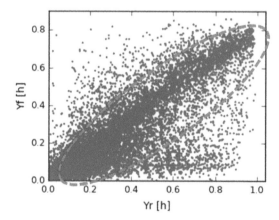

**Figure 1.** Scatterplot of system yield $Y_f$ versus reference yield $Y_R$ of a commercial rooftop PV system of 2.45 kWp capacity. The reference yield is calculated from GHI, DHI and the HDKR solar model [9,10]. The oval red shape emphasizes the fact that a strong majority of the measurements follow a linear trend, hence the higher density of measurement points inside that area.

### 2.3. Neighboring PV Systems Used as Reference

#### 2.3.1. Shade in One of the Systems

In case of neighboring systems, both systems, the reference one and the studied one, can be affected by partial shading, which will create more noise in the scatterplot. Moreover, if the distance of the systems is large, any local, small clouds can affect the linearity of the measurement. However, this effect will be smaller in case of nearby systems and it is absent in case of systems on the same rooftop.

#### 2.3.2. Difference in Energy Production

Furthermore, due to different reasons (wiring, panel and inverter brand, age) one of the compared systems may produce less energy, in a non-linear way. Such an example is presented in Figure 2, where two different panels (same micro inverter, same capacity, different ages and wired panels) are compared. Except for the shadow, between 14:00 and 15:30, where panel one (blue line) is generating much less energy, the production curve of panel one is lower than the curve of panel two (green curve) and this difference is getting larger at higher energies. In this case, clustering of the inliers and the outliers can give two different values of energy loss (with respect to the panel which is used as reference) to the owner: (a) due to shade and (b) due to the system. In that case the owner can decide if it is economically profitable to replace the panel or remove the shade, or both. Furthermore, in case of no replacement, it will be very useful to know the production relationship of the monitored PV systems, in order to detect future malfunctions.

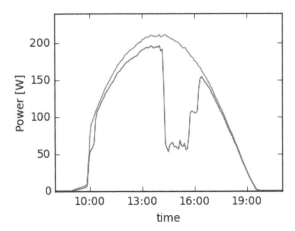

**Figure 2.** Comparison of *Pac* output of two panels with the same micro inverter, same capacity, but different brand. Except for the shadow, the production curve of panel one is lower than the curve of panel two (green curve) and this difference is getting bigger for higher power outputs.

## 3. Methodology

### 3.1. Data Preparation

The proposed method is operating with two different sets of data. One is the power output (either DC or AC) of the studied PV system (referred to as *studied PV* from now on) and the other the reference data. The reference data could be the tilted irradiance measurements or the power output (either DC or AC) of a neighboring PV system, with same tilt and orientation (referred to as reference data).

Both *studied PV* and *reference data* should be preprocessed into a specific form before the algorithm can be applied, in order to create a linear relationship in the scatterplot "studied PV vs. reference data". The process depends on the type of reference data and it is presented in Table 2.

**Table 2.** The mandatory preprocessing of each data set. $Y_f$ and $Y_R$ are calculated from Equations (2) and (3). Depending on the available reference data, the data are processed and used as shown in the table.

| Data | Reference Data | GTI | Neighboring PV System | |
| | | | Same Capacity | Different Capacity |
|---|---|---|---|---|
| Studied PV | | $Y_f$ | $P_{studied}$ (DC or AC) | $Y_{f,studied}$ |
| Reference data | | $Y_R$ | $P_{ref}$ (DC or AC) | $Y_{f,ref}$ |

For all the cases of data, the difference (error $\varepsilon$) between the studied PV and the reference is calculated, since its role is pivotal for the later steps of the process.

$$\varepsilon = Y_f - Y_R, \text{ if reference data is GTI}$$

or

$$\varepsilon = P_{studied} - P_{ref}, \text{ if reference data is PV system with same capacity} \tag{4}$$

or

$$\varepsilon = Y_{f,studied} - Y_{f,ref}, \text{ if reference data is PV system with different capacity}$$

*3.2. Scope of the Algorithm*

The scope of the algorithm is to define the maximum and lower thresholds of the error, where the measurements within these thresholds will be characterized as inliers, thus the measurements that fit to the linear regression curve. These are measurements where the studied PV system is operating normally. The measurements outside these thresholds will be characterized as outliers, thus the moments where the PV system is malfunctioning.

As we assume that a high density of measurement points in the "*Studied PV* vs. *Reference data*" plot must be around the linear regression line, the thresholds must be set such that the density of measurements points is decreasing.

*3.3. Description of the Algorithm*

3.3.1. First Step—Inliers Determination Using Ran.Sa.C.

In the first step, the iterative method Ran.Sa.C. (Random Sample Consensus) [33] is applied on the scatterplot "*Studied PV* vs. *Reference data*", which clusters the measurements in two groups, the inliers (the ones that following a linearity, i.e., normal operation) and outliers.

If the data are limited, such as data for only a few days with hourly resolution, no further processing is needed. The inliers from Ran.Sa.C. are used directly for the calculation of the *PR* and the outliers are further studied in order to determine the cause of energy loss and its total impact on the energy production of the system.

However, if the sample is large, such as data for long periods at minutely time resolution, extra processing may be needed, since Ran.Sa.C. could be misloaded and measurements could have been denoted as inliers while they do not exactly fit in the linear relationship. In this case, Ran.Sa.C. is used as a first cleaning of measurements to identify clear outliers.

3.3.2. Second Step—Data Clustering and Polynomial Regression

From this step the focus of the analysis is on the calculated error between the studied PV and the reference (calculated with Equation (4)). Furthermore, the analysis involves only the inliers as calculated in step 1.

The inliers from step 1 are clustered into groups, based on the actual value of the reference data. Typically fifteen groups are used to cover the range of the reference data. For each group, a histogram of the errors is calculated. Subsequently, a polynomial linear regression is performed:

$$f = poly(\varepsilon) \tag{5}$$

where $f$ is the frequency and $\varepsilon$ the value of the error.

In the plot of Figure 3 the histogram of one group is presented, as an example, together with the estimated polynomial linear regression between the errors and their frequency (red points).

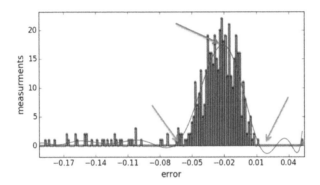

**Figure 3.** Example of a histogram of one group of measurements, with $Y_R$ from 0.4 to 0.5. Green bars and red dots represent the data, and the blue line the polynomial fit.

### 3.3.3. Third Step—Determine the Thresholds of Each Group

Based on the assumption described in Section 3.2, the algorithm is focusing on the maximum (depicted by a red arrow in Figure 3) of $poly(\varepsilon)$, the $\varepsilon_{max}$. As thresholds the two local minimums on the left and right ($\varepsilon_{left}$ and $\varepsilon_{right}$) of the global maximum (blue arrows on Figure 3) are used. If $poly(\varepsilon l/r) < 0$ then as threshold is set the $\varepsilon$ where $poly(\varepsilon) = 0$, with $\varepsilon$ in ($\varepsilon_{left}$, $\varepsilon_{max}$) or $\varepsilon$ in ($\varepsilon_{max}$, $\varepsilon_{right}$).

The logic behind this is that as the polynomial curve is moving away (left or right) from the maximum, the frequency of the errors will rapidly reduce, thus $poly(\varepsilon)$ will also rapidly reduce and will form a local minimum. If the reduction is high, the $poly(\varepsilon)$ will take values lower than zero. That means that the frequency of the errors dramatically changes, as well the density of the measurements in the initial "*Studied PV* vs. *Reference data*" scatterplot.

### 3.3.4. Fourth Step—Normalization and Connection of All Limits

During this step the relationship of each defined threshold ($\varepsilon_{left}$, $\varepsilon_{right}$) and the global maximum ($\varepsilon_{max}$) of all power groups versus the reference power is studied, see Figure 4.

The clustering of the sample in groups during step 2, based on reference power, is random and it depends on the sample size. Furthermore, since one value of $\varepsilon$ will represent a group, with reference power from $P_n$ to $P_{n+1}$ then the plots of $\varepsilon_{max}$, $\varepsilon_{left}$ and $\varepsilon_{right}$ versus reference will be stair plots, as it is clear from Figure 4a. In order to use these data in practice, polynomial fits are made, to obtain continuous functions for each error. An example is presented in Figure 4b.

The polynomial fits of the $\varepsilon_{left}$ and $\varepsilon_{right}$ versus reference data ($\varepsilon_{left}^{Poly}(ref)$ and $\varepsilon_{right}^{Poly}(ref)$) is the most important output of the algorithm since they define the relationship between the studied and the reference data, for any value of the reference data. That is to say, for any measurement of a studied sample, the application of its reference data using $\varepsilon_{left}^{Poly}(ref)$ and $\varepsilon_{right}^{Poly}(ref)$ defines the maximum and the lower allowed value of its error, in order to be characterized as inlier or outlier. Furthermore, polynomials from historical data, can be applied to new measurements and characterize them as inliers and outliers.

The polynomial fit of $\varepsilon_{max}$ versus the reference ($\varepsilon_{max}^{Poly}(ref)$) provides the information about the most frequent value of the error, for the respective value of the reference data.

Provided that the relationship of the error $\varepsilon$ and $Y_f$, $Y_R$ is for any single measurement defined as $\varepsilon = Y_f - Y_R$ (Equation (4)), for any single measurement the $\varepsilon_{max}$, $\varepsilon_{left}$ and $\varepsilon_{right}$ are calculated by the application of the reference data on the respective polynomial fits, $\varepsilon_{max}^{Poly}(ref)$, $\varepsilon_{left}^{Poly}(ref)$ and $\varepsilon_{right}^{Poly}(ref)$. Thus:

$$\varepsilon_i^{Poly}(Y_R) = Y_{f,i} - Y_R, \text{ for } i \text{ in } [max, left, right] \tag{6}$$

Rewriting Equation (6) with respect to $Y_f$:

$$\begin{aligned} Y_{f,max}(Y_R) &= \varepsilon_{max}^{Poly}(Y_R) - Y_R \\ Y_{f,left}(Y_R) &= \varepsilon_{left}^{Poly}(Y_R) - Y_R \\ Y_{f,right}(Y_R) &= \varepsilon_{right}^{Poly}(Y_R) - Y_R \end{aligned} \tag{7}$$

in which $Y_{f,\,left}(Y_R)$, $Y_{f,\,right}(Y_R)$ are the polynomial functions that return the thresholds for which $Y_f$ is inlier, for any respective value of $Y_R$. In case that the $Y_f$ is constantly lower than the $Y_R$ (case of PV system vs. solar radiation data) then the right threshold represents the maximum value of $Y_f$ and the left the minimum in order to be characterized as inlier. On the other hand, $Y_{f,\,max}(Y_R)$ is the polynomial which returns the most frequent value of $Y_f$ for any respective value of $Y_R$. In Figure 4c, the Equation (7) is plotted. Any measurement of the scatterplot between green and blue lines is characterized as inlier, while outside as outlier, with the majority of the measurements to be on the red line.

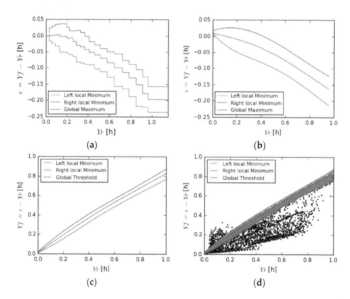

**Figure 4.** (a) The defined thresholds, after step 3, where for each power group a single $\varepsilon$ value is defined and (b) after step 4, where a polynomial regression is applied to obtain a continuous function. Step 4 is applied separately to each threshold: left and right as defined in step 3 in red and blue, and the maximum in red. In (c), polynomial fits of (b) are solved with respect to $Y_f$ and plotted vs. $Y_R$, in order to point out that any measurement of the plot inside blue and red line is characterized as inlier and in (d) all the measurements are plotted. Within blue and red they are inliers (grey) and outside of them, outliers (black).

### 3.3.5. Fifth Step—Pplication of the Limits to the Data

Finally, the polynomial regressions of the thresholds, calculated during step four are applied to the data. Thus, for every measurement, the thresholds are calculated, based on its reference value and if the error ($Y_f$ can be used as well) is within these thresholds the measurement is defined as inlier, otherwise as outlier.

For better understanding an example is presented in Figure 4b, where the blue and red lines are the thresholds of $Y_f$, based on the Equations (6) and (7). Every measurement inside these lines is characterized as inlier (gray), while outside as outlier (black).

### 3.4. Calculation of Energy Loss

After the clustering in inliers and outliers, the amount of lost energy due to the outliers can be calculated. In this paper, energy loss is defined as the energy that each outlier would have produced if it would follow the normal relationship between the studied data and the reference (which is defined by the polynomial fits calculated in Section 3.3.4). The calculation is applied only if it is known to the users that the outliers are due to a malfunction on the system, thus if the proposed algorithm is used for the second purpose, as described in Section 1.3.

If $E_{hyp\ Studied}$ is the hypothetical energy production of the studied PV, for a single outlier, in case that it was inlier and $E_{studied\ PV}$ the produced energy, then for a single outlier:

$$E_{loss} = E_{hyp\ Studied} - E_{studied\ PV} \tag{8}$$

And for all outliers:

$$\sum_{n=1}^{N} E_{loss}^{i} = \sum_{n=1}^{N} \left( E_{hyp\ studied}^{i} - E_{studied\ PV}^{i} \right) \tag{9}$$

where $i$ the number of outliers.

In this paper, for the hypothetical energy production three scenarios are chosen:

1. Error is equal to the higher frequent error (global maximums), thus the most probable value
2. Error of outliers is slightly higher (105%) than the smaller threshold ($\varepsilon_{left}$)
3. Error is slightly lower (95%) than the higher threshold ($\varepsilon_{right}$)

From the first scenario the most probable energy loss is calculated, since it is assumed that all the outliers would have the same error $\varepsilon$ with the most frequent one. From the second and third scenarios the lower and the higher possible energy losses are calculated respectively, under the assumption that all outliers would have the same lower or higher $Y_f$ values in order to be characterized as inliers.

### 4. Application of the Method—Examples

The proposed method is applied in different examples and their scatterplots are presented below. The first three examples are from data from the experimental facility of SEAC (Solar Energy Application Center, Eindhoven, The Netherlands). The facility contains three PV systems, with identical panel structure (six panels in two rows, one front, one back, same tilt and orientation) and different inverter technology [22,34]. The system can be seen in Figure 5. The system on the right-hand side consist of 6 micro inverters (265 W each), the system in the middle consist of a series connection of the panels and a standard string inverter of 1.5 kW. Finally, the system on the left consist of 6 power optimizers connected in parallel (boost DC/DC) and a central inverter of 1.5 kW especially made for the power optimizer system. In front of each system a pole is placed (same dimension for every system) in order to create an artificial shadow on the front rows of each system during the day which is equal for all systems. Furthermore, two pyranometers are available for the measurement of the tilted irradiance. The initial purpose of the system was to study the performance of different inverter technologies (string inverter, power optimizer and micro inverter) under shading conditions [22,34].

For these cases it is known that the strong majority of the outliers in the data is caused by shadow, since the reference data are measured through the most accurate methods (pyranometer and neighboring PV system) as defined in Sections 2.1 and 2.3. Thus the algorithm will applied for purpose 2 as defined in Section 1.3. Moreover, the energy loss can be calculated since it is known that the outliers are due to a malfunction (shadow).

The percentage of lost energy compared to the actual production is presented, according to:

$$E_{loss}(\%) = E_{loss}/E_{produced} \times 100\% \tag{10}$$

where $E_{loss}$ is the energy loss as calculated from Equation (9).

For each of the first two examples, the studied PV system is compared to the same reference data, two cases are used and two different plots are presented: the shaded system/panel versus (a) GTI obtained by an in-plane pyranometer, and (b) the average production of the 3 panels with power optimizers, in the back row of the system, which are shaded partially during late afternoon, different hours than the shaded panels. In the third example, two different panels with same micro-inverters are compared and monitored each other.

In the fourth example, a PV system from a house in The Netherlands is compared with the tilted irradiance, obtained by the application of the HDKR solar model [9,10,29] on satellite data from MeteoSat. The system consists of a 2.5 kWp inverter and has panel capacity of 2.45 kWp. This example is different than the other three, since the reference data will contain a large amount of input anomalies and any presence of shadow is unknown. Thus the algorithm will be used for purpose 1 of Section 1.3 and energy loss cannot be calculated, since it is not known if the outliers are due to data input anomalies (thus any energy loss calculation is pointless since there are false measurements) or for any other reason, for instance shadow.

**Figure 5.** The experimental facility of SEAC.

*4.1. Shaded Panel with Power Optimizer*

In the first example the algorithm is applied to the DC power output of a panel with power optimizer. Figure 5a,c shows the scatterplots "$Y_f$ vs. $Y_R$" and "$P_{DC\_shaded}$ vs. $P_{DC\_reference}$", respectively, while at the right plots, Figure 5b,d show a zoom (red box in Figure 5a,c) for more details. The time resolution is 5 min and the studied period is 116 days, from beginning of July 2015 until end of October. The panel is shaded for a part of the day (approximately from 10:30 am to 12:20 pm).

In the scatterplots b and d of the Figure 6, it is obvious how the algorithm is detecting as inliers the areas with the higher density of measurements. Due to the large number of the measurements, this is not obvious in the scatterplots of the whole sample, where the inliers and outliers seem to have the same density. However, in the zoomed areas, in plots b,d differences in density are clearer.

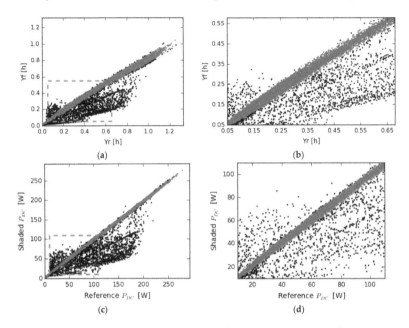

**Figure 6.** Example of the application of the algorithm for monitoring a panel with power optimizer (Shaded $P_{DC}$). Two different types of data are used: GTI measured by an in-plane pyranometer (**a,b**) and the average production of the 3 panels with power optimizers (**c,d**). In plots (**b,d**) the marked red square of plots (**a,c**) is presented.

Differences in the classification of inliers and outliers between the two cases are observed since the reference data are different. In case of the pyranometer as reference data (Figure 6a,b), the threshold of the inliers is broader for higher values of irradiance, due to the accuracy of the pyranometer and the reduction of the efficiency of the panels in higher temperatures. However, these differences are very small, since for the studied period, 95% of the measurements have the same classification (both inliers or outliers) and only 3% different in the two cases of the reference data.

As mentioned at the end of Section 1.2.1, the plot of "$Y_f$ vs. $Y_R$" is not strongly linear, especially at higher values of energy due to the reduction of the efficiency of the panels at higher temperatures. Thus the use of the fitted linear regression equation ($Y_f = a + bY_R$) is not so accurate, as the use of the polynomial fits (Section 3.3.4), where the two polynomials (left and right in Figure 4) will define the upper and lower threshold for the inliers and the third (global maximum in Figure 4) the most possible position of the inliers. By contrast, in the "$P_{DC\_shaded}$ vs. $P_{DC\_reference}$" the relationship is very strongly linear, because both panels have the same efficiency reduction at higher temperatures. In this case the polynomial fits are almost a straight line and the fitted linear regression equation could be used as well, in order to describe the performance relationship between the compared panels.

The performance ratio ($PR_{DC}$ in this case) is calculated, according to Section 1.1 and Equation (1). As reference data the pyranometer is used (Figure 6c). The $PR_{DC}$ of the panel, with the use of all data is 81.7% while using only the inliers (green markers in Figure 6c) the $PR_{DC}$ is 86.8%. That means that the real performance of the panel is 86.8%, however due to presence of the shadow (external reason, not malfunction inside the system) it drops to 81.7%.

The energy loss due to the shadow is calculated using Equations (9) and (10) presented in Table 3, for each of the scenarios and for both cases of the pyranometer and neighboring panels as reference data. Clearly, the panel produces less energy due to the shadow (outliers) and depending the reference data, the calculated energy loss is slightly different. As mentioned above, these differences are due to the accuracy of the pyranometer, and the fact that in case of the pyranometer as reference data the threshold of the inliers is broader for higher values of irradiance, explains the larger difference between the scenarios.

**Table 3.** The energy loss due to the shadow, for each reference data and for each scenario. According to the scenario, that the error of each outlier should be equal to the most frequent error, the panel is producing 6.7% or 6.1% (depending on the reference data taken for comparison) less energy.

| Reference Data | Error | | |
|---|---|---|---|
| | Left | Most Frequent | Right |
| Pyranometer | 4.9% | 6.7% | 8.1% |
| Neighboring PV | 5.4% | 6.1% | 6.8% |

### 4.2. PV System with String Inverter

In this case the algorithm is applied to the PV system with string inverter, where three panels in the front row are heavily shaded during most of the day. The data resolution is 5 min and the studied period is 133 days, from beginning of July until end of October. In Figure 7, the results of applying the algorithm are presented. Figure 7a,c show the complete data sample, while Figure 6b,d show a smaller portion for better understanding.

The results from the study of this system show that the duration of the shadow is clearly different. While the panel in Section 4.1 was shaded only a part of the day, this system is shaded during most of the day. Furthermore, the shaded panels are shaded one at a time and only a few moments two panels are shaded at the same time. Due to this fact, the values of the errors are much smaller and the measurements with high error are rare.

In Figure 7a the clustering threshold of the algorithm is almost impossible to be discerned. In the more detailed observation (Figure 7b) the threshold is clearer, since the density of the green marks is reduced (where the yellow arrows are pointing), and the marks after the reduction are clustered as outliers (black colored). A thin white line between the arrow heads can be seen, reflecting absent data points, which in fact shows where the algorithm separates the inliers from the outliers.

However, in Figure 7c,d, where the average power of the neighboring panels is used as reference data, the results are more clear and detailed. Furthermore, three distinct different lines of outliers can be observed (highlighted by the yellow circle). This demonstrates that the use of neighboring panels for performance evaluations can show much more detail and can be more accurate than the use of a pyranometer as reference data.

These differences can be explained by the accuracy of the pyranometer and the reduction of the panels efficiency at higher temperatures. Similar to example 1, for higher irradiances, where the temperature is higher as well, the efficiency of the panels is reducing. Thus the scatterplot of the energy production versus the solar irradiance shows more scatter for higher irradiances and the relationship is not strongly linear. Thus a polynomial fit is more accurate to describe the relationship of the $Y_f$ versus the $Y_R$. By contrast, the relationship with the neighboring panel is considerably stronger linear, since its efficiency is reducing at higher temperatures as well.

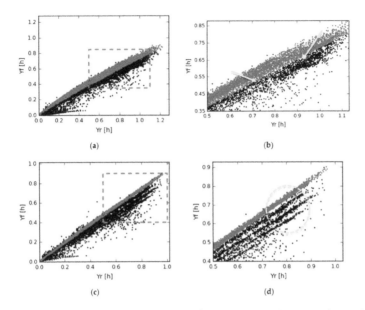

**Figure 7.** Example of the application of the algorithm for monitoring a PV system with string inverter ($Y_f$). Two different types of data are used, the GTI measured by an on-plane pyranometer (**a**,**b**) and the average production of the 3 panels with power optimizers (**c**,**d**). In plots (**b**,**d**) the marked square of plots (**a**,**c**) is presented, for more detailed observation and better understanding of the plots. Yellow arrows in plot b are pointing to the area where the density of marks is changing and the algorithm sets a threshold between the inliers and outliers. The yellow cycle in plot d points to the detail of the secondary linear behaviors, which are caused when a panel (different for each line) is shaded.

### 4.3. Systems with Same Capacity, Different Production and Different Shadows

In this example, two different panels with same micro inverters are monitored. Panel with micro inverter 3 (Micro 3 from now on) is shaded by the placed pole, while the other panel (Micro 6 from now on) is shaded by the corner of the rooftop in the late afternoon. Each panel is used as reference for the other, thus the algorithm runs two times, one for each panel as studied and one as reference. The results are presented in Figure 8.

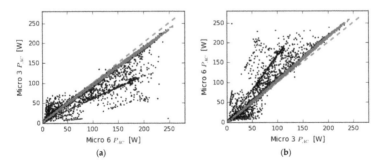

**Figure 8.** Two different panels with micro inverters are compared to each other. The one is heavily shaded and less productive (micro 3) than the other, which is less shaded and newer, thus more efficient (micro 6). The red line corresponds to identical performance of the two systems (45°) and, is shown for better comparison of the performances. (**a**) $P_{AC}$ of system micro 3 versus micro 6; (**b**) $P_{AC}$ of system micro 6 versus micro 3.

The differences between the figures-cases are negligible since 96% of the measurements have the same classification in both cases. The major difference is in the angle of the linear regression lines of the two plots, where in plot (a) is lower than (b), thus system 6 is functioning better than system 3. It is clear in the plots that the system 3 is shaded much more than system 6. Thus in this example 3 values for energy loss can be calculated:

1.  Energy loss of system 3 due to shadow
2.  Energy loss of system 6 due to shadow
3.  Energy loss of system 3 due to the older panel

Values 1 and 2 are calculated through Equation (5), similarly to examples 1 and 2. System 6 is producing due to the shadow (percentwise according to Equation (6)) from 0.2 to 0.4% less energy and system 3 from 2.5 to 3.5%. As mentioned above, system 3 is affected much more by shade than system 6. Energy loss of system 6 is negligible since it is affected by shadow by the very end of the day.

Furthermore, system 3 has an older and less efficient panel. In order to calculate this energy loss, only the inliers are used as the panel is not disturbed by other reasons (shade in this case). Thus, system 3 is producing 95.6% less energy than system 6, due to age difference between the panels, since both micro inverters and the wirings are the same and no other differences have been observed in DC to AC conversion during the operation of the system [22,34].

*4.4. System of a Regular House in The Netherlands, Monitored with MeteoSat Data*

In the last example a PV system with system capacity 2.45 kW and inverter capacity 2.5 kW is compared with solar radiation, determined from images taken with the satellite MeteoSat 10 [25]. The measurements of MeteoSat 10 consist of 15 min time resolution data of GHI and DHI and these are converted to GTI with the use of the HDKR solar model [9,10] see Figure 9a. The results of applying the proposed method are presented in Figure 9b.

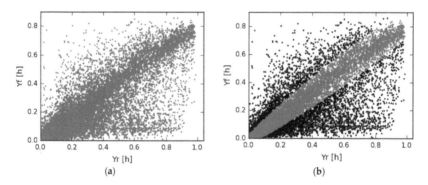

**Figure 9.** (a) System Yield of a residential PV system versus the reference yield obtained by satellite data and solar models; (b) The plot after the application of the algorithm, where it is clustered to inliers (green) and outliers (black).

In this example, 75% of all measurements are classified as inliers (green markers in Figure 9b) and 25% as outliers. About 1/3 of the outliers are above the linear regression line. These are mostly due to spatial measuring faults. About 2/3 of the outliers are below the line, and these are due to the presence of a local shadow in addition to spatial measuring faults.

The Performance Ratio of this system from the total sample of measurement is determined to be 71.3%. By taking into account only the inliers the real *PR* is 79.4%.

Clearly in this example the proposed algorithm is used for purpose 1, as explained in Section 1.3, thus in order to filter irradiance data with large input anomalies and make them suitable for accurate performance analysis for residential systems where the only available data is the power output.

The two-thirds of the outliers which are below the line could be better studied and malfunction detection techniques could be applied in order to determine if there are malfunctions or irradiance data input anomalies. For instance, if DC power is available as well, a DC to AC conversion study on the outliers could detect a malfunction on the inverter. A study on voltage (if it is available) can detect the presence of shadow [22]. If further data on the DC side are available, fault detection and classification methods that have been proposed by Akram and Lotfifard [35] or in [24], a study which also describes detection of ground faults.

## 5. Conclusions

In conclusion, this paper describes the development of a new method of cluster analysis and its application for the analysis of PV performance. The method itself can be applied to any pair of data sets which tend to fit to a linear relationship, and allows to distinguish the data between those which are following the linear relationship and those which are not.

The application of the developed algorithm for monitoring and performance evaluation of PV systems is based on the fact that the produced power of a PV system, with the application of the correct formulas) tends to be linear with irradiance in the plane of array (GTI). As the fit is not strongly linear the proposed method aims to replace this linear fit with three polynomial fits. Two are providing the lower and higher value of production (as system yield, $Y_f$) in order for the measurement to be characterized as inlier or outlier, for any value of the irradiance (as reference yield, $Y_R$). The third polynomial is providing the position of the majority of the inliers in the scatterplot of the compared data.

The proposed algorithm is used for two different purposes in the field of PV monitoring. For each monitored PV system, the purpose depends strongly on the available reference data and any additional information about the system (known shadow or malfunction). Presuming that the monitored PV system is a residential one, on a rooftop, and the only available reference data is tilted irradiance obtained by satellite or local weather station data after application of solar models, then the method is applied in order to detect and exclude data input anomalies (purpose 1). If the reference data is obtained by a tilted pyranometer or from a neighboring PV system then the method is applied in order to detect any malfunction that causes changing energy loss (shadow in the given examples) and will calculate the energy loss due to the malfunction as well (purpose 2). Furthermore, in that case the performance of the PV system, the "real PR", for the moments that the system is not malfunctioning is calculated. Thus, if the *PR* is between 0.8 and 0.7 the system is performing well but it can be improved and if the *PR* is lower than 70% then it is suffering from a constant energy loss malfunction. On the other hand, if the number of the outliers is increasing then a new shadow or another changing energy loss malfunction may be affecting the system.

Unfortunately the case of the neighboring PV systems is rare, except for systems with micro inverters and power optimizers. Its linear relationship with the studied PV is much stronger and detailed compared to pyranometer measurements as reference data, as the pyranometer accuracy is limited to ~2.5% and calibration is needed, ideally every two years according to surveys [36] and pyranometer manufacturers [37,38].

As a suggestion, the outliers could be further studied for the detection and classification of any malfunctions. For instance, the hourly occurrence of outliers should be studied and used for detection of any shadow created by any obstacle around the studied PV system. Other possible studies are depending on the case and the available data.

As a final suggestion, the proposed method can be used for validation of any model that simulates PV performance. The simulated and measured data should follow a strong linear relationship (similar to the comparison of two PV systems, see Figures 6c, 7c and 8) and the algorithm can detect from the measurements if the system is malfunctioning or if the model may have any weaknesses at specific moments. Especially in case of building integrated PV (BIPV), where the calculation of the tilted irradiance is more complicated due to the existence of different surfaces around the building, the combination of the proposed algorithm with energy performance models, such as proposed in [39] could lead to monitoring and identification of outliers.

**Acknowledgments:** The authors gratefully acknowledge fruitful discussions with Panagiotis Moraitis (UU) Chris Tzikas (ECN), Tom Lemmens (Eneco), Jasper Müller (Eneco) and Fonger Ypma (Eneco). This work is partly financially supported by the Netherlands Enterprise Agency (RVO) within the framework of the Dutch Topsector Energy (project Automatic Malfunction Detection for Improvement of solar PV yield, AMDIS).

**Author Contributions:** Odysseas Tsafarakis conceived and designed the method, performed the validation, and analyzed the data; Odysseas Tsafarakis, Kostas Sinapis and Wilfried G. J. H. M. van Sark discussed results and wrote the paper; Wilfried G. J. H. M. van Sark conceived the project.

**Conflicts of Interest:** The authors declare no conflict of interest. The funding organization had no role in the design of the study; in the collection, analyses, or interpretation of data; in the writing of the manuscript, and in the decision to publish the results".

## References

1. Centraal Bureau voor de Statistiek (CBS). Bijgeplaatst Vermogen Zonnestroom Bijgesteld. Available online: https://www.cbs.nl/nl-nl/achtergrond/2015/51/bijgeplaatst-vermogen-zonnestroom-bijgesteld (accessed on 17 December 2015).
2. Reich, N.H.; Mueller, B.; Armbruster, A.; Van Sark, W.G.J.H.M.; Kiefer, K.; Reise, C. Performance ratio revisited: Is PR > 90% realistic ? *Prog. Photovolt.* **2012**, *20*, 717–726. [CrossRef]
3. Silvestre, S.; Chouder, A.; Karatepe, E. Automatic fault detection in grid connected PV systems. *Sol. Energy* **2013**, *94*, 119–127. [CrossRef]
4. Firth, S.K.; Lomas, K.J.; Rees, S.J. A simple model of PV system performance and its use in fault detection. *Sol. Energy* **2010**, *84*, 624–635. [CrossRef]
5. Platon, R.; Martel, J.; Woodruff, N.; Chau, T.Y. Online Fault Detection in PV Systems. *IEEE Trans. Sustain. Energy* **2015**, *6*, 1200–1207. [CrossRef]
6. Eke, R.; Senturk, A. Monitoring the performance of single and triple junction amorphous silicon modules in two building integrated photovoltaic (BIPV) installations. *Appl. Energy* **2013**, *109*, 154–162. [CrossRef]
7. Woyte, A.; Richter, M.; Moser, D.; Green, M.; Mau, S.; Beyer, H.G. *Analytical Monitoring of Grid-Connected Photovoltaic Systems*; NET Ltd.: St. Ursen, Switzerland, 2014; Volume 13.
8. Perez, R.; Seals, R.; Ineichen, P.; Stewart, R.; Menicucci, D. A new simplified version of the Perez diffuse irradiance model for tilted surfaces. *Sol. Energy* **1987**, *39*, 221–231. [CrossRef]
9. Hay, J.E. Calculating solar radiation for inclined surfaces: Practical approaches. *Renew. Energy* **1993**, *3*, 373–380. [CrossRef]
10. Davies, J.A.; McKay, D.C. Evaluation of selected models for estimating solar radiation on horizontal surfaces. *Sol. Energy* **1989**, *43*, 153–168. [CrossRef]
11. Olmo, F.J.; Vida, J.; Foyo, I.; Castro-Diez, Y.; Alados-Arboledas, L. Prediction of global irradiance on inclined surfaces from horizontal global irradiance. *Energy* **1999**, *24*, 689–704. [CrossRef]
12. Tsafarakis, O.; Moraitis, P.; Kausika, B.B.; Van Der Velde, H.; Hart'T, S.; de Vries, A.; de Rijk, P.; De Jong, M.M.; Van Leeuwen, H.P.; van Sark, W. Three years experience in a Dutch public awareness campaign on photovoltaic system performance. *IET Renew. Power Gener.* **2017**, *11*, 1229–1233. [CrossRef]
13. Leloux, J.; Narvarte, L.; Pereira, A.D.; Leader, W.P.; Madrid, R.; SENES, C.; de Navarra, P. *Analysis of the State of the Art of PV Systems in Europe*; Universidad Politécnica de Madrid: Madrid, Spain, 2015.
14. Taylor, J.; Leloux, J.; Everard, A.M.; Briggs, J.; Buckley, A.; Solar, S.; Building, H.; Road, H.; Sheffield, S. *Monitoring Thousands of Distributed PV Systems in the UK: Energy Production and Performance*; Universidad Politécnica de Madrid: Madrid, Spain, 2011.
15. Leloux, J.; Narvarte, L.; Trebosc, D. Review of the performance of residential PV systems in France. *Renew. Sustain. Energy Rev.* **2012**, *16*, 1369–1376. [CrossRef]
16. Leloux, J.; Narvarte, L.; Trebosc, D. Review of the performance of residential PV systems in Belgium. *Renew. Sustain. Energy Rev.* **2012**, *16*, 178–184. [CrossRef]

17. Leloux, J.; Taylor, J.; Moretón Villagrá, R.; Narvarte Fernández, L.; Trebosc, D.; Desportes, A. Monitoring 30,000 PV Systems in Europe: Performance, Faults, and State of the Art. In Proceedings of the 31st European PV Solar Energy Conference and Exhibition, Hamburg, Germany, 14–18 September 2015; Volume 153, pp. 1574–1582.

18. Mallor, F.; León, T.; de Boeck, L.; van Gulck, S.; Meulders, M.; Van Der Meerssche, B. A method for detecting malfunctions in PV solar panels based on electricity production monitoring. In Proceedings of the 31st European PV Solar Energy Conference and Exhibition, Hamburg, Germany, 14–18 September 2015; Volume 153, pp. 51–63.

19. IEC 61724. *Photovoltaic System Performance Monitoring—Guidelines for Measurement, Data Exchange and Analysis*, 10th ed.; International Electrotechnical Commission: Geneva, Switzerland, 1998.

20. Van Sark, W.; Hart, S.; de Jong, M.; de Rijk, P.; Moraitis, P.; Kausika, B.B.; van der Velde, H. "Counting the Sun"—A Dutch Public Awareness Campaign on Pv Performance. In Proceedings of the 29th European Photovoltaic Solar Energy Conference and Exhibition, Amsterdam, The Netherlands, 22–26 September 2014; Volume 2014, pp. 3545–3548.

21. Tsafarakis, O.; van Sark, W.G.J.H.M. Development of a data analysis methodology to assess PV system performance. In Proceedings of the 29th European Photovoltaic Solar Energy Conference, Amsterdam, The Netherlands, 22–26 September 2014; pp. 2908–2910.

22. Sinapis, K.; Tzikas, C.; Litjens, G.; Van Den Donker, M.; Folkerts, W.; Van Sark, W.G.J.H.M.; Smets, A. A comprehensive study on partial shading response of c-Si modules and yield modeling of string inverter and module level power electronics. *Sol. Energy* **2016**, *135*, 731–741. [CrossRef]

23. Alam, M.; Johnson, J. PV faults: Overview, modeling, prevention and detection techniques. In Proceedings of the 2013 IEEE 14th Workshop on Control and Modeling for Power Electronics (COMPEL), Salt Lake City, UT, USA, 23–26 June 2013; pp. 2–9.

24. Alam, M.K.; Khan, F.; Member, S.; Johnson, J.; Flicker, J. A Comprehensive Review of Catastrophic Faults in PV Arrays: Types, Detection, and Mitigation Techniques. *IEEE J. Photovolt.* **2015**, *5*, 982–997. [CrossRef]

25. Eumetsat. Meteosat Second Generation (MSG) Provides Images of the Full Earth Disc, and Data for Weather Forecasts. Available online: https://www.eumetsat.int/website/home/Satellites/CurrentSatellites/Meteosat/index.html (accessed on 16 April 2018).

26. Cucumo, M.; De Rosa, A.; Ferraro, V.; Kaliakatsos, D.; Marinelli, V. Experimental testing of models for the estimation of hourly solar radiation on vertical surfaces at Arcavacata di Rende. *Sol. Energy* **2007**, *81*, 692–695. [CrossRef]

27. Gueymard, C.A. Direct and indirect uncertainties in the prediction of tilted irradiance for solar engineering applications. *Sol. Energy* **2009**, *83*, 432–444. [CrossRef]

28. Yang, D.; Dong, Z.; Nobre, A.; Khoo, Y.S.; Jirutitijaroen, P.; Walsh, W.M. Evaluation of transposition and decomposition models for converting global solar irradiance from tilted surface to horizontal in tropical regions. *Sol. Energy* **2013**, *97*, 369–387. [CrossRef]

29. Ineichen, P.; Perez, R.R.; Seal, R.D.; Maxwell, E.L.; Zalenka, A. Dynamic global-to-direct irradiance conversion models. *ASHRAE Trans.* **1992**, *98*, 354–369.

30. Maxwell, E.L. *A Quasi-Physical Model for Converting Hourly Global Horizontal to Direct Normal Insolation*; Solar Energy Research Institute: Golden, CO, USA, 1987.

31. Erbs, D.G.; Klein, S.A.; Duffle, J.A. Estimation of the diffuse radiation fraction for hourly, daily and monthly-average global radiation. *Sol. Energy* **1982**, *28*, 293–302. [CrossRef]

32. Aler, R.; Galván, I.M.; Ruiz-arias, J.A.; Gueymard, C.A. Improving the separation of direct and diffuse solar radiation components using machine learning by gradient boosting. *Sol. Energy* **2017**, *150*, 558–569. [CrossRef]

33. Fischler, M.A.; Bolles, R.C. Random Sample Consensus: A Paradigm for Model Fitting with Applications to Image Analysis and Automated Cartography. *Commun. ACM* **1981**, *24*, 381–395. [CrossRef]

34. Sinapis, K.; Litjens, G.; Van Den Donker, M.; Folkerts, W.; Van Sark, W. Outdoor characterization and comparison of string and MLPE under clear and partially shaded conditions. *Energy Sci. Eng.* **2015**, *15*, 510–519. [CrossRef]

35. Akram, M.N.; Lotfifard, S. Modeling and Health Monitoring of DC Side of Photovoltaic Array. *IEEE Trans. Sustain. Energy* **2015**, *6*, 1245–1253. [CrossRef]

36. Chohfi, R.E. *Calibration and Installation of a Pyranometer*; Department of Geography, University of California: Los Angeles, CA, USA, 2017.

37. Clive, L. Kipp & Zonen Pyranometer & Pyrheliometer Calibration Frequency. Kipp & Zonen Website. Available online: http://www.kippzonen.com/Download/553/Kipp-Zonen-Pyranometer-Pyrheliometer-Calibration-Frequency?ShowInfo=true (accessed on 16 April 2018).
38. Gengenbach, M. What Is the Calibration Frequency of a Pyranometer? Gengenbach Messtechnik Website. Available online: http://www.rg-messtechnik.de/faq-pyranometer.php (accessed on 16 April 2018).
39. Costanzo, V.; Yao, R.; Essah, E.; Shao, L.; Shahrestani, M.; Oliveira, A.C.; Araz, M.; Hepbasli, A.; Biyik, E. A method of strategic evaluation of energy performance of Building Integrated Photovoltaic in the urban context. *J. Clean. Prod.* **2018**, *184*, 82–91. [CrossRef]

*Article*

# Non-Destructive Failure Detection and Visualization of Artificially and Naturally Aged PV Modules

Gabriele C. Eder [1,*], Yuliya Voronko [1], Christina Hirschl [2], Rita Ebner [3], Gusztáv Újvári [3] and Wolfgang Mühleisen [2]

1   OFI Austrian Research Institute for Chemistry and Technology, Arsenal Object 213, Franz-Grill-Str. 5, 1030 Vienna, Austria; yuliya.voronko@ofi.at
2   CTR Carinthian Tech Research AG, Europastr.12, 9524 Villach, Austria; christina.hirschl@ctr.at (C.H.); wolfgang.muehleisen@ctr.at (W.M.)
3   Center for Energy, AIT Austrian Institute of Technology, Giefinggasse 2, A-1210 Vienna, Austria; rita.ebner@ait.ac.at (R.E.); gusztav.ujvari@ait.ac.at (G.Ú.)
*   Correspondence: gabriele.eder@ofi.at; Tel.: +43-1-7981601-250

Received: 28 March 2018; Accepted: 23 April 2018; Published: 25 April 2018

**Abstract:** Several series of six-cell photovoltaic test-modules—intact and with deliberately generated failures (micro-cracks, cell cracks, glass breakage and connection defects)—were artificially and naturally aged. They were exposed to various stress conditions (temperature, humidity and irradiation) in different climate chambers in order to identify (i) the stress-induced effects; (ii) the potential propagation of the failures and (iii) their influence on the performance. For comparison, one set of test-modules was also aged in an outdoor test site. All photovoltaic (PV) modules were thoroughly electrically characterized by electroluminescence and performance measurements before and after the accelerated ageing and the outdoor test. In addition, the formation of fluorescence effects in the encapsulation of the test modules in the course of the accelerated ageing tests was followed over time using UV-fluorescence imaging measurements. It was found that the performance of PV test modules with mechanical module failures was rather unaffected upon storage under various stress conditions. However, numerous micro-cracks led to a higher rate of degradation. The polymeric encapsulate of the PV modules showed the build-up of distinctive fluorescence effects with increasing lifetime as the encapsulant material degraded under the influence of climatic stress factors (mainly irradiation by sunlight and elevated temperature) by forming fluorophores. The induction period for the fluorescence effects of the polymeric encapsulant to be detectable was ~1 year of outdoor weathering (in middle Europe) and 300 h of artificial irradiation (with 1000 W/m$^2$ artificial sunlight 300–2500 nm). In the presence of irradiation, oxygen—which permeated into the module through the polymeric backsheet—bleached the fluorescence of the encapsulant top layer between the cells, above cell cracks and micro-cracks. Thus, UV-F imaging is a perfect tool for on-site detection of module failures connected with a mechanical rupture of solar cells.

**Keywords:** failure detection; ageing and degradation of PV-modules; performance analysis; UV-fluorescence imaging

## 1. Introduction

Photovoltaic systems (PV) have developed into one of the most promising key technologies within renewable energy supply systems. Starting as a niche market of small scale, special purpose applications more than 30 years ago, PV has turned into a mainstream electricity source with a cumulative global capacity of 303 GW in 2016 according to the Trend Report of the International Energy Agency (IEA) [1] and the Global Market Report of Solar Power Europe [2]. Electricity generation by photovoltaic systems is strongly gaining in importance with utility scale solar plants already being cost-competitive with wind energy and also fossil fuels and nuclear energy [2]. However, there are still some hurdles to overcome to strengthen the role of PV systems in a global electricity market [2,3] and to let PV develop into a technology which can contribute to the obtainment of local and global sustainable goals [4,5]. The growing PV production rates suggest that new sectors like PV recycling will be essential in the world's transition to a sustainable, economically viable and increasingly renewables-based energy future (as described in detail in the joined International Renewable Energy Agency (IRENA)/International Energy Agency (IEA) Photovoltaic Power Systems (PVPS) report on End-of-Life Management of PV panels [4]), [6].

Induced by the enhanced competition in the PV market over the last decade, the PV industry has experienced tremendous pressure to reduce production costs. At the same time, there is high demand for improving the module's efficiency, reliability and long-term performance [7,8] as well as its sustainability [4–6]. In order to be competitive in the market, a warranty of operational lifetimes of 25+ years at a maximum total yield loss of 20% has to be granted [7–10]. Upon improper handling, transport and/or installation of PV modules, failures due to mechanical impact, such as cracked cells, disruptions in the electric connection system or glass breakage, can occur (an elaborated review of failure modes of PV modules is given in [11]), [12]. Furthermore, extreme stress imposed on installed modules by storm events, heavy snow loads or hail storms can also cause such failures [13–17]. Thus, one of the major concerns, especially for operation and maintenance (O & M) companies, is to forecast how mechanically-induced module failures develop over time and how this will affect yield under different stress conditions in the future [18,19]. However, a recent comprehensive report of the IEA PVPS Task13 on the "Assessment of Photovoltaic Module Failures in the Field" [20] came to the preliminary conclusion that no strong correlations between observed failure occurrences and impacts on climatic zones exist. An extended data analysis with larger data sets is planned for the future. Some failure modes like potential induced degradation, failure of bypass diodes and also cell cracks seem to be independent of climatic zones [20].

In the Austrian flagship research project, INFINITY, the topics of in-field failure detection and failure propagation were also addressed [17,21,22]. The objective of the work presented here is to investigate the detectability of various mechanically-induced failures (glass crack, solar cell micro-cracks and defects in the cell connection system) and their potential propagation under artificial stress conditions (enhanced temperature, humidity and irradiation) as well as under outdoor weathering conditions. The effect of the failures on the module's performance is analysed and visualized by current–voltage (IV) and electroluminescence (EL) measurements. Furthermore, the detectability of such failures by the non-destructive characterisation tool, ultraviolet fluorescence imaging (UV-F), is investigated. With this experimental approach, we aim to see (i) the effects that deliberately-generated failures (micro-cracks, cell cracks, glass breakage, and defects in the cell interconnection) have on the developing UV-fluorescence (UV-F) patterns and (ii) whether a propagation of such failures occurs with storage time. This will allow the deduction of additional information from UV-F images of failure modules in the field and will facilitate their interpretation.

## 2. Materials and Methods

### 2.1. Experimental Approach

For this investigation, identical 6-cell PV modules (polycrystalline Si-cells, ethylene vinyl acetate (EVA)-encapsulation, non-gastight polyethylene terephthalate (PET)-based backsheet and glass frontsheet) were produced. Modules were manufactured (i) without failures (ii) with a defect in the cell connection system (one cell connector was deliberately cut), or (iii) with micro-cracks, introduced in the solar cells by means of mechanical impact. Two test modules of each module type (i–iii) were stored as follows:

- under accelerated ageing conditions, such as damp heat (DH in accordance with IEC61215 [23]) at 85 °C, 85% relative Humidity [r.H.] for 1000 h and 2000 h,

- under artificial irradiation (I) at 1000 W/m$^2$ of simulated sunlight with metal halide lamps (300–2500 nm) at a chamber temperature of 50 °C and 40% r.H. for 1000 h,
- under outdoor (OD) conditions (in middle Europe, Vienna and Austria) for 1.5 years (see Figure 1).

All samples were characterized in detail (i) before and after light stabilization with 20 kWh and (iI) after the aging procedure with power and EL measurements. In addition, the effects of the failures on the formed fluorescence patterns of the polymeric encapsulant were monitored by measuring the UV-F of the test modules every 100 h during the accelerated aging tests (indoor) and every month during outdoor storage.

**Figure 1.** Outdoor test stand (Austrian Institute of Technology (AIT), Vienna, Austria).

## 2.2. Characterisation Methods

### 2.2.1. Electrical Measurements

By means of IV measurements, the open circuit voltage ($U_{OC}$), short circuit current ($I_{sc}$), series resistance ($R_{SER}$), shunt resistance ($R_{SH}$), maximum power point ($P_{MPP}$), voltage at $P_{MPP}$ ($U_{MP}$), current at $P_{MPP}$ ($I_{MP}$), and fill factor (FF) were determined. The IV measurements of all test modules before and after the accelerated ageing test or following natural weathering for 1.5 years were performed in accordance with IEC61215 [9,23]. The electrical performance was measured under laboratory standard conditions (25 °C; 1000 W/m$^2$ with air mass (AM) 1.5 spectral distribution) using a PASAN HighLIGHT VLMT A+A+A+ flasher. The HighLIGHT VLMT (Very Large Module Tester) is designed to flash a surface of 3 m × 3 m. The accuracy of the power measurement is set by the quality and reliability of the module tester, with the used system being classified as A+A+A+ (according to the IEC Standard IEC 60904-9 [24]) based on three parameters: a spectral irradiance distribution of ≤12.5%, a non-uniformity of irradiance of 1% and a pulse instability (long term) of ≤1%.

### 2.2.2. Electroluminescence Measurements

EL measurements take advantage of the radiative interband recombination of excited charge carriers in solar cells. For EL investigations, the module is operated as a light emitting diode. The emitted radiation due to recombination effects can be detected with a sensitive Si-CCD-camera. The wavelength window of the Si-CCD camera is 300 to 1100 nm. The solar cells are supplied with a defined external excitation current (current applied ≤ short circuit current ($I_{sc}$) of the cell or module) while the camera takes an image of the emitted photons. Damaged areas of a solar module appear dark or radiate less than areas without defects. EL has proven to be a useful tool for investigating electrical inhomogeneities caused by intrinsic defects (e.g., grain boundaries, dislocations, shunts, or other process failures) and extrinsic defects (e.g., cell cracks, corrosion, or interrupted contacts) [25–27].

To determine the influences of defects (e.g., shunts), the EL behaviour of the test modules was investigated with different current densities: 10% and 100% of $I_{SC}$. When applying a low current density (10% of $I_{sc}$ of the module), the conductivity of shunts was very high. When applying higher current densities (~$I_{SC}$), the conductivity of the positive-negative (p-n)-junction increased compared to the shunt conductivity, and shunts were less influential on the EL intensity distribution. Thus, with low current densities, the material properties can be investigated, and with high current densities, the properties of the electrical contacts can be investigated [28].

### 2.2.3. UV-Fluorescence Measurements

Fluorescence is a form of luminescence and stands for the physical effect of the emission of light by a material that has absorbed light or other electromagnetic radiation. The emitted light (e.g., in the visible region) has a longer wavelength than the absorbed radiation (e.g., UV light). A fluorophore is thus a fluorescent chemical compound that can re-emit light upon light excitation and mostly contains several combined aromatic groups or other plane or cyclic molecules with several π bonds. Typical fluorophores are degradation products of polymers and/or additives with chromophoric/fluorophoric groups. The fluorescence of materials can be extinguished by "photobleaching" effects which lead to a decrease in fluorescence due to reaction processes with, for example, oxygen [11,29,30]. The first description of fluoresence effects in PV encapsulants was performed in 1993 by Pern [31,32]. The use of this effect as a diagnosis tool for polymeric encapsulant processing and degradation was also reported be Pern et al. one year later [30]. The first to observe the UV fluorescing effects of PV encapsulants caused by ageing-induced polymer degradation was Röder et al. in 2008 [33–38], and the first to use these UV-F effects and the formation of specific patterns of fluorescing and bleached parts within the module to detect cell breakage of aged PV modules in the field was Köntges et al. in 2012 [14,39–41].

In this study, UV-Fluorescence measurements were performed in a dark environment by illumination of the PV modules with UV light and detection of the fluorescing light in the visible region by a photographic camera system (Olympus OM-D, equipped with high pass filter to cut off the UV irradiation). Excitation with UV light was performed with a self-made UV lamp [42] consisting of 3 power-tunable light emitting diode (LED) arrays with an emission maximum at 365 nm and a low pass filter to cut off all visible light. The power supply was a modified DC/DC converter with controllable and piecewise constant voltage/constant current characteristics, sourced by a 12-cell, lithium-polymer accumulator with a capacity of 5000 mAh. This characterization method is non-destructive, non-invasive, easy to handle and fast (an exposure time of 30 s is sufficient to achieve a well contrasted UV fluorescence image of a module) [17,22].

## 3. Results

In order to see (i) the effects that deliberately-generated failures (micro-cracks, cell cracks, glass breakage, and defects in the cell interconnection) have on developing UV fluorescence patterns and (ii) to determine whether propagation of such failures occurs with storage time, reference modules and defective test modules were treated in the same way (as described above).

### 3.1. Test Modules without Failures

As expected, the reference modules (test modules without failures) did not show relevant degradation effects, as determined by performance and EL measurements, irrespective of the stress conditions applied. The power of the modules after light stabilization was 24.3 ± 0.1 W under standard test conditions (STC) conditions (see Table 1) and changed only slightly upon accelerated and natural ageing. While irradiation with artificial sunlight for 1000 h caused a slight decrease in the power output (−1%) at the maximum power point ($P_{MPP}$), storage under DH for 1000 h and OD storage for 1.5 years resulted in slight increases in the $P_{MPP}$ of +0.3% and +0.7%, respectively.

**Table 1.** Power output in [W] at the maximum power point ($P_{MPP}$) (measured under standard test conditions) of the test modules without defects, before and after ageing.

| Module Number | Original [1] | After 1000 h Artificial Irradiation (I) | After 1000 h Damp Heat (DH) | After 1.5 Years Outdoors (OD) |
|---|---|---|---|---|
| 1-01 | 24.39 | 24.19 | | |
| 1-06 | 24.45 | 24.17 | | |
| 1-03 | 24.29 | | 24.33 | |
| 1-04 | 24.27 | | 24.37 | |
| 1-02 | 24.44 | | | 24.61 |
| 1-20 | 24.32 | | | 24.47 |

[1] Values after light stabilization with 20 kWh.

The original modules did not show any specific fluorescence behavior after manufacturing and light stabilization (see Figure 2a). Upon subsequent DH storage, however, the polymeric encapsulation showed fluorescing effects (Figure 2b). These effects were related to the ingress of water vapour into the polymeric encapsulant and were not homogeneously distributed over the whole module. The fluorescing effects were stronger above the backsheet and the rims of the cells. The broadness of the fluorescing rims increased with increasing storage time (related to water vapour permeating into the encapsulant via the polymeric backsheet).

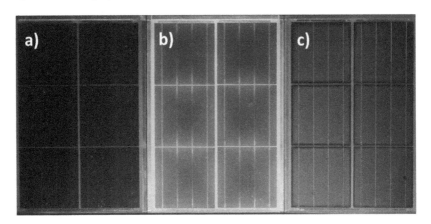

**Figure 2.** UV-F images of three test modules: (**a**) one in the original state (stabilized); (**b**) one after storage under DH -conditions for 2000 h; and (**c**) one after exposition to artificial sunlight (1000 W/m$^2$) for 1000 h.

In contrast, upon irradiation with artificial sunlight, fluorescing groups were formed within the front encapsulant layer of the module (Figure 2c). The UV-fluorescence was evenly distributed in the polymeric encapsulation and increased in intensity with increasing irradiation doses. In our test setup, after 300 h of irradiation in the accelerated ageing test and after 12 months of outdoor storage, UV-F effects could clearly be detected in the encapsulation of the test modules. Between and at the rims of the cells, the fluorescence was extinguished by oxygen (bleaching) permeating into the encapsulation via the polymeric backsheet. The broadness of the extinct area at the cell rims increased with increasing weathering time.

One testmodule showed a glass crack which occurred at the beginning of the irradiation test, most probably caused by thermal tension (untempered glass was used for all test modules). It was first noticed after 100 h storage time and increased in length, reaching its final magnitude after 400 h [8]. It was observed that the fluorescence formed homogeneously in the polymeric encapsulation over the cells and disappeared beneath the glass breakage after 300 h of further irradiation. The EL image showed that the glass breakage (induced by thermal stress in the first phase of the weathering test) did not cause a crack in the underlying cells. The disappearance of the fluorescence beneath the glass breakage is attributed to an oxygen bleaching effect. The incoming oxygen can interact/react with the activated and fluorescing sites of the polymer, leading to a decrease in its fluorescence intensity. The glass breakage had no impact on the performance of the test module after the 1000 h irradiation test.

### 3.2. Test Modules with Cell Cracks/Micro-Cracks

The test module with deliberately induced cell cracks/micro-cracks did not show relevant differences in the power output ($P_{MPP}$, measured under STC conditions) compared to the intact reference modules before ageing (see results in Table 2). Upon aging under irradiation, however, decreases in $P_{MPP}$ by −1.6% and −2.8% were observed, while storage under DH and outdoor conditions resulted in an increased electrical output of +1.1%and +1.9%, respectively.

It has to be noted at this point that it was very difficult to generate a comparable number and size of (micro)cracks in the six test modules. As visualized in the EL images (see Figure 3), there were modules with few (*) and numerous (**) micro-cracks and others with cell cracks (+), and it was expected that their effects on degradation would be quite different.

**Figure 3.** Electroluminescence (EL) images at 10% short circuit current ($I_{SC}$) of trhee test modules with (**a**) few (*) or (**b**) numerous (**) micro-cracks or (**c**) with a cell crack (+).

**Table 2.** Power output in [W] at $P_{MPP}$ (measured under STC conditions) of the test modules with few * and numerous ** micro-cracks or a cell crack + before and after ageing.

| Module Number | Original [1] | After 1000 h I | After 1000 h DH | After 1.5 Years OD |
|---|---|---|---|---|
| 1-13 * | 24.40 | 24.01 | | |
| 1-16 ** | 24.40 | 23.72 | | |
| 1-14 + | 24.42 | | 24.39 | |
| 1-15 * | 24.58 | | 24.53 | |
| 1-09 + | 24.32 | | | 24.77 |
| 1-19 ** | 23.73 | | | 23.99 |

[1] Values after light stabilization with 20 kWh.

One test module (no. 1-14), which showed a cell crack in the EL image (Figure 3c), supplied UV-fluorescence images upon DH storage, as shown in Figure 4. In addition to the fluorescence formed with increasing storage time above the backsheet and the rims of the cells, fluorescence was also observed above the cell crack making it detectable for the observer. It has to be noted that neither the performance (−0.1%) nor the EL image of this test module were changed drastically upon artificial ageing for 1000 h DH. With one module (no. 1-09), the cell crack increased in length following outdoor storage for 1.5 years by a factor of 2, and the performance increased slightly (+1.1%).

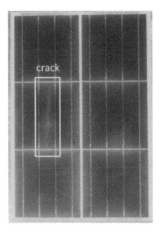

**Figure 4.** UV-F image of test module no. 1-14 after 2000 h DH showing a cell crack.

Upon irradiation with artificial sunlight and natural sunlight (see Figure 5b,c), the evolving fluorescence pattern of the test modules with micro-cracks clearly differed from those of the test modules without failures (see Figure 2).

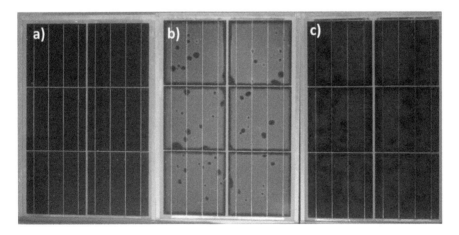

**Figure 5.** UV-F images of three test modules with micro-cracks: (**a**) one in the original state (stabilized); (**b**) one after exposition to artificial sunlight (1000 W/m$^2$) for 1000 h; and (**c**) one after natural storage for 1.5 years.

The UV-F images taken every 100 h of accelerated weathering again showed a continuous increase in the intensity of fluorescence above the cells with unequivocally detectable intensities starting from 300 h. However, there were numerous dark spots where no fluorescence was generated (Figure 5b). In comparison with the EL image of that module, a clear correlation with the positions of the micro-cracks in the cells could be obtained. At the positions of the micro-cracks in the cells, the fluorescence was extinguished by oxygen (bleaching) permeating into the front encapsulation via the polymeric backsheet and cell cracks. In the outdoor weathering, the effect of UV-F in the front encapsulant was clearly detectable after 1 year showing extinctions (dark spots) at the locations of the micro-cracks in the cells (see. Figure 5c; confirmed by parallel EL measurements).

*3.3. Measurement of Test-Modules with Failures in the Connection System*

The EL images of the test modules with defects in the connection system (interruption of one connection ribbon of one cell in the six-cell test module) all showed—as expected—a darker region of one cell, with the rest of the affected cell being brighter (see Figure 6). This effect was visible in the EL image in the original state, after accelerated ageing (under DH or irradiation conditions) and outdoor testing. The electrical performances of these test modules ($P_{MPP}$, measured under STC- conditions) before and after ageing are summarized in Table 3. The power output of the original modules was only ~1% lower than that for the intact reference modules. This effect was due to a higher $R_{SER}$.

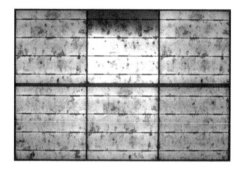

**Figure 6.** EL image at 100% $I_{SC}$ of a test module with a defective cell interconnection in the original state.

**Table 3.** Power output in [W] at $P_{MPP}$ (measured under STC conditions) of the test modules with defects in the cell interconnection system before and after ageing.

| Module Number | Original [1] | After 1000 h I | After 1000 h DH | After 1.5 Years OD |
|:---:|:---:|:---:|:---:|:---:|
| 2-05 | 24.19 | 23.90 | | |
| 2-06 | 24.26 | 23.94 | | |
| 2-03 | 24.12 | | 24.13 | |
| 2-04 | 24.13 | | 24.15 | |
| 2-09 | 23.99 | | | 24.13 |
| 2-10 | 24.07 | | | 24.24 |

[1] Values after light stabilization with 20 kWh.

Upon ageing, the test modules with a deliberately-generated connection defect behaved comparably to the intact modules (decrease in $P_{MPP}$ after artificial irradiation by −1.2%, stable after DH storage, slight increase upon OD testing by +0.7%), and no stress-induced propagation of the failures could be observed in the test program applied. The UV-F images of the test modules with defects in the cell interconnectors were identical to those of the intact reference modules, irrespective of the stress treatment (see Figure 7).

This comparison clearly shows that the UV-F pattern which appears under outdoor conditions (Figure 7b) resembles the fluorescence pattern observed after artificial aging (Figure 7c). However, the intensity of the UV-F above the cells is lower and the extinct area at the cell rims is broader, indicating a pronounced oxygen ingress.

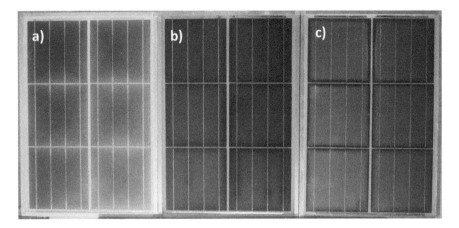

**Figure 7.** UV-F images of three test modules with interconnector defects: (**a**) one after DH storage for 2000 h; (**b**) one after natural weathering outdoors for 1.5 years; and (**c**) one after indoor exposition to artificial sunlight (1000 W/m$^2$) for 1000 h.

## 4. Discussion

The results presented show a good robustness of the PV modules against stress, even when defects such as micro-cracks, cell cracks, glass breakage or interruptions in the cell interconnections are present. Hardly any propagation of the module failures in the accelerated ageing tests (indoors) and the outdoor exposure test could be detected. It was also noticed that the length of the micro-cracks (as visualized by the EL-images) did not (or only minor) increase during the time span of the accelerated (indoor) or natural ageing (outdoor) tests performed within the work presented. These effects were also found by Buerhop-Lutz et al. in indoor tests [15] and Jaeckel et al. in a 15-year outdoor test [16] and are in accordance with a recent study by Dimish et al. on the impact of cracks on photovoltaic power performance [43].

However, the presence of numerous micro-cracks led to a higher rate of degradation; the power loss upon irradiation (1000 W/m$^2$ for 1000 h) was twice as high as that for the intact reference module under the same stress conditions. Defects in the cell interconnection of test modules caused a higher series resistance, leading to a slightly lower power output by ~1%. The relative change in power upon accelerated ageing, though, was not affected by that failure. Outdoor weathering for 1.5 years did not lead to any degradation effects, irrespective of the failures deliberately incorporated in the test modules.

In the unaged state, the polymeric encapsulates of the PV modules (ethylene-vinyl acetate, EVA) did not show distinctive fluorescence. The EVA polymer itself has no fluorescing groups in its original state but UV-excitable fluorophores/chromophores can be formed from the peroxidic curing agent/crosslinker during curing, either as decomposition product [29] or by degradation of stabilizers by peroxides [44,45]. Furthermore, constituents of the encapsulating material, like impurities or additives, also influence the UV-F of the encapsulating polymer [46–48]. With an increasing lifetime, the encapsulant material degraded due to the interaction with irradiation through artificial or natural sunlight and the elevated temperature from forming fluorophores. The fluorescing effect was increasing with exposure time as it is directly dependent on the concentration of fluorophores built [1]. Partially activated and degraded polymer chains show fluorescence effects [48]. As the encapsulant not only comprises the polymeric base material itself but also of a mixture of additives, its composition has an influence on the development of the fluorescence (as studied in detailed by Peike et al. [46,48]), [29].

Our own results as well as detailed studies by Schlothauer et al. [47,49] confirm that in the photobleached areas, the polymeric encapsulant is chemically changed due to the oxidative reaction of the polymer, leading to chain scission and consequently, to changes in the viscoelastic mechanical properties and thermomechanical properties of the polymer.

Comparable fluorescence patterns of the front encapsulant of the test modules were observed following artificial and natural sunlight weathering. However, during outdoor exposure, the fluorescence effects were clearly detectable in the field (in darkness) with a UV-F lamp after ~1 year. This is in good agreement with findings reported in the literature regarding an induction period for the visible detection of UV-fluorescence [25]. Köntges et al. reported [11] that the module should have been exposed to UV dose of approximately 80 kWh/m$^2$, which correlates to about 1.5 years of outdoor exposure in Germany, to get a sufficient fluorescence signal. The formation of the fluorescing groups in the polymer requires irradiation and the absence of oxygen; an elevated temperature accelerates the process. Thus, in the artificial weathering test (1000 W/m$^2$ of simulated sunlight with metal halide lamps (300–2500 nm) performed at a chamber temperature of 50 °C with 40% r.H. for 1000 h), UV-F patterns were already detectable after 300 h.

The abovementioned fluorescence patterns formed in the polymeric encapsulant upon weathering/irradiation can become extinct in the presence of oxygen and irradiation. The permeation of oxygen into the encapsulant via glass breakage or a polymeric backsheet acts as bleaching agent for fluorescence. As the oxygen molecules can interact/react with the activated and fluorescing sites of the polymer, their presence leads to an irreversible decrease in the fluorescence intensity. Bleaching needs irradiation besides oxygen ("photobleaching") [11,29,30]. The width of the bleached rim around the cell and the bleached area above the cell crack increased with time and were correlated with the permeation rate of oxygen into the front encapsulant [47,50,51].

This clearly indicates that UV-F imaging is a suitable tool, not only for detecting cell cracks in operating PV modules but also to allow for estimation of the timescale at when these cell cracks were generated. This may be of special interest to allow one to distinguish between installation- or transport-induced cell cracks and those which are generated during operation by, for example, a heavy hail storm [17]. Furthermore, the temperature-induced degradation effects of the encapsulant lead to increased fluorescing light intensities and make hot spots or overheated module parts visible in the UV-F images [11].

The fluorescence effects observed upon DH storage (high temperature, high humidity, no irradiation) above permeable backsheets were correlated with the ingress of water vapour [34–36,51,52] and its interaction with the polymeric encapsulant and had a different origin from the irradiation-induced fluorescence effects [22,49–51]. The intensity of this type of fluorescence is lower, and fluorescence spectroscopic measurements have shown that the absorption maximum of elevated temperature and water induced fluorescence lies at a lower wavelength than for irradiation induced fluorescence, as described in detail by Röder and Schlothauer in several scientific presentations [33,36,45,50,53].

## 5. Conclusions

The performances of PV modules with mechanical module failures, such as micro-cracks, cell cracks, glass breakage or breakage of interconnectors were found to be rather unaffected upon storage under various stress conditions. In most cases, no propagation of the module failures in the accelerated ageing tests and the outdoor exposure test could be detected. However, the presence of numerous micro-cracks led to a higher rate of degradation; the power loss upon irradiation was twice as high as for the intact reference module upon accelerated ageing.

UV-F imaging is a perfect tool for on-site detection of module failures connected with a mechanical rupture of the solar cells and additionally, gives information about the timescale of the formation of the cracks. The induction period for the fluorescence effects of the polymeric encapsulant is ~1 year of outdoor weathering (in middle Europe) and 300 h of artificial irradiation (with 1000 W/m$^2$ artificial sunlight 300–2500 nm). Characteristic fluorescence patterns and types are formed upon temperature and humidity impact (DH storage) and irradiation stress (artificial weathering and outdoor exposure).

Mechanical failures of the modules, such as cell cracks and glass breakage have strong influences on the local extinction of the fluorescence (photo bleaching).

**Author Contributions:** All authors are participants of the research project INFINITY.

**Acknowledgments:** This work was conducted as part of the Austrian "Energy Research Program" project INFINITY. This project is funded by the Austrian Climate and Energy Fond and the Austrian Research Promotion Agency (FFG) which are gratefully acknowledged.

**Conflicts of Interest:** The authors declare no conflict of interest.

## References

1.  Masson, G.; Kaizuka, I.; International Energy Agency; Photovoltaic Power Systems Programme (Eds.) *Report IEA PVPS T1-32:2017: Trends 2017 in Photovoltaic Applications*; IRENA and IEA PVPS: St. Ursen, Switzerland, 2017; ISBN 978-3-906042-68-8.
2.  Solar Power Europe. Global Market Outlook 2017–2021. M. Schmela, SolarPower Europe. 2017. Available online: www.solarpowereurope.org (accessed on 1 May 2017).
3.  Liu, J.; Long, Y.; Song, X. A Study on the Conduction Mechanism and Evaluation of the Comprehensive Efficiency of Photovoltaic Power Generation in China. *Energies* **2017**, *10*, 723. [CrossRef]
4.  Weckend, S.; Wade, A.; Heath, G. (Eds.) *End-of-Life Management: Solar Photovoltaic Panels*; Joined Report IRENA/IEA-PVPS; IEA-PVPS Report Number: T12-06:2016; IRENA and IEA PVPS: St. Ursen, Switzerland, 2016; ISBN 978-92-95111-98-1.
5.  Gómez-Lorente, D.; Rabaza, O.; Aznar-Dols, F.; Mercado-Vargas, M.J. Economic and Environmental Study of Wineries Powered by Grid-Connected Photovoltaic Systems in Spain. *Energies* **2017**, *10*, 222. [CrossRef]
6.  D'Adamo, I.; Miliacca, M.; Rosa, P. Economic Feasibility for Recycling of Waste Crystalline Silicon Photovoltaic Modules. *Int. J. Photoenergy* **2017**, *2017*, 4184676. [CrossRef]
7.  Nordmann, T.; Clavadetscher, L.; van Sark, W.G.J.H.M.; Green, M.; International Energy Agency; Photovoltaic Power Systems Programme (Eds.) *Report IEA-PVPS T13-05-2014: Analysis of Long-Term Performance of PV Systems*; IRENA and IEA PVPS: St. Ursen, Switzerland, 2014; ISBN 978-3-906042-21-3.
8.  Report of the EU-Project Solar Bankability: Technical Bankability Guidelines: Recommendations to Enhance Technical Quality of Existing and New PV Investments, EC Grant Agreement Number: No 649997; von Armansperg, M., Oechslin, D., Schweneke, M., Eds. Available online: https://www.tuv.com/content-media-files/master-content/services/products/p06-solar/solar-downloadpage/solar-bankability_d4.3_technical-bankability-guidelines.pdf (accessed on 1 February 2017).
9.  Wohlgemuth, J. Standards for PV Modules and Components—Recent Developments and Challenges. In Proceedings of the 27th European PVSEC, Frankfurt, Germany, 24–28 September 2012.
10. Dunlop, E.D. Lifetime performance of crystalline silicon PV modules. In Proceedings of the 3rd World Conference on Photovoltaic Energy Conversion, Osaka, Japan, 11–18 May 2003.
11. Köntges, M.; Kurtz, S.; Packard, C.; Jahn, U.; Berger, K.A.; Kato, K.; Friesen, T.; Liu, H.; Van Iseghem, M.; International Energy Agency; et al. (Eds.) *Report IEA-PVPS T13-01:2014: Review of Failures of PV Modules*; IRENA and IEA PVPS: St. Ursen, Switzerland, 2014; ISBN 978-3-906042-16-9.
12. Köntges, M.; Kunze, I.; Kajari-Schröder, S.; Breitenmoser, X.; Bjørneklett, B. Quantifying the Risk of Power Loss in PV Modules Due to Micro Cracks. In Proceedings of the 25th European PV SEC, 4BO.9.4, Valencia, Spain, 6–10 September 2010; pp. 3745–3752. [CrossRef]
13. Köntges, M.; Kunze, I.; Kajari-Schröder, S.; Breitenmoser, X.; Bjørneklett, B. The risk of power loss in crystalline silicon based photovoltaic modules due to microcracks. *Sol. Energy Mater. Sol. Cells* **2011**, *5*, 1131–1137. [CrossRef]
14. Köntges, M.; Kajari-Schröder, S.; Kunze, I. Crack Statistic for Wafer-Based Silicon Solar Cell Modules in the Field Measured by UV Fluorescence. *IEEE J. Photovolt.* **2013**, *3*, 95–101. [CrossRef]
15. Buerhop-Lutz, C.; Winkler, T.; Fecher, F.W.; Bemm, A.; Hauch, J.; Camus, C.; Brabec, C.J. Performance Analysis of Pre-Cracked PV-Modules at Realistic Loading Conditions. In Proceedings of the 33rd European PV-SEC, 5CO.8.2, Amsterdam, The Netherlands, 25–29 September 2017; pp. 1451–1456. [CrossRef]

16. Jaeckel, B.; Franke, T.; Arp, J. Long Term Statistics on Micro Cracks and Their Impact on Performance. In Proceedings of the 33rd European PV-SEC, 5CO5.6, Amsterdam, The Netherlands, 25–29 September 2017; pp. 1396–1401. [CrossRef]

17. Mühleisen, W.; Eder, G.C.; Voronko, Y.; Spielberger, M.; Sonnleitner, H.; Knöbl, K.; Ebner, R.; Ujvari, G.; Hirschl, C. Outdoor detection and visualization of hailstorm damages of photovoltaic plants. *Renew. Energy* **2018**, *118*, 138–145. [CrossRef]

18. Köntges, M.; Siebert, M.; Illing, R.; Wegert, F. Influence of Photovoltaic-Module Handling on Solar Cell Cracking. In Proceedings of the 30th European PV SEC, 5BO12.6, Kyoto, Japan, 23–27 November 2014; pp. 2276–2282. [CrossRef]

19. Mathiak, G.; Pohl, L.; Sommer, J.; Fritzsche, U.; Herrmann, W.; Reil, F.; Althaus, J. PV Module Damages Caused by Hail Impact—Field Experience and Lab Tests. In Proceedings of the 31th European PV SEC, 5DO.12.5, Hamburg, Germany, 14–18 September 2015; pp. 1915–1919. [CrossRef]

20. Köntges, M.; Oreski, G.; Jahn, U.; Herz, M.; Hacke, P.; Weiss, K.-A.; Razongles, G.; Paggi, M.; Parlevliet, D.; Tanahashi, T.; et al. (Eds.) *Report IEA-PVPS T13-09:2017: Assessment of Photovoltaic Module Failures in the Field*; IRENA and IEA PVPS: St. Ursen, Switzerland, 2017; ISBN 978-3-906042-54-1.

21. Hirschl, C.; Mühleisen, W.; Brantegger, G.; Neumaier, L.; Spielberger, M.; Sonnleitner, H.; Kubicek, B.; Ujvari, G.; Ebner, R.; Schwark, M.; et al. Scientific and economic comparison of outdoor evaluation methods for photovoltaic power plants. *Renew. Energy* **2018**. submitted.

22. Hirschl, C.; Eder, G.C.; Neumaier, L.; Mühleisen, W.; Voronko, Y.; Ebner, R.; Kubicek, B.; Berger, K.A. Long term development of Photovoltaic module failures during accelerated ageing tests. In Proceedings of the 33rd European PV-SEC, 5BV.4.72, Amsterdam, The Netherlands, 25–29 September 2017; pp. 1709–1712. [CrossRef]

23. International Standard IEC61215. *Terrestrial Photovoltaic (PV) Modules—Design Qualification and Type Approval—Part 1: Test Requirements and—Part 2: Test Procedures*, 1.0 ed.; TC 82—Solar Photovoltaic Energy Systems; IEC: Geneva, Switzerland, 9 March 2016.

24. International Standards IEC 60904-9:2007. *Photovoltaic Devices—Part 9: Solar Simulator Performance Requirements*, 2.0 ed.; TC 82—Solar Photovoltaic Energy Systems; IEC: Geneva, Switzerland, 16 October 2007.

25. Wang, G.; Gong, H.; Zhu, J. Failure analysis of dark cells detected by electroluminescence (EL). In Proceedings of the 28th European PVSEC 2014, 4AV.5.1, Paris, France, 30 September–4 October 2013; pp. 3173–3179. [CrossRef]

26. Köntges, M.; Siebert, M.; Hinken, D.; Eitner, U.; Bothe, K.; Potthof, T. Quantitative analysis of PV-modules by electroluminescence images for quality control. In Proceedings of the 24th European PVSEC, 4CO.2.3, Hamburg, Germany, 21–25 September 2009; pp. 3226–3231. [CrossRef]

27. Ebner, R.; Zamini, S.; Újvári, G. Defect Analysis of Different Photovoltaic Modules Using Electroluminescene (EL) and Infrared (IR)-Thermography. In Proceedings of the 25th European PVSEC 2010, 1DV.2.8, Valencia, Spain, 6–10 September 2010. [CrossRef]

28. Crozier, J.L. Identifying voltage dependent features in photovoltaic modules using electroluminescence imaging. In Proceedings of the 29th EU-PVSEC, Amsterdam, The Netherlands, 22–26 September 2014.

29. Pern, F.J. Factors that affect the EVA encapsulant discoloration rate upon accelerated exposure. *Sol. Energy Mater. Sol. Cells* **1996**, *41–42*, 587–615. [CrossRef]

30. Pern, F.J.; Glick, S.H. Fluorescence analysis as a diagnostic tool for polymer encapsulation processing and degradation. *AIP Conf. Proc.* **1994**, *306*, 573–585. [CrossRef]

31. Pern, F.J. Polymer encapsulants characterized by fluorescence analysis before and after degradation. In Proceedings of the Conference Record of the 23rd IEEE Photovoltaic Specialists Conference, Louisville, KY, USA, 10–14 May 1993. [CrossRef]

32. Czanderna, A.W.; Pern, F.J. Encapsulation of PV modules using ethylene vinyl acetate copolymer as a pottant: A critical review. *Sol. Energy Mater. Sol. Cells* **1996**, *43*, 101–181. [CrossRef]

33. Röder, B.; Ermilov, E.A.; Philipp, D.; Köhl, M. Observation of polymer degradation processes in photovoltaic modules via luminescence detection. In *Reliability of Photovoltaic Cells, Modules, Components, and Systems*; SPIE—The International Society for Optical Engineering: Bellingham, WA, USA, 2008; Volume 7048.

34. Schlothauer, J.; Jungwirth, S.; Röder, B.; Köhl, M. Fluorescence imaging: A powerful tool for the investigation of polymer degradation in PV modules. *Photovolt. Int.* **2010**, *10*, 149–154.

35. Jungwirth, S.; Röder, B.; Weiss, K.A.; Köhl, M. The influence of different back sheet materials on EVA degradation in photovoltaic modules investigated by luminescence detection. In *SPIE Proceedings—Reliability of Photovoltaic Cells, Modules, Components, and Systems III*; Dhere, N.G., Wohlgemuth, J.H., Lynn, K., Eds.; SPIE: Bellingham, WA, USA, 2010; Volume 7773, ISBN 9780819482693.
36. Schlothauer, J.C.; Jungwirth, S.; Röder, B.; Köhl, M. Fluorescence imaging: A powerful tool for the investigation of polymer degradation in PV modules. In Proceedings of the 37th IEEE Photovoltaic Specialists Conference (PVSC), Seattle, WA, USA, 19–24 June 2011; pp. 003606–003608. [CrossRef]
37. Röder, B.; Jungwirth, S.; Braune, M.; Philipp, D.; Köhl, M. The influence of different ageing factors on polymer degradation in photovoltaic modules investigated by luminescence detection. In *SPIE Proceedings—Reliability of Photovoltaic Cells, Modules, Components, and Systems III*; Dhere, N.G., Wohlgemuth, J.H., Lynn, K., Eds.; SPIE: Bellingham, WA, USA, 2010; Volume 7773, ISBN 9780819482693.
38. Röder, B.; Schlothauer, J.; Jungwirth, S.; Köhl, M. Investigation of outdoor degradation of photovoltaic modules by luminescence and electroluminescence. In *SPIE Proceedings—Reliability of Photovoltaic Cells, Modules, Components, and Systems III*; Dhere, N.G., Wohlgemuth, J.H., Lynn, K., Eds.; SPIE: Bellingham, WA, USA, 2010; Volume 7773, ISBN 9780819482693.
39. Morlier, A.; Siebert, M.; Kunze, I.; Mathiak, G.; Köntges, M. Detecting photovoltaic module failures in the field during daytime with ultraviolet fluorescence module inspection. In Proceedings of the 44th IEEE PVSC, Washington, DC, USA, 25–30 June 2017.
40. Morlier, A.; Köntges, M.; Siebert, M.; Kunze, I. UV fluorescence imaging as fast inspection method for PV modules in the field. In Proceedings of the 14th IEA PVPS Task 13 Meeting, Bolzano, Italy, 3–8 April 2016.
41. Köntges, M.; Kajari-Schröder, S.; Kunze, I. Cells Cracks Measured by UV Fluorescence in the Field. In Proceedings of the 27th European PVSEC, 4CO.11.4, Frankfurt, Germany, 24–28 September 2012; pp. 3033–3040. [CrossRef]
42. Knöbl, K. (University of Applied Sciences, Technicum, Vienna, Austria). Personal communication, 2016.
43. Dhimish, M.; Holmes, V.; Mehrdadi, B.; Dales, M. The impact of cracks on photovoltaic power performance. *J. Sci. Adv. Mater. Dev.* **2017**, *2*, 199–209. [CrossRef]
44. Morlier, A.; Köntges, M.; Blankemeyer, S.; Kunze, I. Contact-free determination of ethylene vinyl acetate crosslinking in PV modules with fluorescence emission. *Energy Procedia* **2014**, *55*, 348–355. [CrossRef]
45. Schlothauer, J.C.; Ralaiarisoa, R.M.; Morlier, A.; Köntges, M.; Röder, B. Determination of the cross-linking degree of commercial ethylene-vinyl-acetate polymer by luminescence spectroscopy. *J. Polym. Res.* **2014**, *21*, 457. [CrossRef]
46. Peike, C.; Purschke, L.; Weiß, K.A.; Köhl, M.; Kempe, M. Towards the origin of photochemical EVA discoloration. In Proceedings of the 39th IEEE Photovoltaic Specialists Conference, PVSC 2013, Tampa, FL, USA, 16–21 June 2013; pp. 1579–1584. [CrossRef]
47. Schlothauer, J.C.; Grabmayer, K.; Hintersteiner, I.; Wallner, G.M.; Röder, B. Non-destructive 2D-luminescence detection of EVA in aged PV modules: Correlation to calorimetric properties, additive distribution and a clue to aging parameters. *Sol. Energy Mater. Sol. Cells* **2017**, *159*, 307–317. [CrossRef]
48. Beinert, A.; Peike, C.; Dürr, I.; Kempe, M.D.; Reiter, G.; Weiß, K.A. The influence of the additive composition on degradation induced changes in Polyethylene-co-vinyl-acetate during photochemical aging. In Proceedings of the 29th European PVSEC, 5BV.4.6, Amsterdam, The Netherlands, 22–26 September 2014; pp. 3126–3132. [CrossRef]
49. Schlothauer, J.C.; Grabmayer, K.; Wallner, G.M.; Röder, B. Correlations of spatially resolved photoluminescence and viscoelastic mechanical properties of encapsulating EVA in differently aged PV modules. *Prog. Photovolt.* **2016**, *24*, 855–870. [CrossRef]
50. Schlothauer, J.; Jungwirth, S.; Köhl, M.; Röder, B. Degradation of the encapsulant polymer in outdoor weathered photovoltaic modules: Spatially resolved inspection of EVA ageing by fluorescence and correlation to electroluminescence. *Sol. Energy Mater. Sol. Cells* **2012**, *102*, 75–85. [CrossRef]
51. Eder, G.C.; Voronko, Y.; Grillberger, P.; Kubicek, B.; Knöblm, K. UV-Fluorescence measurements as tool for the detection of degradation effects in PV-modules. In Proceedings of the 8th European Weathering Symposium, Vienna, Austria, 20–22 September 2017; Gesellschaft für Umweltsimulation e.V. GUS: Pfinztal, Germany, 2017.
52. Peike, C.; Kaltenbach, T.; Weiß, K.A.; Koehl, M. Indoor vs. outdoor aging—Polymer degradation in PV modules investigated by Raman spectroscopy. In *Reliability of Photovoltaic Cells, Modules, Components,*

*and Systems V*; SPIE—The International Society for Optical Engineering: Bellingham, WA, USA, 2012; Volume 8472. [CrossRef]

53. Röder, B.; Schlothauer, J.; Köhl, M. Luminescence spectroscopy as powerful tool for non destructive inspection of PV module encapsulants. In Proceedings of the 5th Sophia Workshop PV Module Reliability, Loughborough, UK, 16–17 April 2015.

*Article*

# A Novel Improved Cuckoo Search Algorithm for Parameter Estimation of Photovoltaic (PV) Models

Tong Kang [1,*], Jiangang Yao [1], Min Jin [2], Shengjie Yang [3] and ThanhLong Duong [4]

[1] College of Electrical and Information Engineering, Hunan University, Changsha 410082, China; yaojiangang@126.com
[2] College of Computer Science and Electronic Engineering, Hunan University, Changsha 410082, China; jinmin@hnu.edu.cn
[3] College of Computer and Information Engineering, Hunan University of Commerce, Changsha 410205, China; yangsj16@hotmail.com
[4] Department of Electrical Engineering, Industrial University of Ho Chi Minh City, Ho Chi Minh City 700000, Vietnam; thanhlong802003@yahoo.com
* Correspondence: kangtong126@126.com; Tel.: +86-731-8882-4089

Received: 30 March 2018; Accepted: 23 April 2018; Published: 25 April 2018

**Abstract:** Parameter estimation of photovoltaic (PV) models from experimental current versus voltage (I-V) characteristic curves acts a pivotal part in the modeling a PV system and optimizing its performance. Although many methods have been proposed for solving this PV model parameter estimation problem, it is still challenging to determine highly accurate and reliable solutions. In this paper, this problem is firstly transformed into an optimization problem, and an objective function (OF) is formulated to quantify the overall difference between the experimental and simulated current data. And then, to enhance the performance of original cuckoo search algorithm (CSA), a novel improved cuckoo search algorithm (ImCSA) is proposed, by combining three strategies with CSA. In ImCSA, a quasi-opposition based learning (QOBL) scheme is employed in the population initialization step of CSA. Moreover, a dynamic adaptation strategy is developed and introduced for the step size without Lévy flight step in original CSA. A dynamic adjustment mechanism for the fraction probability ($P_a$) is proposed to achieve better tradeoff between the exploration and exploitation to increase searching ability. Afterwards, the proposed ImCSA is used for solving the problem of estimating parameters of PV models based on experimental I-V data. Finally, the proposed ImCSA has been demonstrated on the parameter identification of various PV models, i.e., single diode model (SDM), double diode model (DDM) and PV module model (PMM). Experimental results indicate that the proposed ImCSA outperforms the original CSA and its superior performance in comparison with other state-of-the-art algorithms, and they also show that our proposed ImCSA is capable of finding the best values of parameters for the PV models in such effective way for giving the best possible approximation to the experimental I-V data of real PV cells and modules. Therefore, the proposed ImCSA can be considered as a promising alternative to accurately and reliably estimate parameters of PV models.

**Keywords:** photovoltaic modeling; parameter estimation; optimization problem; metaheuristic; opposition-based learning; quasi-opposition based learning; improved cuckoo search algorithm

---

## 1. Introduction

In recent years, several reasons such as gradually depleting fossil fuel resources, environmental protection concerns, and political issues have resulted in a high demand for electrical energy [1]. Thus, the conflict between the vigorously increasing power demands and scarcity of fossil resource is becoming more and more serious, promoting the development of renewable energy resources, especially solar energy [2,3]. Since solar energy is emission-free, freely available, and easy to install, the use of solar energy via photovoltaic (PV) systems has attracted great attention all over the world [4,5]. Lately reported by the Photovoltaic Power Systems Programme of the International Energy Agency (IEA PVPS) [6], the global solar PV capacity at the end of 2016 amounted to about 300 GW, with a 50% growth bringing the additional installed solar PV capacity worldwide to at least 75 GW. Three countries, namely China, USA and Japan represented the largest solar PV markets in 2015 as well as 2016, in which there was a 75% increase in newly installed solar PV capacity. Meanwhile, the Asia Pacific region installed more than 66% of the global solar PV capacity in 2016, where China (with at least 34 GW installed) ranked first. Many countries were increasing their installed PV capacity during 2016, which is still going on. With dozens of countries developing solar PV now, and much more to come, the globalization of PV is now a reality. So far, no other single energy technology has shown such a distributed set-up and modularity as PV systems [7]. However, in PV systems, solar PV cells or modules are applied for harnessing the Sun's energy and turn it into electricity. In particular the solar PV cell/module is the most important part of a PV system [8]. Therefore, with regard to the modeling a PV system and optimizing its performance, an accurate modeling of PV cells or modules is necessary.

The modeling of PV cells or modules consists of three major processes: choice of proper electrical circuit models, the expression of mathematical models and precise estimation of values of parameters for them. Although various equivalent electronic circuit models were proposed years ago, in practice, the SDM and the DDM are two most commonly adopted models [9–11]. For the mathematical model, the I-V characteristic that describes PV cell/module behavior is taken into account, and the current equation of PV model is an implicit transcendental equation [10]. Therefore, under the circumstances, a precise parameter estimation of such models is extremely essential and hard work and has drawn much attention recently [11].

Various approaches have been proposed for solving the PV models parameter identification problem, mainly classified into three categories: analytical methods, numerical methods and metaheuristic methods. In analytical methods, the Lambert W-function- based method was applied for estimating solar cells' parameters in [12]. In [13], a novel technique based on Taylor's series expansion was presented to obtain the explicit single-diode model of solar cells. Although analytical methods are simple and can provide rapid solution, they are not flexible and especially, making approximations in them often reduces accuracy. In numerical methods, the Newton-based method was proposed to obtain the parameters of solar cell [14]. In [15,16], the Gauss–Seidel-based method was used to identify the parameters for a SDM of a PV module. The Levenberg-Marquardt (LM) algorithm was employed for estimating five parameters of the SDM of PV modules in [17]. Although numerical methods can offer accurate results, their accuracy relies on the selection of the initial values. Moreover, they may easily trap into local optima. In [18], a new strategy based on the reduced forms of the five-parameter model was proposed for solving the problem of identification of the five unknown parameters from the experimental I-V data of the PV panel. Using the reduced forms, the dimension of the search space can be reduced from five unknown parameters to two. Moreover, the original nonconvex optimization problem can be transformed into a convex optimization problem and any kind of deterministic approach can easily and efficiently find the solution. The capabilities of the proposed reduced forms were verified on two case studies. Comparison results showed the high performances of the novel techniques based on reduced forms. The metaheuristic methods have been widely used for the PV models parameter estimation problem [19–34]. Such methods include genetic algorithm (GA) [19], chaos particle swarm optimization (CPSO) [20], pattern search (PS) [21,22], simulated annealing (SA) [23], harmony search (HS) [1], artificial bee swarm optimization (ABSO) [24], $R_{cr}$-IJADE [25], mutative-scale parallel chaos optimization algorithm (MPCOA) [26], biogeography-based optimization algorithm with mutation strategies (BBO-M) [27], artificial bee colony (ABC) [2], modified artificial bee colony (MABC) [28], improved artificial bee colony (IABC) [29], chaotic asexual reproduction optimization (CARO) [4], EHA-NMS [30], generalized oppositional teaching learning based optimization (GOTLBO) [10], self-adaptive teaching-learning-based optimization (SATLBO) [31], improved JAYA (IJAYA) [32], modified simplified swarm optimization (MSSO) [33], chaotic improved artificial bee colony (CIABC) [11], and teaching-learning-based artificial bee colony (TLABC) [34]. These metaheuristic methods are very flexible and can achieve satisfied results, however, in the light of "no free lunch" (NFL) theorem, there is no single metaheuristic method best suited for all optimization problems [35]. That is to say, a particular algorithm provides best results for a set of problem, while the same algorithm may give the worst performance on a different set of problems. Therefore, searching for the new and most accurate and reliable metaheuristic method for solving PV models parameter estimation problem is still ongoing and always appreciated.

Recently, a new metaheuristic algorithm called cuckoo search algorithm (CSA) is developed by Yang and Deb [36] inspired from the obligate brood parasitic behavior of some cuckoo species and some birds' Lévy flight characteristic. It has a simple structure, a few control parameters and is easy for users to implement [37]. The CSA uses a control parameter called fraction probability or discovery rate, $P_a$ to balance the global exploration and local exploitation [38]. Thus, the CSA has attracted great attention of researchers and been successfully employed in various problems from different fields [38,39] compared with a variety of optimization algorithms. However, the original CSA suffers from some drawbacks, which have been improved in this study. Firstly, the CSA uses random initialization cuckoo population of host nests, which decreases the global exploration ability, and causes the convergence of original CSA to deteriorate and results in being easy to trap into local optimum, especially when tackling the problem of dimensional increasing. Secondly, the Lévy flight step size in original CSA needs initializing fixed value for both step size scaling factor, $\alpha$ and distribution factor, $\beta$ parameters, which cannot be amended in the next iterations. It is important but difficult to tune proper values of such parameters of the Lévy flight step size for the provided problems. In addition, no strategy is used to control over the step size during the process of iteration while obtaining global optimization in original CSA. Thirdly, the original CSA uses fixed value for fraction probability, $P_a$. Thus, an ideal value of $P_a$ needs to be carefully tuned for a given problem, which is not trouble free. The fixed value of such parameter still lacks an appropriate balance between the global search ability and local search capability of original CSA. Hence, it is necessary to overcome these drawbacks and enhance the performance of the original CSA.

Opposition-based learning (OBL) recently introduced by Tizhoosh [40], is a new scheme for machine intelligence and applied for speeding up various optimization algorithms' convergence and improving the accuracy of their solutions [41], which has attracted a lot of research attention in recent years [42]. The major concept of OBL is the simultaneous consideration of a guess and its corresponding opposite guess which is closer to the global optimum for finding out a better candidate answer to given problems. Nevertheless, recently, researchers introduced the QOBL and established that a quasi-opposite number is more likely to be closer to the solution than an opposite number [43,44]. Thus, the idea of QOBL has also been successfully used to reinforce several global optimization algorithms like DE, GA, PSO, and BBO [42,44].

For addressing the aforementioned drawbacks of original CSA and improving its performance, in this paper, a novel improved cuckoo search algorithm (ImCSA) is proposed, by combining three strategies with original CSA. Firstly, a strategy called QOBL scheme is employed in the population initialization step of CSA to accelerate its convergence and enhance its solution accuracy. Secondly, a dynamic adaptation strategy is developed and introduced for the step size without Lévy flight step in original CSA, which makes the step size with zero parameter initialization adaptively change according to the individual nest's fitness value over the course of the iteration and the current iteration number. This strategy is useful for optimization with a faster rate. Thirdly, a dynamic adjustment mechanism for the fraction probability or discovery rate ($P_a$) is proposed for providing better tradeoff between the exploration and exploitation to increase searching ability. This paper focuses on the PV models parameter estimation problem. In this paper, this problem is firstly transformed into an optimization problem, and an OF is formulated to quantify the overall difference between the experimental and simulated current data. And then, a novel improved version of CSA called ImCSA is proposed and employed to solve the problem of estimating the parameters of PV models based on measured I-V data from the real PV cells/modules. Finally, the proposed ImCSA has been demonstrated on the various PV models, i.e., SDM, DDM and PMM. The main contributions of this article are summarized as follows:

- A new improved variant of CSA, known as ImCSA, is proposed for solving the PV models parameter estimation problem based on experimental I-V data.
- A novel improved CSA, named as ImCSA, by combining three strategies with original CSA to enhance its performance is proposed. First, a QOBL scheme is used in the population initialization step of original CSA. Then, a dynamic adaptation strategy is developed and introduced for the step size without Lévy flight step in original CSA. Finally, a dynamic adjustment mechanism for the fraction probability, $P_a$ is proposed to provide better balance between the global exploration and local exploitation to increase searching ability. The proposed ImCSA is a global optimization method and could be applied to other real-world problems.
- The proposed ImCSA is able to seek out the best parameter values for PV models in such effective way for giving the best possible approximation to the experimental I-V data of real PV cells and modules. Compared with original CSA and other different methods used in recent literature, the superior performance of the ImCSA is confirmed. Therefore, the proposed ImCSA can serve as a potential alternative to accurately and reliably identify PV models parameters.

The remainder of the article is arranged as follows: Section 2 introduces the PV models in this study. The proposed mathematical problem formulation for parameter estimation of PV models is also presented. The original CSA is given in Section 3. The proposed ImCSA and its application for the PV models parameter estimation problem were described in Section 4. Section 5 demonstrates the experimental results and discussion. Section 6 summarizes the conclusions.

## 2. Photovoltaic (PV) Modeling and Problem Formulation

This section firstly describes the modeling of PV cells and modules. Then, the objective function for the problem is detailed.

### 2.1. PV Cell Model

In the literature, various circuit models have been employed for describing the electrical behavior of PV cells, but in practice, only two widely used models, namely, SDM and DDM, are suitable for electrical engineering applications [4,11,24,26]. These two models will be concisely presented in the following subsections.

#### 2.1.1. Single Diode Model

The SDM is the most normally adopted in the researches for describing the static I-V characteristic of a PV cell due to its simplicity and accuracy [32]. The equivalent circuit of SDM is illustrated in Figure 1a. This model comprises a photo generated current source in parallel with a diode, a series resistor to denote the ohmic losses related to load current and a shunt resistor to present the leakage current. Thus, in term of Kirchhoff's current law (KCL), the PV cell terminal current, $I_t$, can be expressed by:

$$I_t = I_{ph} - I_d - I_{sh} \tag{1}$$

where $I_{ph}$ denotes the photo generated current, $I_d$ denotes the diode current, and $I_{sh}$ denotes the shunt resistor current, respectively. Additionally, in term of Shockley equation, $I_d$ is computed by:

$$I_d = I_{sd}[\exp(q(V_t + I_t R_s)/akT) - 1] \tag{2}$$

where $I_{sd}$ is the reverse saturation current of diode, $V_t$ is the cell terminal voltage, $R_s$ is the series resistance, $a$ is the diode ideality factor, $k$ is the Boltzmann constant ($1.380 \times 10^{-23}$ J/K), $q$ is the electronic charge ($1.602 \times 10^{-19}$ C), and $T$ is the PV cell absolute temperature in Kelvin, respectively. Moreover, using Kirchhoff's voltage law (KVL), $I_{sh}$ is obtained as:

$$I_{sh} = (V_t + I_t R_s)/R_{sh} \tag{3}$$

where $R_{sh}$ is the shunt resistance. Therefore, by substituting from Equations (2) and (3) into Equation (1), the I-V relationship of the SDM can be rewritten as follows [2,11]:

$$I_t = I_{ph} - I_{sd}[\exp{(q(V_t + I_t R_s)/akT)} - 1] - (V_t + I_t R_s)/R_{sh} \tag{4}$$

Consequently, for this SDM, there are five unknown parameters, namely, $I_{ph}$, $I_{sd}$, $a$, $R_s$, and $R_{sh}$ that can be estimated based on experimental I-V data. Accurate estimations of these parameters are vital to reflect the PV cell characteristics closer to the real characteristics, and this can be achieved by an optimization technique.

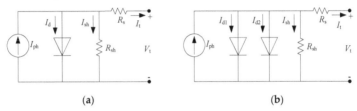

(a)                      (b)

**Figure 1.** Equivalent circuits of a PV cell: (**a**) The SDM; (**b**) The DDM.

### 2.1.2. Double Diode Model

The DDM is the second most widely used circuit model in practice for PV cells. Although the SDM is known to provide a satisfactory approximation to the characteristic of a practical PV cell, the effect of recombination current loss in the depletion region need to be taken into account for making the model more realistic and achieving higher degrees of accuracy. The equivalent circuit of DDM is shown in Figure 1b. This model includes two diodes in parallel with the photo generated current source, a series resistance and a shunt resistance. Hence, by applying KCL, $I_t$ can be expressed by:

$$I_t = I_{ph} - I_{d1} - I_{d2} - I_{sh} \tag{5}$$

where $I_{d1}$ denotes the first diode current, and $I_{d2}$ denotes the second diode current, respectively. In addition, according to the Shockley equation, $I_{d1}$ and $I_{d2}$ are given as follows:

$$I_{d1} = I_{sd1}[\exp{(q(V_t + I_t R_s)/a_1 kT)} - 1] \tag{6}$$

$$I_{d2} = I_{sd2}[\exp{(q(V_t + I_t R_s)/a_2 kT)} - 1] \tag{7}$$

where $I_{sd1}$ and $I_{sd2}$ represent the diffusion and saturation currents, respectively. $a_1$ and $a_2$ stand for the diode ideality factors. Thereby, like the SDM, the I-V relationship of the DDM is finally computed by [11,34]:

$$\begin{aligned} I_t = I_{ph} &- I_{sd1}[\exp{(q(V_t + I_t R_s)/a_1 kT)} - 1] \\ &- I_{sd2}[\exp{(q(V_t + I_t R_s)/a_2 kT)} - 1] \\ &- (V_t + I_t R_s)/R_{sh} \end{aligned} \tag{8}$$

Obviously, from Equation (8), seven unknown parameters, namely, $I_{ph}$, $I_{sd1}$, $I_{sd2}$, $a_1$, $a_2$, $R_s$, and $R_{sh}$ need to be identified based on the given I-V data from a real PV cell. Therefore, this is a crucial task in PV systems to accurately estimate such values of parameters for ensure a better performance of a practical PV cell.

*2.2. PV Module Model*

The PMM that comprises of several PV cells interconnected in series and/or in parallel to raise the level of output voltage and/or current [4,21–23,26]. The equivalent circuit model of a PV module (based on SDM) is depicted in Figure 2. Therefore, the Equation (4) of SDM is directly employed to express the I-V relationship of a PMM as follows:

$$I_t = I_{ph}N_p - I_{sd}N_p\left[\exp\left(q(V_t + I_tR_sN_s/N_p)/aN_skT\right) - 1\right] - (V_t + I_tR_sN_s/N_p)/R_{sh}N_s/N_p \quad (9)$$

where $N_s$ and $N_p$ denote the number of PV cells in series and parallel, respectively.

Considering the concision, Equation (9) is also rewritten as:

$$I_t = I_{phm} - I_{sdm}\left[\exp\left(q(V_t + I_tR_{sm})/a_mkT\right) - 1\right] - (V_t + I_tR_{sm})/R_{shm} \quad (10)$$

where $I_{phm} = I_{ph}N_p$, $I_{sdm} = I_{sd}N_p$, $a_m = aN_s$, $R_{sm} = R_sN_s/N_p$, and $R_{shm} = R_{sh}N_s/N_p$, respectively.

Considering this PV module model based on SDM, five unknown parameters, namely, $I_{phm}$, $I_{sdm}$, $a_m$, $R_{sm}$, and $R_{shm}$ must be estimated based on the given I-V data of real PV modules. Similarly, an accurate identification of these parameters is critical to optimizing the performance of a PV module.

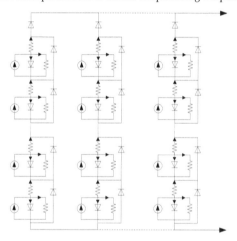

**Figure 2.** Equivalent circuit model of a PV module.

*2.3. Objective Function*

The main intention of mathematical modeling of PV models is to precisely estimate the values of unknown parameters that characterize several models, especially the aforementioned PV models such as SDM, DDM and PMM, based on measured I-V data from real PV cells and PV modules. However, estimation of the most optimal values of unknown parameters is a difficult and challenge problem since the characteristic current equations describing the PV models are implicit, nonlinear and transcendental [10]. Thus, this PV models parameter estimation problem can be transformed into an optimization problem, in which the aim is to minimize the difference between the experimental I-V data and the I-V data from model computed by taking into consideration a specific set of estimated parameters. This difference also called error function can be defined by rewriting the Equations (4), (8) and (10) in their homogeneous forms for SDM, DDM and PMM respectively as follows:

$$\begin{cases} e_{SDM}(V_t, I_t, \theta) = I_{ph} - I_{sd}\left[\exp\left(q(V_t + I_tR_s)/akT\right) - 1\right] - (V_t + I_tR_s)/R_{sh} - I_t \\ \theta = [I_{ph}, I_{sd}, a, R_s, R_{sh}] \end{cases} \quad (11)$$

$$\begin{cases} e_{DDM}(V_t, I_t, \theta) = I_{ph} - I_{sd1}[\exp(q(V_t + I_t R_s)/a_1 kT) - 1] - I_{sd2}[\exp(q(V_t + I_t R_s)/a_2 kT) - 1] \\ \qquad - (V_t + I_t R_s)/R_{sh} - I_t \\ \theta = [I_{ph}, I_{sd1}, I_{sd2}, a_1, a_2, R_s, R_{sh}] \end{cases} \tag{12}$$

$$\begin{cases} e_{PMM}(V_t, I_t, \theta) = I_{phm} - I_{sdm}[\exp(q(V_t + I_t R_{sm})/a_m kT) - 1] - (V_t + I_t R_{sm})/R_{shm} - I_t \\ \theta = [I_{phm}, I_{sdm}, a_m, R_{sm}, R_{shm}] \end{cases} \tag{13}$$

where $e(V_t, I_t, \theta)$ is the error function which means the difference between the simulated current using model determined by estimated parameters and experimental current from a PV cell and module, and computed for each pair of the measured data. $\theta$ is the solution vector which includes the several unknown parameters of PV models to be identified, where $\theta = [I_{ph}, I_{sd}, a, R_s, R_{sh}]$ is for the SDM, $\theta = [I_{ph}, I_{sd1}, I_{sd2}, a_1, a_2, R_s, R_{sh}]$ is for the DDM and $\theta = [I_{phm}, I_{sdm}, a_m, R_{sm}, R_{shm}]$ is for the PMM, respectively.

Hence, considering that defining an OF is necessary for the optimization problem, we adopt the root mean square error (RMSE) as the OF in our study to quantify the overall difference between the simulated and experimental current data. And this OF has been widely used in the literature [1,2,4,10,11,29], which is formulated as follows:

$$\text{Min OF}(\theta) = \text{Min RMSE}(\theta) = \text{Min} \sqrt{\frac{1}{N} \sum_{i=1}^{N} (e_i(V_t, I_t, \theta))^2} \tag{14}$$

where $N$ is the number of measured data points.

Therefore, in our study, the PV models parameter estimation is an optimization process that minimizes the OF($\theta$) by successively regulating the model parameters solution vector $\theta$ within the specified search interval. Obviously, the smaller value of the OF, the better the solution is and that is, the more precise the parameter values estimated from the model. Moreover, it is significant that any decrease occurs in the OF value, representing an improvement in the knowledge about the real values of the parameters [24].

## 3. The Original CSA

The CSA is a recent metaheuristic algorithm proposed by Yang and Deb [36]. The main idea behind CSA is the combination of the cuckoo bird's obligate brood parasitic behaviour and some insects' Lévy flights characteristics. To simply give a description of the original CSA, cuckoo search is based on the following three idealized rules [36,38,39]:

- One egg is laid by each cuckoo at a time and dumps its egg into any chosen nest randomly.
- Nests with the best quality eggs are maintained to the forthcoming generations.
- The fraction probability ($P_a$) of the host birds discovering cuckoo's egg lies within probability range $P_a \in [0, 1]$. The available host nest is fixed.

Combining cuckoo search based on three idealized rules with Lévy flight phenomenon, the CSA can be easily formed. In CSA, a fraction probability or discovery rate, $P_a$ is used to balance the global search ability and local search capability. The local search can be expressed by:

$$X_i^{t+1} = X_i^t + \alpha s \otimes H(P_a - \varepsilon) \otimes (X_j^t - X_k^t) \tag{15}$$

where $X_i$, $X_j$ and $X_k$ are three different solutions; $\alpha > 0$ is the step size scaling factor; $s$ is the step size; $\otimes$ means entry-wise multiplications; $H(\cdot)$ is a Heaviside function and $\varepsilon$ is a random number uniformly distributed. On the other hand, the global search is conducted by using Lévy flights as:

$$X_i^{t+1} = X_i^t + \alpha L(s, \lambda), \qquad L(s, \lambda) = \frac{\lambda \Gamma(\lambda) \sin(\pi \lambda / 2)}{\pi} \frac{1}{s^{1+\lambda}}, \qquad (s \gg s_0 > 0) \tag{16}$$

where $\Gamma(\cdot)$ is a Gamma function and expressed by:

$$\Gamma(z) = \int_0^\infty t^{z-1} e^{-t} dt \tag{17}$$

in a special case when $z = n$ is an integer, then we have $\Gamma(n) = (n-1)!$.

The Lévy flight essentially provides a random walk whose random step length is drawn from a Lévy distribution as:

$$Le'vy \sim \frac{1}{s^{\lambda+1}}, \quad (0 < \lambda \le 2) \tag{18}$$

which has an infinite variance with an infinite mean.

In Mantegna's algorithm, the step length $s$ is calculated as [39]:

$$s = \frac{u}{|v|^{\frac{1}{\beta}}} \tag{19}$$

where $u$ and $v$ are normally distributed stochastic variables as:

$$\begin{cases} u \sim N(0, \sigma_u^2) \\ v \sim N(0, \sigma_v^2) \end{cases} \tag{20}$$

and $\sigma_u$, $\sigma_v$ represent the standard deviations and are given by:

$$\begin{cases} \sigma_u = \left\{ \dfrac{\Gamma(1+\beta)\sin(\pi\beta/2)}{\Gamma[(1+\beta)/2]\beta 2^{(\beta-1)/2}} \right\}^{1/\beta} \\ \sigma_v = 1 \end{cases} \tag{21}$$

where $\beta$ is the distribution factor ($0.3 \le \beta \le 1.99$).

Hence, the pseudocode of the original CSA is presented in Algorithm 1.

---

**Algorithm 1: Pseudocode of the original CSA**

---

1.     Randomly initialize $n$ host nests within specified range as $\theta_i$ ($i = 1, \dots, n$)
2.     Compute fitness value $f_i$ ($i = 1, \dots, n$)
3.     Determine the global best nest with the best fitness value
4.     for $It = 1:It_{max}$
5.         Randomly generate a new solution (say $C_i$) using Lévy flights
6.         Compute its fitness value $f_{tr}$
7.         Randomly choose a solution (say $\theta_m$) from current $n$ solutions
8.         if ($f_{tr} < f_m$) then
9.             $\theta_m = C_i$
10.        $f_m = f_{tr}$
11.       end if
12.       Drop several worst nests via probability ($P_a$) and build new ones
13.       Keep the best solutions
14.       Rank and seek out the current global best nest
15.     end for
16.   Postprocess results and visualization

---

The major procedure of original CSA can be presented as follows:

1.     Randomly initialize $n$ host nests within specified range:

$$\theta_i = (\theta_{i1}, \theta_{i2}, \cdots, \theta_{ij})^T \quad i = 1, 2, \cdots, n \quad j = 1, 2, \cdots, d \tag{22}$$

where $\theta_i$ denotes the *i*th nest; $\theta_{ij}$ denotes the *j*th element of the *i*th nest; *d* denotes the dimension. Set the value of discovery rate $P_a \in [0, 1]$. Set the maximum number of iterations $It_{max}$.

2.  Compute fitness value $f_i$ ($i = 1, \ldots, n$), select the best value of each nest $\theta best_i$ and the global best nest *Gbest*, memorize fitness values and the best fitness value.

3.  Randomly generate a new solution using Lévy flights. As aforementioned, the new solution is given by:

$$\theta_i^{new} = \theta best_i + rand_1 \times S_i^{new} \times (\theta best_i - Gbest) \tag{23}$$

where $rand_1$ is a random number drawn from a normal distribution and the step size $S_i^{new}$ is determined by:

$$S_i^{new} = \alpha \times \frac{u}{|v|^{1/\beta}} \tag{24}$$

where $\alpha$ is the step size scaling factor and set to 0.01; $\beta$ is the distribution factor and set to 1.5; *u* and *v* are two normally distributed stochastic variables ($u \sim N(0, \sigma_u^2)$ and $v \sim N(0, \sigma_v^2)$) with respective the $\sigma_u$ and $\sigma_v$ aforementioned in (21).

4.  Compute the fitness values of the new solutions, decide the newly $\theta best_i$ and *Gbest* via comparing the memorized fitness values in Step 2 with newly computed ones, update $\theta best_i$ and *Gbest*, and memorize fitness values and the best fitness value.

5.  Drop several worst nests via probability ($P_a$) and build new solution. Due to this action, the new solution can be calculated by:

$$\theta_i^{disc} = \theta best_i + C \times \Delta\theta_i^{disc} \tag{25}$$

where *C* is the updated coefficient resolved by $P_a$ and given by:

$$C = \begin{cases} 1 & if \ rand_2 < P_a \\ 0 & otherwise \end{cases} \tag{26}$$

and the increased value $\Delta\theta_i^{disc}$ is computed by:

$$\Delta\theta_i^{disc} = rand_3 \times [randp_1(\theta best_i) - randp_2(\theta best_i)] \tag{27}$$

where $rand_2$ and $rand_3$ are random numbers drawn from normal distributions; $randp_1(\theta best_i)$ and $randp_2(\theta best_i)$ are the random perturbation for positions of nests in $\theta best_i$.

6.  Compute the fitness values of the new solutions, decide the newly $\theta best_i$ and *Gbest* via comparing the computed fitness values of these new solutions with memorized fitness values in Step 4, update $\theta best_i$ and *Gbest*, memorize fitness values and the best fitness value.

7.  If the predefined maximum number of iterations $It_{max}$ is reached, stop the calculation and display the results, else go to Step 3.

## 4. The Proposed Novel Improved Cuckoo Search Algorithm (ImCSA) and Its Application

In this section, the novel improved cuckoo search algorithm (ImCSA) is firstly proposed, by combining three strategies with CSA to enhance the performance of the original CSA. Then, we present the procedure of employing the proposed ImCSA to solve the problem of PV models parameter estimation.

### 4.1. Proposed ImCSA

The ImCSA is proposed in this subsection. Three main strategies as improvements of the original CSA exist in the ImCSA. First, a QOBL scheme is employed in the population initialization step of CSA to accelerate its convergence and enhance its solution accuracy. Second, a dynamic adaptation strategy is developed and introduced for the step size without Lévy flight step in original CSA, which makes the step size with zero parameter initialization adaptively change according to the individual nest's fitness value over the course of the iteration and the current iteration number. This strategy is useful for optimization with a faster rate. Third, a dynamic adjustment mechanism for the fraction probability or discovery rate ($P_a$) is proposed to provide better tradeoff between the exploration and exploitation to increase searching ability. These three main strategies in the ImCSA are elucidated in the following subsections and the implementation of the proposed ImCSA is finally described.

### 4.1.1. Quasi-Opposition Based Learning Scheme for the Population Initialization

As mentioned in Section 3, the original CSA adopts random initialization cuckoo population of host nests. This random initialization population method decreases the global exploration ability, which causes the convergence of original CSA to deteriorate and results in being easy to fall into local optimal solution. Here, to overcome this drawback, a strategy called QOBL scheme is introduced to accelerate convergence rate and enhance the solutions quality of CSA.

The OBL recently introduced by Tizhoosh [40], is a new scheme for machine intelligence and applied for speeding up various optimization algorithms' convergence and improving the accuracy of their solutions [41]. The major concept of OBL is the simultaneous consideration of a guess and its corresponding opposite guess which is closer to the global optimum for finding out a better candidate answer to given problems.

In general, all population-based optimization algorithms start with some initial solutions and try to improve them toward some optimal solution(s). The process of searching stops when several predefined criteria are satisfied. We usually start with random estimations for the absence of a priori knowledge or information about the solution. Researchers have established that an opposite candidate solution has a higher probability of being closer to the global optimum than a random candidate solution [41]. Hence, starting with the closer of the two guesses has the potential to speed up convergence and improve solution's accuracy. Recently, researchers introduced QOBL [43,44] and established that a quasi-opposite number is more likely to be closer to the solution than an opposite number.

In order to easily explain OBL and QOBL, we need to define some concepts clearly. The opposite number and opposite point adopted for OBL are defined by [41]:

1.  Opposite number: Let $X \in R$ be a real number defined on a certain interval: $X \in [a, b]$. The opposite number $\breve{X}^o$ is defined by:

$$\breve{X}^o = a + b - X \tag{28}$$

2.  Opposite point: Let $P = (X_1, X_2, \cdots, X_n)$ be a point in n-dimensional space, where $X_1, X_2, \cdots, X_n \in R$ and $X_i \in [a_i, b_i] \forall i \in \{1, 2, \cdots, n\}$. The opposite point $\breve{P}^o = (\breve{X}^o_1, \breve{X}^o_2, \cdots, \breve{X}^o_n)$ is completely defined by its components $\breve{X}^o_1, \breve{X}^o_2, \cdots, \breve{X}^o_n$ where:

$$\breve{X}^o_i = a_i + b_i - X_i \tag{29}$$

Here, the quasi-opposite number and quasi-opposite point adopted for QOBL are defined by [43]:

1.  Quasi-opposite number: Let $X \in R$ be a real number defined on a certain interval: $X \in [a, b]$. The quasi-opposite number $\breve{X}^{qo}$ is defined by:

$$\breve{X}^{qo} = rand((a+b)/2, \breve{X}^o) \tag{30}$$

where $\breve{X}^o$ is the opposite number of $X$; $rand((a+b)/2, \breve{X}^o)$ is a random number uniformly distributed between $(a+b)/2$ and $\breve{X}^o$.

2. Quasi-opposite point: Let $P = (X_1, X_2, \cdots, X_n)$ be a point in n-dimensional space, where $X_1, X_2, \cdots, X_n \in R$ and $X_i \in [a_i, b_i] \forall i \in \{1, 2, \cdots, n\}$. The quasi-opposite point $\overset{\smile qo}{P} = (\overset{\smile qo}{X}_1, \overset{\smile qo}{X}_2, \cdots, \overset{\smile qo}{X}_n)$ is completely defined by its components $\overset{\smile qo}{X}_1, \overset{\smile qo}{X}_2, \cdots, \overset{\smile qo}{X}_n$ where:

$$\overset{\smile qo}{X} = rand((a+b)/2, \overset{\smile 0}{X})\qquad (31)$$

where $\overset{\smile 0}{X}_i$ is the opposite point of $X_i$; $rand((a_i + b_i)/2, \overset{\smile 0}{X}_i)$ is a random point uniformly distributed between $(a+b)/2$ and $\overset{\smile 0}{X}_i$.

Overall, in our paper, for improving the performance of original CSA, the QOBL scheme is chosen and employed in the population initialization step of the original CSA. By considering a guess and its corresponding quasi-opposite guess simultaneously, the QOBL scheme leads to searching of search space more thoroughly, which can provide a faster rate of convergence and a higher probability of seeking candidate solutions closer to the global optimum.

### 4.1.2. Dynamic Adaptation Strategy for the Step Size

Accordingly, in CSA, the global exploration phase for generation of new eggs is governed by Lévy flight based random walks and one has to define the Lévy flight step size. However, in the literature [38,39], the Lévy flight step size needs initializing fixed value for both step size scaling factor, $\alpha$ and distribution factor, $\beta$ parameters, which cannot be amended in the next iterations. Moreover, the characteristics of the next generation nests are decided by step size scaling factor, $\alpha$ and fraction probability, $P_a$ in original CSA. On one hand, if the fixed value of $\alpha$ is set too large, the iterations of algorithm will considerably increase while the rate of convergence cannot be guaranteed. Consequently, the host nest will fly beyond boundaries, out of search space, which will affect the accuracy of solution. On the other hand, though a small value of $\alpha$ leads to a high speed convergence rate, it may be unable to seek out global optimum.

Hence, it is crucial and difficult to choose an appropriate value of the step size scaling factor $\alpha$ of the Lévy flight step size for a given problem. Additionally, there is no strategy to control over the step size during the process of iteration while obtaining global optimization by using an original CSA. In order to overcome these drawbacks, we ignore the parameters. Here, a dynamic adaptation strategy is developed and introduced for the step size without Lévy flight step in original CSA. In this sense, the step size $S_i^{new}$ can be modeled as follows:

$$S_i^{new} = \left(\frac{1}{It}\right)^{\left|\frac{Bestf(It)-f_i(It)}{Bestf(It)-Worstf(It)}\right|}\qquad (32)$$

where $It$ is the current iteration number; $Bestf(It)$ is the best fitness value in the iteration $It$; $f_i(It)$ is the fitness value of $i$th nest in the iteration $It$; $Wortf(It)$ is the worst fitness value in the iteration $It$.

Quite evidently, as can be seen from Equation (32), the step size is now with zero parameter to be initialized, which not only relies on the current iteration number but also relies on the fitness value of individual nest in the search space. It is obvious that the step size is automatically determined during the iterative search process of the algorithm and adaptively changed according to the individual nest's fitness value over the course of the iteration and the current iteration number. Therefore, though the step size is large at the beginning, when the number of the iteration increases, the step size decreases. That is to say, the step size is very small, when the algorithm reaches to the global optimum. Thus, in our study, the dynamic adaptation strategy for the step size without Lévy flight step in original CSA has been investigated and is beneficial to optimization with a faster rate and higher quality solutions.

### 4.1.3. Dynamic Adjustment Mechanism for the Fraction Probability

As a matter of fact, considering the search process, the original CSA uses a combination of global explorative Lévy flight based random walk and local exploitative random walk which is controlled by fraction probability or discovery rate, $P_a$. From the viewpoint of fraction probability $P_a$, the large value of $P_a$ leads to increase the diversity of solutions and inhibit premature convergence, while the small value of $P_a$ will increase search accuracy but slow down the search process.

However, the original CSA uses fixed value for fraction probability, $P_a$. Thus, an ideal value of $P_a$ needs to be carefully tuned for a given problem, which is not trouble free. The fixed value of $P_a$ still lacks an appropriate balance between the global search ability and local search capability of original CSA. To overcome this problem and improve the search ability, in this paper, a dynamic adjustment mechanism is introduced into the original CSA to realize the dynamic control of the fraction probability or discovery rate, $P_a$, which is calculated as follows:

$$P_a = P_{a,max} - (P_{a,max} - P_{a,min}) \times \frac{It}{It_{max}} \tag{33}$$

where $P_{a,max}$ is the maximum fraction probability and equal to 0.25; $P_{a,min}$ is the minimum fraction probability and equal to 0.01; $It$ and $It_{max}$ are the current iteration number and the maximum number of iterations, respectively.

### 4.1.4. Implementation of the Proposed ImCSA

In this paper, for further enhancing the performance of CSA, a novel ImCSA is proposed based on three strategies detailed above. First, a QOBL scheme is used in the population initialization step of original CSA. Then, a dynamic adaptation strategy is developed and introduced for the step size without Lévy flight step in original CSA. Finally, a dynamic adjustment mechanism for the fraction probability, $P_a$ is proposed to achieve better tradeoff between the global exploration and local exploitation to increase searching ability. In addition, the proposed ImCSA has a simple structure and is thus easy for user to implement, which is the same as that of original CSA. The implementation processes of the proposed ImCSA can be presented as the pseudocode listed in Algorithm 2. Newly added/extended code segments are highlighted in bold.

---

**Algorithm 2: Pseudocode of the proposed ImCSA**

---

**/\* QOBL scheme for the population initialization \*/**
**1. Generate uniformly distributed initial $n$ host nests $N_0$**
**2. for $i$ = 1:$n$ //$n$: Host nests size**
**3.     for $j$ = 1:$d$ //$d$: Problem dimension**
**4.         $ON_{0i,j} = a_j + b_j - N_{0i,j}$ //$ON_0$: Opposite of initial host nests $N_0$; $[a_j, b_j]$: Range of the $j$th variable**
**5.         $M_{i,j} = (a_j + b_j)/2$ //$M_{i,j}$: Middle point**
**6.         if ($N_{0i,j} < M_{i,j}$)**
**7.             $QON_{0i,j} = M_{i,j} + (ON_{0i,j} - M_{i,j}) \times rand(0,1)$ //$QON_0$: Quasi-opposite of initial host nests $N_0$**
**//$rand(0,1)$: A random number uniformly generated**

---

| | |
|---|---|
| 8. | else |
| 9. | $QON_{0i,j} = ON_{0i,j} + (M_{i,j} - ON_{0i,j}) \times rand(0,1)$ |
| 10. | end if |
| 11. | end for |

12. end for

13. Choose $n$ fittest nests from set of $\{N_0, QON_0\}$ as initial host nests $N_0$

/* End of QOBL scheme for the population initialization */

14. Compute fitness value $f_i$ $(i = 1, \dots, n)$

15. Determine the global best nest with the best fitness value

16. for $It = 1 : It_{max}$

/* Dynamic adaptation strategy for the step size */

| | |
|---|---|
| 17. | Find the best fitness value $Bestf(It)$ and the worst fitness value $Wortf(It)$ in the iteration $It$ |
| 18. | Randomly generate a new solution (say $C_i$) using Equations (23) and (32) |

/* End of dynamic adaptation strategy for the step size */

| | |
|---|---|
| 19. | Compute its fitness value $f_{tr}$ |
| 20. | Randomly choose a solution (say $\theta_m$) from current $n$ solutions |
| 21. | if $(f_{tr} < f_m)$ then |
| 22. | $\theta_m = C_i$ |
| 23. | $f_m = f_{tr}$ |
| 24. | end if |

/* Dynamic adjustment mechanism for the fraction probability */

| | |
|---|---|
| 25. | Calculate the dynamic adjustment fraction probability ($P_a$) using Equation (33) |

/* End of dynamic adjustment mechanism for the fraction probability */

| | |
|---|---|
| 26. | Drop several worst nests via probability ($P_a$) and build new ones |
| 27. | Keep the best solutions |
| 28. | Rank and seek out the current global best nest $Gbest$ |

29. end for

30. Postprocess results and visualization

## 4.2. Procedure of the Proposed ImCSA-based PV Models Parameter Estimation

This subsection describes the major procedures of employing the proposed ImCSA for solving the PV models parameter estimation problem based on experimental I-V data of real PV cells and modules. The successive steps can be detailed below:

1. Read the $N$ measured I-V data values of $V_t$ and $I_t$ and set associated parameters of the proposed ImCSA such as the host nests size $n$, the dynamic adjustment fraction probability $P_a$ amount within the domain of $[P_{a,min} \ P_{a,max}]$, the number of variables to be optimized $d$, and $It_{max}$.

2. Initialize $n$ host nests $N_0$ considering the variables to be optimized (the unknown parameters of solar cell models, where the parameters solution vector $\theta = [I_{ph}, I_{sd}, a, R_s, R_{sh}]$ is for the SDM, $\theta = [I_{ph}, I_{sd1}, I_{sd2}, a_1, a_2, R_s, R_{sh}]$ is for the DDM, and $\theta = [I_{phm}, I_{sdm}, a_m, R_{sm}, R_{shm}]$ is for the PMM, respectively). The solution vector $\theta$ is randomly generated within the specified range which is widely used in the literature [4,25,31,32,34].

3. Create quasi-opposite of initial host nests ($QON_0$) using Equation (31).

4. Evaluate the OF for the initial host nests ($N_0$) and quasi-opposite of initial host nests ($QON_0$) according to the OF($\theta$) in Equation (14).

5. Select $n$ (host nests size) fittest nests from the initial host nests ($N_0$) and quasi-opposite of initial host nests ($QON_0$) as initial host nests ($N_0$).

6.   Evaluate the OF values for *n* host nests $N_0$, select the best value of each nest $\theta best_i$ and the global best nest *Gbest* which is corresponding to the best $OF(\theta)$, memorize objective values and the best objective value.

7.   Find the best objective value $Bestf(It)$ and the worst objective value $Wortf(It)$ in the current iteration number *It*.

8.   Randomly generate a new solution using Equations (23) and (32)

9.   Compute the OF values of the new solutions, decide the newly $\theta best_i$ and *Gbest* via comparing the memorized objective values in Step 6 with newly computed ones, update $\theta best_i$ and *Gbest*, and memorize objective values and the best objective value.

10.  Calculate the dynamic adjustment fraction probability ($P_a$) using Equation (33)

11.  Drop several worst nests with a dynamic control of the fraction probability or discovery rate, $P_a$ and build new solution. Due to this action, the new solution can be calculated using Equations (25)–(27).

12.  Compute the OF values of the new solutions, decide the newly $\theta best_i$ and *Gbest* via comparing the computed $OF(\theta)$ in Equation (14) of these new solutions with memorized objective values in Step 9, update $\theta best_i$ and *Gbest*, memorize objective values and the best objective value.

13.  If the predefined maximum number of iterations $It_{max}$ is reached, terminate the computation and display the results (the best solution vector $\theta$ and the corresponding objective value OF (RMSE)), else go to Step 7.

The flowchart of the procedure of employing the proposed ImCSA for solving the PV models parameter estimation problem is depicted in Figure 3.

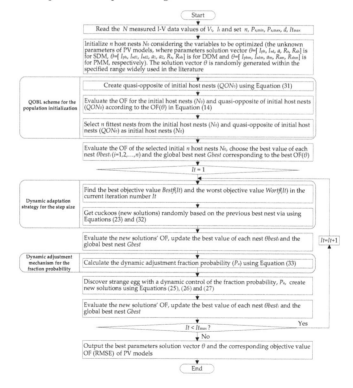

**Figure 3.** The flowchart of the procedure of employing the proposed ImCSA for solving the PV models parameter estimation problem.

## 5. Experimental Results and Discussion

This section is to fully evaluate the performance of proposed ImCSA for parameter estimation of various PV models, i.e., SDM, DDM and PMM. Two datasets of experiments, namely, benchmark datasets of a standard PV cell and a standard PV module, and real datasets of PV panels are used in the following subsections. First, the benchmark datasets of a standard PV cell and a standard PV module are chosen to verify the effectiveness of proposed ImCSA and compare with the results reported in literature. The benchmark datasets are acquired from [14], where the experimental I-V data are measured using a 57 mm diameter commercial RTC (the R.T.C. Company, Paris, France) France silicon solar cell (under a 1000 W/m$^2$ irradiance and 33 °C temperature) and a PV module named Photowatt-PWP201 module consisting of 36 polycrystalline silicon cells in series (under a 1000 W/m$^2$ at 45 °C). These two data sets of experimental I-V data have been widely used as the benchmark datasets to test and compare the performance of diverse methods [1,2,4,11,31,32,34] developed for parameter estimation of SDM, DDM and PMM. And then, in addition to the benchmark datasets, the real datasets of two recent reported PV panels are also chosen to further establish the ability of proposed ImCSA for parameter estimation under a real implementation. The real datasets of PV panels are gotten from [45], where the experimental I-V data of two PV panels, namely, polycrystalline STP6-120/36 panel and monocrystalline STM6-40/36 panel are measured by setting up a simple load scanning experiment. Both PV panels consist of 36 cells in series, while operating at 55 °C and 51 °C, respectively.

All the programs are executed using MATLAB in a computer with an Intel(R) Core(TM) i5-2415M @ 2.30 GHz CPU processor, 4 GB RAM and Windows 7 system. The parameters for the original CSA are set as follows: the population size $n = 25$, the fraction probability $P_a = 0.25$, the step size scaling factor $\alpha = 0.01$, the distribution factor $\beta = 1.5$. For the proposed ImCSA, the parameters are given by: the population size $n = 25$, the maximum and minimum fraction probability $P_{a,max}$ and $P_{a,min}$ are 0.25 and 0.01 respectively. The maximum number of iterations $It_{max}$ is set to 1500 for SDM, 8000 for DDM and 1000 for PMM. In addition, all experiments are performed for 30 independent runs and the best result is presented at each case.

### 5.1. Results on Benchmark Datasets

#### 5.1.1. Case Study 1: Single Diode Model

In this case, there are five unknown parameters that need to be estimated for the SDM. The range of each parameter used in the literature [1,2,4,11,24] are set as follows: $I_{ph}$ (A) $\in$ [0, 1], $I_{sd}$ ($\mu$A) $\in$ [0, 1], $a \in$ [1, 2], $R_s$ ($\Omega$) $\in$ [0, 0.5], $R_{sh}$ ($\Omega$) $\in$ [0, 100]. The experimental data measured from RTC France silicon solar cell at 33 °C contain 26 pairs of voltage and current values used the same as in the literature [1,2,4,11,14,24]. These data are cited to obtain the optimal parameters vector $\theta$ for the SDM of RTC France silicon solar cell by the proposed ImCSA.

Table 1 tabulates the statistics of the OF (RMSE) values for the SDM of RTC France silicon solar cell computed using the ImCSA and CSA. Table 1 shows that the ImCSA performs better than CSA in all terms of the best, mean, median, worst and standard deviation (Std) of the OF (RMSE) values in all 30 independent runs. Moreover, the best OF (RMSE) value quantifies the best accuracy, the mean OF (RMSE) value quantifies the average accuracy, and the standard deviation (Std) of the OF (RMSE) value indicates the reliability of the parameter estimation methods, respectively. From Table 1 it can be found that the ImCSA achieves the best, mean, median, and worst of the OF (RMSE) values as low as $9.860219 \times 10^{-4}$. Especially, the ImCSA obtains a Std of $2.987589 \times 10^{-12}$, which is obviously far better than that calculated by CSA as shown in Table 1. These results indicate that the proposed ImCSA really enhances the performance of original CSA and is more accurate and reliable than CSA. Furthermore, the convergence performance for the best run of the proposed ImCSA for parameter estimation of the SDM of RTC France silicon solar cell is represented in Figure 4. It can be seen from Figure 4 that the ImCSA fastly converges to a comparatively stable OF value in less than 300 iterations.

**Table 1.** Statistics of the OF (RMSE) values for the SDM of RTC France silicon solar cell using the proposed ImCSA and CSA.

| Algorithm | OF (RMSE) | | | | |
|---|---|---|---|---|---|
| | **Best** | **Mean** | **Median** | **Worst** | **Std** |
| ImCSA | $9.860219 \times 10^{-4}$ | $9.860219 \times 10^{-4}$ | $9.860219 \times 10^{-4}$ | $9.860219 \times 10^{-4}$ | $2.987589 \times 10^{-12}$ |
| CSA | $9.860227 \times 10^{-4}$ | $9.894848 \times 10^{-4}$ | $9.865435 \times 10^{-4}$ | $1.031010 \times 10^{-3}$ | $8.570571 \times 10^{-6}$ |

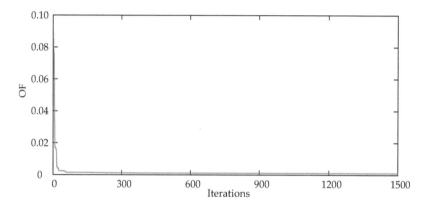

**Figure 4.** Convergence characteristic of the proposed ImCSA for parameter estimation of the SDM of RTC France silicon solar cell.

Table 2 summarizes the optimal parameters such as $I_{ph}$, $I_{sd}$, $a$, $R_s$, $R_{sh}$ values and the corresponding objective value of OF (RMSE) for the SDM achieved by the ImCSA compared with those by CSA and several other parameter estimation methods such as TLABC [34], CIABC [11], MSSO [33], IJAYA [32], SATLBO [31], GOTLBO [10], EHA-NMS [30], CARO [4], IABC [29], MABC [28], ABC [2], BBO-M [27], $R_{cr}$-IJADE [25], ABSO [24], HS [1], PS [21], CPSO [20], and GA [1]. These approaches are selected for comparison here due to their good performance in estimating parameters for the SDM of the PV cell reported in the recent literature. From the OF (RMSE) values in Table 2, it is apparent that the proposed ImCSA, together with the TLABC, CIABC, SATLBO, EHA-NMS, and $R_{cr}$-IJADE obtain the best OF (RMSE) value ($9.8602 \times 10^{-4}$), and CSA gets the second best OF (RMSE) value ($9.86023 \times 10^{-4}$), followed by IJAYA, MSSO, MABC, ABC, BBO-M, CARO, GOTLBO, ABSO, HS, IABC, CPSO, PS and GA, which indicates that the proposed ImCSA improves the performance of the original CSA. Consequently, the optimal parameters values sought out via the proposed ImCSA are closer to the real ones for the SDM of the solar cell, thus the parameters estimated by ImCSA are accurate.

To make a further investigation on the quality of the parameters estimated by the proposed ImCSA, these estimated parameters values of $I_{ph}$, $I_{sd}$, $a$, $R_s$ and $R_{sh}$ are put into the SDM in Equation (4) to reconstruct the calculated current data and calculated power data at experimental voltage point. The experimental data (voltage and current), the calculated data and the individual absolute error (IAE) between experimental and calculated data are listed in Table 3. Table 3 (columns 5 and 7) and the last line of Table 3 show that both the *IAE* and their sum are so small, which gives concrete evidence that the parameter values estimated by the ImCSA are very precise. The I-V and P-V (power versus voltage) characteristics of the best model parameters estimated by the ImCSA and the experimental data are illustrated in Figure 5. Figure 5 shows that the calculated data of SDM are in excellent accordance with the experimental data almost in all data points, which further demonstrates the optimal parameters values estimated by the ImCSA are very precise.

**Table 2.** Comparison among various parameter estimation algorithms for the SDM of RTC France silicon solar cell.

| Algorithm | $I_{ph}$ (A) | $I_{sd}$ (µA) | $a$ | $R_s$ (Ω) | $R_{sh}$ (Ω) | OF (RMSE) |
|---|---|---|---|---|---|---|
| ImCSA | 0.760776 | 0.323021 | 1.481718 | 0.036377 | 53.718524 | $9.8602 \times 10^{-4}$ |
| CSA | 0.760776 | 0.322821 | 1.481656 | 0.036380 | 53.696699 | $9.86023 \times 10^{-4}$ |
| TLABC [34] | 0.76078 | 0.32302 | 1.48118 | 0.03638 | 53.71636 | $9.8602 \times 10^{-4}$ |
| CIABC [11] | 0.760776 | 0.32302 | 1.48102 | 0.036377 | 53.71867 | $9.8602 \times 10^{-4}$ |
| MSSO [33] | 0.760777 | 0.323564 | 1.481244 | 0.036370 | 53.742465 | $9.8607 \times 10^{-4}$ |
| IJAYA [32] | 0.7608 | 0.3228 | 1.4811 | 0.0364 | 53.7595 | $9.8603 \times 10^{-4}$ |
| SATLBO [31] | 0.7608 | 0.32315 | 1.48123 | 0.03638 | 53.7256 | $9.8602 \times 10^{-4}$ |
| GOTLBO [10] | 0.760780 | 0.331552 | 1.483820 | 0.036265 | 54.115426 | $9.87442 \times 10^{-4}$ |
| EHA-NMS [30] | 0.760776 | 0.323021 | 1.481184 | 0.036377 | 53.718521 | $9.8602 \times 10^{-4}$ |
| CARO [4] | 0.76079 | 0.31724 | 1.48168 | 0.03644 | 53.0893 | $9.8665 \times 10^{-4}$ |
| IABC [29] | 0.7599 | 0.33243 | 1.4842 | 0.0363 | 54.4610 | $10.000 \times 10^{-4}$ |
| MABC [28] | 0.760779 | 0.321323 | 1.481385 | 0.036389 | 53.39999 | $9.861 \times 10^{-4}$ |
| ABC [2] | 0.7608 | 0.3251 | 1.4817 | 0.0364 | 53.6433 | $9.862 \times 10^{-4}$ |
| BBO-M [27] | 0.76078 | 0.31874 | 1.47984 | 0.03642 | 53.36227 | $9.8634 \times 10^{-4}$ |
| $R_{cr}$-IJADE [25] | 0.760776 | 0.323021 | 1.481184 | 0.036377 | 53.718526 | $9.8602 \times 10^{-4}$ |
| ABSO [24] | 0.76080 | 0.30623 | 1.47583 | 0.03659 | 52.2903 | $9.9124 \times 10^{-4}$ |
| HS [1] | 0.76070 | 0.30495 | 1.47538 | 0.03663 | 53.5946 | $9.9510 \times 10^{-4}$ |
| PS [21] | 0.7617 | 0.9980 | 1.6000 | 0.0313 | 64.1026 | $14.94 \times 10^{-3}$ |
| CPSO [20] | 0.7607 | 0.4000 | 1.5033 | 0.0354 | 59.012 | $1.39 \times 10^{-3}$ |
| GA [1] | 0.7619 | 0.8087 | 1.5751 | 0.0299 | 42.3729 | $19.08 \times 10^{-3}$ |

**Table 3.** The calculated results of the proposed ImCSA for the SDM of RTC France silicon solar cell.

| Item | Experimental Data | | Calculated Current Data | | Calculated Power Data | |
|---|---|---|---|---|---|---|
| | V (V) | I (A) | $I_{cal}$ (A) | IAE | $P_{cal}$ (W) | IAE |
| 1 | −0.2057 | 0.7640 | 0.76408764 | 0.00008764 | −0.15717283 | 0.00001803 |
| 2 | −0.1291 | 0.7620 | 0.76266264 | 0.00066264 | −0.09845975 | 0.00008555 |
| 3 | −0.0588 | 0.7605 | 0.76135473 | 0.00085473 | −0.04476766 | 0.00005026 |
| 4 | 0.0057 | 0.7605 | 0.76015423 | 0.00034577 | 0.00433288 | 0.00000197 |
| 5 | 0.0646 | 0.7600 | 0.75905585 | 0.00094415 | 0.04903501 | 0.00006099 |
| 6 | 0.1185 | 0.7590 | 0.75804301 | 0.00095699 | 0.08982810 | 0.00011340 |
| 7 | 0.1678 | 0.7570 | 0.75709159 | 0.00009159 | 0.12703997 | 0.00001537 |
| 8 | 0.2132 | 0.7570 | 0.75614207 | 0.00085793 | 0.16120949 | 0.00018291 |
| 9 | 0.2545 | 0.7555 | 0.75508732 | 0.00041268 | 0.19216972 | 0.00010503 |
| 10 | 0.2924 | 0.7540 | 0.75366447 | 0.00033553 | 0.22037149 | 0.00009811 |
| 11 | 0.3269 | 0.7505 | 0.75138806 | 0.00088806 | 0.24562876 | 0.00029031 |
| 12 | 0.3585 | 0.7465 | 0.74734834 | 0.00084834 | 0.26792438 | 0.00030413 |
| 13 | 0.3873 | 0.7385 | 0.74009688 | 0.00159688 | 0.28663952 | 0.00061847 |
| 14 | 0.4137 | 0.7280 | 0.72739678 | 0.00060322 | 0.30092405 | 0.00024955 |
| 15 | 0.4373 | 0.7065 | 0.70695327 | 0.00045327 | 0.30915067 | 0.00019822 |
| 16 | 0.4590 | 0.6755 | 0.67529489 | 0.00020511 | 0.30996036 | 0.00009414 |
| 17 | 0.4784 | 0.6320 | 0.63088431 | 0.00111569 | 0.30181505 | 0.00053375 |
| 18 | 0.4960 | 0.5730 | 0.57208207 | 0.00091793 | 0.28375271 | 0.00045529 |
| 19 | 0.5119 | 0.4990 | 0.49949164 | 0.00049164 | 0.25568977 | 0.00025167 |
| 20 | 0.5265 | 0.4130 | 0.41349356 | 0.00049356 | 0.21770436 | 0.00025986 |
| 21 | 0.5398 | 0.3165 | 0.31721950 | 0.00071950 | 0.17123509 | 0.00038839 |
| 22 | 0.5521 | 0.2120 | 0.21210317 | 0.00010317 | 0.11710216 | 0.00005696 |
| 23 | 0.5633 | 0.1035 | 0.10272135 | 0.00077865 | 0.05786294 | 0.00043861 |
| 24 | 0.5736 | −0.0100 | −0.00924885 | 0.00075115 | −0.00530514 | 0.00043086 |
| 25 | 0.5833 | −0.1230 | −0.12438136 | 0.00138136 | −0.07255165 | 0.00080575 |
| 26 | 0.5900 | −0.2100 | −0.20919308 | 0.00080692 | −0.12342392 | 0.00047608 |
| Sum of IAE | | | | 0.01770412 | | 0.00658366 |

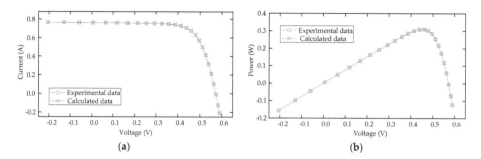

**Figure 5.** Comparisons between the experimental data and calculated data obtained by the proposed ImCSA for the SDM of RTC France silicon solar cell: (**a**) I-V characteristics; (**b**) P-V characteristics.

### 5.1.2. Case Study 2: Double Diode Model

For the DDM in this case, seven unknown parameters need to be estimated. The range of each parameter reported in the literature [1,2,4,11,24] are set as follows: $I_{ph}$ (A) $\in$ [0, 1], $I_{sd1}$ ($\mu$A) $\in$ [0, 1], $I_{sd2}$ ($\mu$A) $\in$ [0, 1], $a_1 \in$ [1, 2], $a_1 \in$ [1, 2], $R_s$ ($\Omega$) $\in$ [0, 0.5], $R_{sh}$ ($\Omega$) $\in$ [0, 100]. The 26 pairs of voltage and current values measured from RTC France silicon solar cell at 33 °C are the same as in Table 3 (columns 2 and 3) from case study 1. Here, the proposed ImCSA is employed to estimate the optimal parameters vector $\theta$ for the DDM of the RTC France silicon solar cell.

Table 4 shows the statistics of the OF (RMSE) values for the DDM of RTC France silicon solar cell obtained by the ImCSA and CSA. Table 4 clearly shows that the ImCSA presents better statistics when compared with CSA. The ImCSA achieves a best OF (RMSE) value of $9.8249 \times 10^{-4}$, which is apparently better than the best OF (RMSE) value achieved by CSA as shown in Table 4. The proposed ImCSA outperforms original CSA in all terms of the best, mean, median, worst and Std of the OF (RMSE) values over 30 independent runs. Moreover, the ImCSA obtains a good Std of $2.8197 \times 10^{-7}$ while CSA obtains a Std of $4.1755 \times 10^{-6}$ as presented in Table 4. These results imply that the proposed ImCSA remarkably enhances the performance of original CSA and is better than original CSA in terms of accuracy and reliability since the best OF (RMSE) value quantifies the best accuracy and the Std of the OF (RMSE) value implies the reliability of parameter estimation methods as aforementioned. In addition, the convergence performance for the best run of the ImCSA for parameter estimation of the DDM of RTC France silicon solar cell is shown in Figure 6. It can be observed from Figure 6 that the objective value becomes relatively stable in less than 1000 iterations.

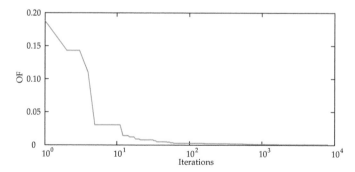

**Figure 6.** Convergence characteristic of the proposed ImCSA for parameter estimation of the DDM of RTC France silicon solar cell.

**Table 4.** Statistics of the OF (RMSE) values for the DDM of RTC France silicon solar cell using the proposed ImCSA and CSA.

| Algorithm | OF (RMSE) | | | | |
|---|---|---|---|---|---|
| | Best | Mean | Median | Worst | Std |
| ImCSA | $9.8249 \times 10^{-4}$ | $9.8258 \times 10^{-4}$ | $9.8249 \times 10^{-4}$ | $9.8396 \times 10^{-4}$ | $2.8197 \times 10^{-7}$ |
| CSA | $9.8292 \times 10^{-4}$ | $9.8626 \times 10^{-4}$ | $9.8535 \times 10^{-4}$ | $1.0056 \times 10^{-3}$ | $4.1755 \times 10^{-6}$ |

Table 5 illustrates the optimal parameters such as $I_{ph}$, $I_{sd1}$, $I_{sd2}$, $a_1$, $a_1$, $R_s$, $R_{sh}$ values and the corresponding objective value of OF (RMSE) for the DDM estimated by the ImCSA compared with those by CSA and several other reported parameter estimation methods such as TLABC [34], CIABC [11], MSSO [33], IJAYA [32], SATLBO [31], GOTLBO [10], EHA-NMS [30], CARO [4], IABC [29], MABC [28], ABC [2], BBO-M [27], $R_{cr}$-IJADE [25], ABSO [24], HS [1], and PS [21]. From the OF (RMSE) values in Table 5, the EHA-NMS and $R_{cr}$-IJADE provide the best OF (RMSE) value ($9.8248 \times 10^{-4}$). The ImCSA achieves the second best OF (RMSE) value ($9.8249 \times 10^{-4}$), which is very close to that of EHA-NMS and $R_{cr}$-IJADE. The other approaches are ranked as CARO, CIABC, BBO-M, MABC, SATLBO, MSSO, CSA, IJAYA, GOTLBO, ABSO, TLABC, ABC, IABC, HS, and PS. These results imply that the proposed ImCSA considerably improves the performance of the original CSA. Consequently, the optimal parameters values determined by the ImCSA are more close to the real ones for the DDM of the PV cell.

To further establish the quality of the parameters estimated by the ImCSA, seven estimated parameters values of $I_{ph}$, $I_{sd1}$, $I_{sd2}$, $a_1$, $a_1$, $R_s$ and $R_{sh}$ are put into Equation (8) to reconstruct the calculated data of DDM of the RTC France silicon solar cell. The calculated data and the experimental data are compared in Table 6 for observation on the accordance between them, and the IAE between experimental and calculated data are also presented in Table 6. It can be seen from Table 6 (columns 5 and 7) and the last line of Table 6 that both the *IAE* and their sum are negligible small and the computed data of DDM are remarkably consistent with the experimental data. Moreover, Figure 7 plots the I-V and P-V characteristics of the best model parameters identified by the proposed ImCSA and the experimental data. It is clear from Figure 7 that the computed data are in good agreement with the experimental data. Cross checking Tables 3 and 5, Figures 6 and 7, we can see that the sum of *IAE* of DDM are smaller than those of SDM, which further validates the optimal parameter values estimated by ImCSA are very precise.

**Table 5.** Comparison among various parameter estimation algorithms for the DDM of RTC France silicon solar cell.

| Algorithm | $I_{ph}$ (A) | $I_{sd1}$ (µA) | $I_{sd2}$ (µA) | $a_1$ | $a_2$ | $R_s$ (Ω) | $R_{sh}$ (Ω) | OF (RMSE) |
|---|---|---|---|---|---|---|---|---|
| ImCSA | 0.760781 | 0.225966 | 0.747309 | 1.451543 | 2.000000 | 0.036740 | 55.482685 | $9.8249 \times 10^{-4}$ |
| CSA | 0.760772 | 0.503010 | 0.255099 | 1.999954 | 1.461682 | 0.036620 | 54.890635 | $9.8292 \times 10^{-4}$ |
| TLABC [34] | 0.76081 | 0.42394 | 0.24011 | 1.9075 | 1.45671 | 0.03667 | 54.66797 | $9.8414 \times 10^{-4}$ |
| CIABC [11] | 0.760781 | 0.227828 | 0.647650 | 1.451623 | 1.988343 | 0.036728 | 55.378261 | $9.8262 \times 10^{-4}$ |
| MSSO [33] | 0.760748 | 0.234925 | 0.671593 | 1.454255 | 1.995305 | 0.036688 | 55.714662 | $9.8281 \times 10^{-4}$ |
| IJAYA [32] | 0.7601 | 0.0050445 | 0.75094 | 1.2186 | 1.6247 | 0.0376 | 77.8519 | $9.8293 \times 10^{-4}$ |
| SATLBO [31] | 0.76078 | 0.25093 | 0.545418 | 1.45982 | 1.99941 | 0.03663 | 55.1170 | $9.82804 \times 10^{-4}$ |
| GOTLBO [10] | 0.760752 | 0.800195 | 0.220462 | 1.999973 | 1.448974 | 0.036783 | 56.075304 | $9.83177 \times 10^{-4}$ |
| EHA-NMS [30] | 0.760781 | 0.225974 | 0.749346 | 1.451017 | 2.000000 | 0.036740 | 55.485441 | $9.8248 \times 10^{-4}$ |
| CARO [4] | 0.76075 | 0.29315 | 0.09098 | 1.47338 | 1.77321 | 0.03641 | 54.3967 | $9.8260 \times 10^{-4}$ |
| IABC [29] | 0.7609 | 0.26900 | 0.28198 | 1.4670 | 1.8722 | 0.0364 | 55.2307 | $10.000 \times 10^{-4}$ |
| MABC [28] | 0.76078 | 0.63069 | 0.241029 | 2.000005 | 1.45685 | 0.036712 | 54.75500 | $9.8276 \times 10^{-4}$ |
| ABC [2] | 0.7608 | 0.0407 | 0.2874 | 1.4495 | 1.4885 | 0.0364 | 53.7804 | $9.861 \times 10^{-4}$ |
| BBO-M [27] | 0.76083 | 0.59115 | 0.24523 | 2.00000 | 1.45798 | 0.03664 | 55.0494 | $9.8272 \times 10^{-4}$ |
| $R_{cr}$-IJADE [25] | 0.760781 | 0.225974 | 0.749347 | 1.451017 | 2.000000 | 0.036740 | 55.485443 | $9.8248 \times 10^{-4}$ |
| ABSO [24] | 0.76078 | 0.26713 | 0.38191 | 1.46512 | 1.98152 | 0.03657 | 54.6219 | $9.8344 \times 10^{-4}$ |
| HS [1] | 0.76176 | 0.12545 | 0.25470 | 1.49439 | 1.49989 | 0.03545 | 46.82696 | $1.26 \times 10^{-3}$ |
| PS [21] | 0.7602 | 0.9889 | 0.0001 | 1.6000 | 1.1920 | 0.0320 | 81.3008 | $15.18 \times 10^{-3}$ |

**Table 6.** The calculated results of the proposed ImCSA for the DDM of RTC France silicon solar cell.

| Item | Experimental Data | | Calculated Current Data | | Calculated Power Data | |
|---|---|---|---|---|---|---|
| | V (V) | I (A) | $I_{cal}$ (A) | IAE | $P_{cal}$ (W) | IAE |
| 1 | −0.2057 | 0.7640 | 0.76398357 | 0.00001643 | −0.15715142 | 0.00000338 |
| 2 | −0.1291 | 0.7620 | 0.76260378 | 0.00060378 | −0.09845215 | 0.00007795 |
| 3 | −0.0588 | 0.7605 | 0.76133716 | 0.00083716 | −0.04476663 | 0.00004923 |
| 4 | 0.0057 | 0.7605 | 0.76017397 | 0.00032603 | 0.00433299 | 0.00000186 |
| 5 | 0.0646 | 0.7600 | 0.75910819 | 0.00089181 | 0.04903839 | 0.00005761 |
| 6 | 0.1185 | 0.7590 | 0.75812190 | 0.00087810 | 0.08983745 | 0.00010405 |
| 7 | 0.1678 | 0.7570 | 0.75718834 | 0.00018834 | 0.12705620 | 0.00003160 |
| 8 | 0.2132 | 0.7570 | 0.75624409 | 0.00075591 | 0.16123124 | 0.00016116 |
| 9 | 0.2545 | 0.7555 | 0.75517755 | 0.00032245 | 0.19219269 | 0.00008206 |
| 10 | 0.2924 | 0.7540 | 0.75372279 | 0.00027721 | 0.22038854 | 0.00008106 |
| 11 | 0.3269 | 0.7505 | 0.75139612 | 0.00089612 | 0.24563139 | 0.00029294 |
| 12 | 0.3585 | 0.7465 | 0.74729625 | 0.00079625 | 0.26790571 | 0.00028546 |
| 13 | 0.3873 | 0.7385 | 0.73999153 | 0.00149153 | 0.28659872 | 0.00057767 |
| 14 | 0.4137 | 0.7280 | 0.72726505 | 0.00073495 | 0.30086955 | 0.00030405 |
| 15 | 0.4373 | 0.7065 | 0.70683595 | 0.00033595 | 0.30909936 | 0.00014691 |
| 16 | 0.4590 | 0.6755 | 0.67523018 | 0.00026982 | 0.30993065 | 0.00012385 |
| 17 | 0.4784 | 0.6320 | 0.63088762 | 0.00111238 | 0.30181664 | 0.00053216 |
| 18 | 0.4960 | 0.5730 | 0.57214020 | 0.00085980 | 0.28378154 | 0.00042646 |
| 19 | 0.5119 | 0.4990 | 0.49957049 | 0.00057049 | 0.25573014 | 0.00029204 |
| 20 | 0.5265 | 0.4130 | 0.41355625 | 0.00055625 | 0.21773737 | 0.00029287 |
| 21 | 0.5398 | 0.3165 | 0.31724205 | 0.00074205 | 0.17124726 | 0.00040056 |
| 22 | 0.5521 | 0.2120 | 0.21208151 | 0.00008151 | 0.11709020 | 0.00004500 |
| 23 | 0.5633 | 0.1035 | 0.10267162 | 0.00082838 | 0.05783492 | 0.00046663 |
| 24 | 0.5736 | −0.0100 | −0.00929718 | 0.00070282 | −0.00533286 | 0.00040314 |
| 25 | 0.5833 | −0.1230 | −0.12439038 | 0.00139038 | −0.07255691 | 0.00081101 |
| 26 | 0.5900 | −0.2100 | −0.20914698 | 0.00085302 | −0.12339672 | 0.00050328 |
| Sum of IAE | | | | 0.01731892 | | 0.00655397 |

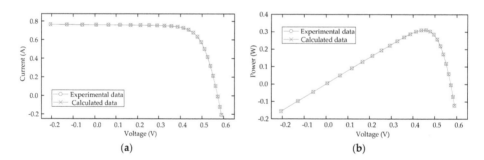

**Figure 7.** Comparisons between the experimental data and calculated data obtained by the proposed ImCSA for the DDM of RTC France silicon solar cell: (**a**) I-V characteristics; (**b**) P-V characteristics.

### 5.1.3. Case Study 3: PV Module Model

In this case, there are five unknown parameters that need to be estimated for the PMM. The range of each parameter used in the literature [4,30–32,34] are set as follows: $I_{phm}$ (A) ∈ [0, 2], $I_{sdm}$ (μA) ∈ [0, 50], $a_m$ ∈ [1, 50], $R_{sm}$ (Ω) ∈ [0, 2], $R_{shm}$ (Ω) ∈ [0, 2000]. The experimental data measured from Photowatt-PWP201 module at 45 °C contain 25 pairs of voltage and current values reported in the literature [4,14,31,32,34]. These data are cited to find the optimal parameters vector θ for the PMM of Photowatt-PWP201 module by the proposed ImCSA. The statistics of the OF (RMSE) values for the PMM of Photowatt-PWP201 module achieved by the ImCSA and CSA are displayed in Table 7. As can be seen in this table, the ImCSA performs better than CSA in terms of all statistical indicators, including the best, mean, median, worst and Std of the OF (RMSE) values over 30 runs. Additionally, the ImCSA achieves the best, mean, and median of the OF (RMSE) values as low as 2.425075 × 10⁻³

as shown in Table 7. Particularly, it can be observed from Table 7 that the ImCSA obtains a Std of $2.915426 \times 10^{-9}$, which is clearly far better than that calculated by CSA. Similar to previous cases, these results prove that the proposed ImCSA is indeed still more accurate and reliable than original CSA and improves the performance of CSA. Furthermore, the convergence performance for the best run of the proposed ImCSA for parameter estimation of the PMM of the Photowatt-PWP201 module is given in Figure 8. It can be found from Figure 8 that the ImCSA rapidly converges to a comparatively stable objective value in less than 100 iterations.

**Table 7.** Statistics of the OF (RMSE) values for the PMM of Photowatt-PWP201 module using the proposed ImCSA and CSA.

| Algorithm | OF (RMSE) | | | | |
|---|---|---|---|---|---|
| | Best | Mean | Median | Worst | Std |
| ImCSA | $2.425075 \times 10^{-3}$ | $2.425075 \times 10^{-3}$ | $2.425075 \times 10^{-3}$ | $2.425091 \times 10^{-3}$ | $2.915426 \times 10^{-3}$ |
| CSA | $2.425082 \times 10^{-3}$ | $2.430857 \times 10^{-3}$ | $2.426771 \times 10^{-3}$ | $2.499628 \times 10^{-3}$ | $1.418512 \times 10^{-5}$ |

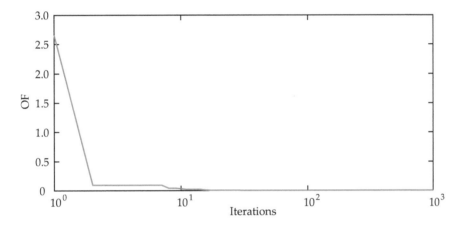

**Figure 8.** Convergence characteristic of the proposed ImCSA for parameter estimation of the PMM of Photowatt-PWP201 module.

Table 8 shows the optimal parameters such as $I_{phm}$, $I_{sdm}$, $a_m$, $R_{sm}$, $R_{shm}$ values and the corresponding objective value of OF (RMSE) for the PMM obtained by the ImCSA in comparisons with those by CSA and some other parameter estimation methods such as TLABC [34], IJAYA [32], SATLBO [31], EHA-NMS [30], CARO [4], MPCOA [26], $R_{cr}$-IJADE [25], SA [23], PS [21], and CPSO [20]. It is obvious from the OF (RMSE) values in Table 8 that the proposed ImCSA, TLABC, IJAYA, SATLBO, EHA-NMS, MPCOA, and $R_{cr}$-IJADE acquire the lowest OF (RMSE) value ($2.425 \times 10^{-3}$), followed by CSA, CARO, SA, CPSO, and PS, which indicates that the proposed ImCSA evidently enhances the performance of the original CSA and the optimal parameters values sought out via the ImCSA are closer to the real ones for the PMM of the PV module.

**Table 8.** Comparison among various parameter estimation algorithms for the PMM of Photowatt-PWP201 module.

| Algorithm | $I_{phm}$ (A) | $I_{sdm}$ (µA) | $a_m$ | $R_{sm}$ (Ω) | $R_{shm}$ (Ω) | OF (RMSE) |
|---|---|---|---|---|---|---|
| ImCSA | 1.030514 | 3.482263 | 48.660397 | 1.201271 | 981.982233 | $2.425 \times 10^{-3}$ |
| CSA | 1.030496 | 3.485411 | 48.663834 | 1.201201 | 984.320163 | $2.42508 \times 10^{-3}$ |
| TLABC [34] | 1.03056 | 3.4715 | 48.63131 | 1.20165 | 972.93567 | $2.425 \times 10^{-3}$ |
| IJAYA [32] | 1.0305 | 3.4703 | 48.6298 | 1.2016 | 977.3752 | $2.425 \times 10^{-3}$ |
| SATLBO [31] | 1.030511 | 3.48271 | 48.6433077 | 1.201263 | 982.40376 | $2.425 \times 10^{-3}$ |
| EHA-NMS [30] | 1.030514 | 3.482263 | 48.642835 | 1.201271 | 981.982256 | $2.425 \times 10^{-3}$ |
| CARO [4] | 1.03185 | 3.28401 | 48.4.363 | 1.20556 | 841.3213 | $2.427 \times 10^{-3}$ |
| MPCOA [26] | 1.03188 | 3.37370 | 48.50646 | 1.20295 | 849.6927 | $2.425 \times 10^{-3}$ |
| $R_{cr}$-IJADE [25] | 1.030514 | 3.482263 | 48.642835 | 1.201271 | 981.982240 | $2.425 \times 10^{-3}$ |
| SA [23] | 1.0331 | 3.6642 | 48.8211 | 1.1989 | 833.3333 | $2.7 \times 10^{-3}$ |
| PS [21] | 1.0313 | 3.1756 | 48.2889 | 1.2053 | 714.2857 | $1.18 \times 10^{-2}$ |
| CPSO [20] | 1.0286 | 8.3010 | 52.2430 | 1.0755 | 1850.1000 | $3.5 \times 10^{-3}$ |

Just like before, for further investigating the quality of the parameters identified by the proposed ImCSA, these identified parameters values of $I_{phm}$, $I_{sdm}$, $a_m$, $R_{sm}$ and $R_{shm}$ are returned to Equation (10) to rebuild the calculated current data and calculated power data at experimental voltage point. Table 9 tabulates the calculated results. From Table 9 (columns 5 and 7) and the last line of Table 9, both the *IAE* and their sum are very tiny, which provides a concrete proof of the ImCSA in accurately estimating the parameters. The I-V and P-V characteristics of the best model parameters estimated by the ImCSA and the experimental data are shown in Figure 9, it can be seen from Figure 9 that the calculated data of PMM match the experimental data nicely, which further demonstrates the high accuracy parameters are achieved again by the proposed ImCSA.

**Table 9.** The calculated results of the proposed ImCSA for the PMM of Photowatt-PWP201 module.

| Item | Experimental Data | | Calculated Current Data | | Calculated Power Data | |
|---|---|---|---|---|---|---|
| | V (V) | I (A) | $I_{cal}$ (A) | IAE | $P_{cal}$ (W) | IAE |
| 1 | 0.1248 | 1.0315 | 1.02912209 | 0.00237791 | 0.12843444 | 0.00029676 |
| 2 | 1.8093 | 1.0300 | 1.02738435 | 0.00261565 | 1.85884651 | 0.00473249 |
| 3 | 3.3511 | 1.0260 | 1.02574214 | 0.00025786 | 3.43736448 | 0.00086412 |
| 4 | 4.7622 | 1.0220 | 1.02410399 | 0.00210399 | 4.87698803 | 0.01001963 |
| 5 | 6.0538 | 1.0180 | 1.02228341 | 0.00428341 | 6.18869931 | 0.02593091 |
| 6 | 7.2364 | 1.0155 | 1.01991740 | 0.00441740 | 7.38053027 | 0.03196607 |
| 7 | 8.3189 | 1.0140 | 1.01635081 | 0.00235081 | 8.45492077 | 0.01955617 |
| 8 | 9.3097 | 1.0100 | 1.01049143 | 0.00049143 | 9.40737206 | 0.00457506 |
| 9 | 10.2163 | 1.0035 | 1.00067876 | 0.00282124 | 10.22323441 | 0.02882264 |
| 10 | 11.0449 | 0.9880 | 0.98465335 | 0.00334665 | 10.87539777 | 0.03696343 |
| 11 | 11.8018 | 0.9630 | 0.95969741 | 0.00330259 | 11.32615687 | 0.03897653 |
| 12 | 12.4929 | 0.9255 | 0.92304875 | 0.00245125 | 11.53155579 | 0.03062316 |
| 13 | 13.1231 | 0.8725 | 0.87258816 | 0.00008816 | 11.45106168 | 0.00115693 |
| 14 | 13.6983 | 0.8075 | 0.80731012 | 0.00018988 | 11.05877623 | 0.00260102 |
| 15 | 14.2221 | 0.7265 | 0.72795782 | 0.00145782 | 10.35308888 | 0.02073323 |
| 16 | 14.6995 | 0.6345 | 0.63646618 | 0.00196618 | 9.35573459 | 0.02890184 |
| 17 | 15.1346 | 0.5345 | 0.53569607 | 0.00119607 | 8.10754576 | 0.01810206 |
| 18 | 15.5311 | 0.4275 | 0.42881615 | 0.00131615 | 6.65998648 | 0.02044123 |
| 19 | 15.8929 | 0.3185 | 0.31866866 | 0.00016866 | 5.06456910 | 0.00268045 |
| 20 | 16.2229 | 0.2085 | 0.20785711 | 0.00064289 | 3.37204517 | 0.01042948 |
| 21 | 16.5241 | 0.1010 | 0.09835421 | 0.00264579 | 1.62521481 | 0.04371929 |
| 22 | 16.7987 | −0.0080 | −0.00816934 | 0.00016934 | −0.13723426 | 0.00284466 |
| 23 | 17.0499 | −0.1110 | −0.11096846 | 0.00003154 | −1.89200116 | 0.00053774 |
| 24 | 17.2793 | −0.2090 | −0.20911762 | 0.00011762 | −3.61340604 | 0.00203234 |
| 25 | 17.4885 | −0.3030 | −0.30202238 | 0.00097762 | −5.28191833 | 0.01709717 |
| Sum of IAE | | | | 0.04178790 | | 0.40460442 |

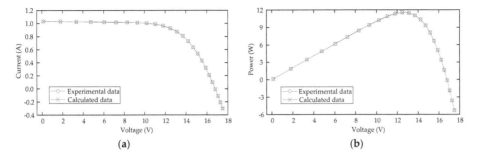

**Figure 9.** Comparisons between the experimental data and calculated data obtained by the proposed ImCSA for the PMM of Photowatt-PWP201 module: (**a**) I-V characteristics; (**b**) P-V characteristics.

### 5.2. Results on Real Datasets of PV Panels

#### 5.2.1. Case Study 1: PV Module Model with Real Dataset of a Polycrystalline Panel

This subsection is to investigate the performance of proposed ImCSA for parameter estimation under a real implementation. A real dataset is considered, where the experimental I–V data from a polycrystalline STP6-120/36 panel at 55 °C [45] contain 22 pairs of voltage and current values. This PV panel consists of 36 polycrystalline cells in series and size of each cell is 156 mm × 156 mm. $V_{OC}$ = 19.21 V, $I_{SC}$ = 7.48 A, $V_M$ = 14.93 V, and $I_M$ = 6.83 A. In this case, there are five unknown parameters needed to be estimated for the PMM of polycrystalline STP6-120/36 panel. The range of each parameter are set as follows: $I_{ph}$ (A) $\in$ [0, 10], $I_{sd}$ (μA) $\in$ [1, 2], $a \in$ [1, 2], $R_s$ (mΩ) $\in$ [0, 10], $R_{sh}$ (Ω) $\in$ [0, 10]. The experimental I–V data are applied for finding optimal parameters vector $\theta$ for the PMM of STP6-120/36 panel by the proposed ImCSA.

Table 10 shows the statistics of the OF (RMSE) values for the PMM of polycrystalline STP6-120/36 panel obtained by the ImCSA and CSA. Evidently, Table 10 shows that all terms of the best, mean, median, worst and Std of the OF (RMSE) values over 30 runs obtained by the ImCSA are smaller than those calculated by CSA. Furthermore, it can be found from Table 10 that the ImCSA provides the best, mean, median, and worst of the OF (RMSE) values as low as $1.5865799 \times 10^{-2}$. In particular, the ImCSA obtains a Std of $4.6901709 \times 10^{-15}$, which is obviously far lower than that calculated by CSA as shown in Table 10. These results give concrete evidence that the ImCSA improves the performance of original CSA and is more accurate and reliable than CSA. In addition, Figure 10 displays the convergence performance for the best run of the ImCSA for parameter estimation of the PMM of polycrystalline STP6-120/36 panel. It can be seen from this figure that the ImCSA can attain a relatively stable OF value in less than 100 iterations, which implies its fast convergence.

**Table 10.** Statistics of the OF (RMSE) values for the PMM of polycrystalline STP6-120/36 panel using the proposed ImCSA and CSA.

| Algorithm | OF (RMSE) | | | | |
|---|---|---|---|---|---|
| | Best | Mean | Median | Worst | Std |
| ImCSA | $1.5865799 \times 10^{-2}$ | $1.5865799 \times 10^{-2}$ | $1.5865799 \times 10^{-2}$ | $1.5865799 \times 10^{-2}$ | $4.6901709 \times 10^{-15}$ |
| CSA | $1.5865806 \times 10^{-2}$ | $1.5869596 \times 10^{-2}$ | $1.5866453 \times 10^{-2}$ | $1.5892796 \times 10^{-2}$ | $6.2673061 \times 10^{-6}$ |

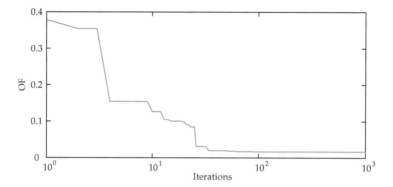

**Figure 10.** Convergence characteristic of the proposed ImCSA for parameter estimation of the PMM of polycrystalline STP6-120/36 panel.

Table 11 illustrates the optimal parameters values and the corresponding objective value of OF (RMSE) for the PMM of polycrystalline STP6-120/36 panel achieved by the ImCSA compared with those by CSA and several other recent parameter estimation methods such as ABC [11], CIABC [11], and Reference [45]. It is obvious from the OF (RMSE) values in Table 11 that the proposed ImCSA obtains the lowest OF (RMSE) value among these methods, followed by CSA, CIABC, Reference [45], and ABC, which implies that the proposed ImCSA enhances the performance of original CSA and outperforms all other algorithms. Consequently, the optimal parameters values found by the proposed ImCSA are closer to the real ones for the PMM of polycrystalline STP6-120/36 panel, whereby the proposed ImCSA achieves the high accuracy parameter values.

For more evaluation on the quality of the parameters estimated by the proposed ImCSA, the estimated parameters values are put into Equation (10) to reconstruct the calculated current data and calculated power data at experimental voltage point. The experimental data, the calculated data and the IAE are listed in Table 12. It can be found from Table 12 (columns 5 and 7) and the last line of Table 12 that both the *IAE* and their sum are very small, which provides positive proof that the high accuracy parameter values identified by the ImCSA. Figure 11 plots the I-V and P-V characteristics of the best model parameters estimated by the ImCSA and the experimental data. It is clear from Figure 11 that the calculated data of the PMM of polycrystalline STP6-120/36 panel are highly in coincidence with the experimental data, which further proves the estimated parameters by the ImCSA are very precise.

**Table 11.** Comparison among various parameter estimation algorithms for the PMM of polycrystalline STP6-120/36 panel.

| Algorithm | $I_{ph}$ (A) | $I_{sd}$ (µA) | $a$ | $R_s$ (mΩ) | $R_{sh}$ (Ω) | OF (RMSE) |
|---|---|---|---|---|---|---|
| ImCSA | 7.482778 | 1.00 | 1.197729 | 5.386970 | 10.00 | 0.015865799 |
| CSA | 7.482777 | 1.00 | 1.197733 | 5.387310 | 10.00 | 0.015865806 |
| ABC [11] | 7.476291 | 1.2 | 1.206992 | 4.91 | 9.70 | 0.019174 |
| CIABC [11] | 7.484126 | 1.29 | 1.214854 | 5.1 | 9.89 | 0.016286553 |
| Reference [45] | 7.4838 | 1.2 | 1.2072 | 4.9 | 9.745 | 0.017879 |

**Table 12.** The calculated results of the proposed ImCSA for the PMM of polycrystalline STP6-120/36 panel.

| Item | Experimental Data | | Calculated Current Data | | Calculated Power Data | |
|---|---|---|---|---|---|---|
| | V (V) | I (A) | $I_{cal}$ (A) | IAE | $P_{cal}$ (W) | IAE |
| 1 | 17.65 | 3.83 | 3.84520015 | 0.01520015 | 67.86778268 | 0.26828268 |
| 2 | 17.41 | 4.29 | 4.27711948 | 0.01288052 | 74.46465022 | 0.22424978 |
| 3 | 17.25 | 4.56 | 4.54504650 | 0.01495350 | 78.40205219 | 0.25794781 |
| 4 | 17.10 | 4.79 | 4.78171108 | 0.00828892 | 81.76725939 | 0.14174061 |
| 5 | 16.90 | 5.07 | 5.07559408 | 0.00559408 | 85.77753992 | 0.09453992 |
| 6 | 16.76 | 5.27 | 5.26678078 | 0.00321922 | 88.27124595 | 0.05395405 |
| 7 | 16.34 | 5.75 | 5.77098920 | 0.02098920 | 94.29796346 | 0.34296346 |
| 8 | 16.08 | 6.00 | 6.03372193 | 0.03372193 | 97.02224861 | 0.54224861 |
| 9 | 15.71 | 6.36 | 6.34833199 | 0.01166801 | 99.73229550 | 0.18330450 |
| 10 | 15.39 | 6.58 | 6.57014416 | 0.00985584 | 101.11451856 | 0.15168144 |
| 11 | 14.93 | 6.83 | 6.81958450 | 0.01041550 | 101.81639658 | 0.15550342 |
| 12 | 14.58 | 6.97 | 6.96396943 | 0.00603057 | 101.53467435 | 0.08792565 |
| 13 | 14.17 | 7.10 | 7.09353516 | 0.00646484 | 100.51539327 | 0.09160673 |
| 14 | 13.59 | 7.23 | 7.22168365 | 0.00831635 | 98.14268079 | 0.11301921 |
| 15 | 13.16 | 7.29 | 7.28648376 | 0.00351624 | 95.89012630 | 0.04627370 |
| 16 | 12.74 | 7.34 | 7.33223712 | 0.00776288 | 93.41270088 | 0.09889912 |
| 17 | 12.36 | 7.37 | 7.36266685 | 0.00733315 | 91.00256226 | 0.09063774 |
| 18 | 11.81 | 7.38 | 7.39363210 | 0.01363210 | 87.31879509 | 0.16099509 |
| 19 | 11.17 | 7.41 | 7.41667187 | 0.00667187 | 82.84422481 | 0.07452481 |
| 20 | 10.32 | 7.44 | 7.43458678 | 0.00541322 | 76.72493553 | 0.05586447 |
| 21 | 9.74 | 7.42 | 7.44205922 | 0.02205922 | 72.48565679 | 0.21485679 |
| 22 | 9.06 | 7.45 | 7.44806806 | 0.00193194 | 67.47949662 | 0.01750338 |
| Sum of IAE | | | | 0.23591924 | | 3.46852297 |

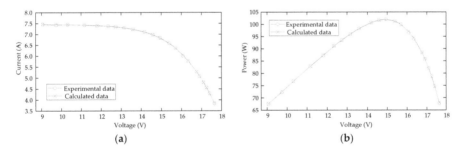

**Figure 11.** Comparisons between the experimental data and calculated data obtained by the proposed ImCSA for the PMM of polycrystalline STP6-120/36 panel: (**a**) I-V characteristics; (**b**) P-V characteristics.

### 5.2.2. Case Study 2: PV Module Model with Real Dataset of a Monocrystalline Panel

In this case, to further verify the performance of proposed ImCSA for parameter estimation under a real implementation of monocrystalline panel. The other real dataset is taken into account, where the experimental I–V data from a monocrystalline STM6-40/36 panel at 51 °C [45] contain 18 pairs of voltage and current values. This PV panel is composed of 36 monocrystalline cells in series and dimension of each cell is 38 mm × 128 mm. $V_{OC}$ = 21.02 V, $I_{SC}$ = 1.663 A, $V_M$ = 16.98 V, and $I_M$ = 1.50 A. There are also five unknown parameters needed to be estimated for the PMM of monocrystalline STM6-40/36 panel in this case. The range of each parameter are set as follows: $I_{ph}$ (A) ∈ [0, 10], $I_{sd}$ (μA) ∈ [0, 2], $a$ ∈ [1, 2], $R_s$ (mΩ) ∈ [0, 10], $R_{sh}$ (Ω) ∈ [0, 20]. The proposed ImCSA is now applied for finding the optimal parameters vector $\theta$ for the PMM of STM6-40/36 panel based on the experimental I-V data.

The statistics of the OF (RMSE) values for the PMM of the monocrystalline STM6-40/36 panel achieved by the ImCSA and CSA are displayed in Table 13. It is notable that the ImCSA performs better than CSA in terms of all statistical indicators, including the best, mean, median, worst and Std of the OF (RMSE) values in all 30 independent runs. Besides, the ImCSA achieves the best, mean, median, and worst of the OF (RMSE) values as low as $1.79436329 \times 10^{-3}$ as tabulated in Table 13. Specially, from this table, it can be observed that the ImCSA obtains a Std of $2.11238634 \times 10^{-14}$, which is markedly smaller than that calculated by CSA. Then, similarly to previous case, these results prove that the proposed ImCSA is indeed still better than original CSA in terms of accuracy and reliability and improves the performance of CSA. Moreover, the convergence performance for the best run of the proposed ImCSA for parameter estimation of the PMM of monocrystalline STM6-40/36 panel is displayed in Figure 12. Figure 12 shows that the objective value achieved by the ImCSA becomes relatively stable in less than 100 iterations, which is an indication of its fast rate.

Table 14 presents the optimal parameters values and the corresponding objective value of OF (RMSE) for the PMM of monocrystalline STM6-40/36 panel estimated by the ImCSA contrasted with those by CSA and several other parameters estimation methods such as ABC [11], CIABC [11], and Reference [45]. From the OF (RMSE) values in Table 14, it is obvious that the ImCSA achieves the best OF (RMSE) value among these methods, followed by CSA, CIABC, ABC, and Reference [45], which indicates that the proposed ImCSA considerably improves the performance of the original CSA and outperforms all other methods. Consequently, the optimal parameters values determined by the ImCSA are more close to the real ones for the PMM of monocrystalline STM6-40/36 panel, thus the parameters estimated by the proposed ImCSA are accurate.

**Table 13.** Statistics of the OF (RMSE) values for the PMM of monocrystalline STM6-40/36 panel using the proposed ImCSA and CSA.

| Algorithm | OF (RMSE) | | | | |
| --- | --- | --- | --- | --- | --- |
| | Best | Mean | Median | Worst | Std |
| ImCSA | $1.79436329 \times 10^{-3}$ | $1.79436329 \times 10^{-3}$ | $1.79436329 \times 10^{-3}$ | $1.79436329 \times 10^{-3}$ | $2.11238634 \times 10^{-14}$ |
| CSA | $1.79436368 \times 10^{-3}$ | $1.79562418 \times 10^{-3}$ | $1.79438763 \times 10^{-3}$ | $1.80652265 \times 10^{-3}$ | $3.06943955 \times 10^{-6}$ |

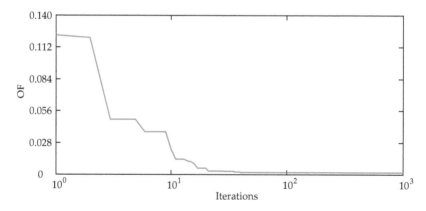

**Figure 12.** Convergence characteristic of the proposed ImCSA for parameter estimation of the PMM of monocrystalline STM6-40/36 panel.

**Table 14.** Comparison among various parameter estimation algorithms for the PMM of monocrystalline STM6-40/36 panel.

| Algorithm | $I_{ph}$ (A) | $I_{sd}$ (µA) | $a$ | $R_s$ (mΩ) | $R_{sh}$ (Ω) | OF (RMSE) |
|---|---|---|---|---|---|---|
| ImCSA | 1.663971 | 2.0000 | 1.533499 | 2.913631 | 15.840511 | 0.00179436329 |
| CSA | 1.663969 | 2.0000 | 1.533497 | 2.912981 | 15.840727 | 0.00179436368 |
| ABC [11] | 1.50 | 1.6644 | 1.4866 | 4.99 | 15.206 | 0.0018379 |
| CIABC [11] | 1.6642 | 1.6760 | 1.4976 | 4.40 | 15.617 | 0.001819 |
| Reference [45] | 1.6635 | 1.4142 | 1.4986 | 4.879 | 15.419 | 0.002181 |

Similarly, to the previous case, for further establishing the quality of the parameters estimated by the ImCSA, five estimated parameters values are back-substituted into Equation (10) to reconstruct the calculated data of PMM of monocrystalline STM6-40/36 panel. The calculated data and the experimental data are compared in Table 15 for observing the accordance between them and the IAE between experimental and calculated data are also listed in Table 15. It can be observed from Table 15 (columns 5 and 7) and the last line of Table 15 that both the *IAE* and their sum are very tiny, which gives concrete evidence that the calculated data of PMM of monocrystalline STM6-40/36 panel are in excellent accordance with the experimental data. Additionally, Figure 13 plots the I-V and P-V characteristics of the best model parameters estimated by the proposed ImCSA and the experimental data. This figure clearly portrays that the calculated data are in close agreement with the experimental data, which further demonstrates the high accuracy parameters are achieved again by the proposed ImCSA. Just like the real implementation of polycrystalline panel, the proposed ImCSA is still able to accurately and reliably estimate the parameters of the PMM of monocrystalline panel.

According to the comparison results mentioned above, it demonstrates that ImCSA can obtain similar or better results contrasted with these methods in literature. Thus, it can be used as an accurate and reliable alternative approach for PV models parameter estimation problem.

**Table 15.** The calculated results of the proposed ImCSA for the PMM of monocrystalline STM6-40/36 panel.

| Item | Experimental Data | | Calculated Current Data | | Calculated Power Data | |
|---|---|---|---|---|---|---|
| | *V* (V) | *I* (A) | $I_{cal}$ (A) | IAE | $P_{cal}$ (W) | IAE |
| 1 | 0.118 | 1.663 | 1.66345723 | 0.00045723 | 0.19628795 | 0.00005395 |
| 2 | 2.237 | 1.661 | 1.65973491 | 0.00126509 | 3.71282700 | 0.00283000 |
| 3 | 5.434 | 1.653 | 1.65406328 | 0.00106328 | 8.98817985 | 0.00577785 |
| 4 | 7.260 | 1.650 | 1.65068943 | 0.00068943 | 11.98400525 | 0.00500525 |
| 5 | 9.680 | 1.645 | 1.64550162 | 0.00050162 | 15.92845565 | 0.00485565 |
| 6 | 11.590 | 1.640 | 1.63922838 | 0.00077162 | 18.99865687 | 0.00894313 |
| 7 | 12.600 | 1.636 | 1.63364948 | 0.00235052 | 20.58398349 | 0.02961651 |
| 8 | 13.370 | 1.629 | 1.62716998 | 0.00183002 | 21.75526261 | 0.02446739 |
| 9 | 14.090 | 1.619 | 1.61814834 | 0.00085166 | 22.79971010 | 0.01199990 |
| 10 | 14.880 | 1.597 | 1.60286544 | 0.00586544 | 23.85063775 | 0.08727775 |
| 11 | 15.590 | 1.581 | 1.58139412 | 0.00039412 | 24.65393434 | 0.00614434 |
| 12 | 16.400 | 1.542 | 1.54224568 | 0.00024568 | 25.29282922 | 0.00402922 |
| 13 | 16.710 | 1.524 | 1.52122273 | 0.00277727 | 25.41963176 | 0.04640824 |
| 14 | 16.980 | 1.500 | 1.49929099 | 0.00070901 | 25.45796106 | 0.01203894 |
| 15 | 17.130 | 1.485 | 1.48541163 | 0.00041163 | 25.44510128 | 0.00705128 |
| 16 | 17.320 | 1.465 | 1.46585878 | 0.00085878 | 25.38867413 | 0.01487413 |
| 17 | 17.910 | 1.388 | 1.38804371 | 0.00004371 | 24.85986286 | 0.00078286 |
| 18 | 19.080 | 1.118 | 1.11802403 | 0.00002403 | 21.33189856 | 0.00045856 |
| Sum of *IAE* | | | | 0.02111015 | | 0.27261495 |

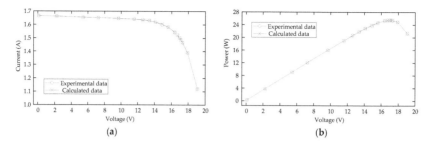

**Figure 13.** Comparisons between the experimental data and calculated data obtained by the proposed ImCSA for the PMM of monocrystalline STM6-40/36 panel: (**a**) I-V characteristics; (**b**) P-V characteristics.

Additionally, in order to verify whether the results achieved by the proposed ImCSA are statistically different from the results obtained by original CSA, the two-sample t-test is conducted, and the corresponding $t$-value, $h$, $CI$, and $p$-value are listed in Table 16. A $t$-value being negative means that the results achieved by the ImCSA are comparatively smaller and vice versa. An $h$ value of one implies that the performances of the two algorithms are statistically different at the 0.05 significance level, whereas value of zero indicates that the performances are not statistically different. The $CI$ is confidence interval. A $p$-value decides the significance level of two algorithms. As can be observed from Table 16, the $t$-values are all negative, the $h$ values are all equal to one, all the $CI$ values are less than zero and do not contain zero and all the $p$-values are less than 0.05, which indicate that the ImCSA significantly outperforms CSA in all case studies from both groups of experiments. Meanwhile, the Wilcoxon rank-sum test is also performed, and the corresponding $z$-value, $h$, and $p$-value are tabulated in Table 17. From Table 17, we can clearly see that the $z$-values are all negative, the $h$ values are all equal to one, and all the $p$-values are extremely less than 0.05, which imply that the ImCSA shows better performance than CSA, in terms of statistical significance. Therefore, the consistent results from both t test and Wilcoxon rank-sum test prove that the proposed ImCSA remarkably enhances the performance of original CSA and is better than CSA and the difference in the results is statistically significant.

**Table 16.** Results of the t test on the data in Tables 1, 4, 7, 10 and 13.

| Comparison | Case Study | t-Value | h | CI | p-Value |
|---|---|---|---|---|---|
| | **Benchmark Datasets** | | | | |
| | Case Study 1 | −2.2130 | 1 | $[-6.5951 \times 10^{-6}, -3.3068 \times 10^{-7}]$ | 0.03084 |
| | Case Study 2 | −4.8129 | 1 | $[-5.2069 \times 10^{-6}, -2.1479 \times 10^{-6}]$ | 0.000011 |
| | Case Study 3 | −2.2325 | 1 | $[-1.0966 \times 10^{-5}, -5.9756 \times 10^{-7}]$ | 0.02946 |
| ImCSA versus CSA | **Real Datasets of PV Panels** | | | | |
| | Case Study 1 | −3.3177 | 1 | $[-6.0867 \times 10^{-6}, -1.5058 \times 10^{-6}]$ | 0.00157 |
| | Case Study 2 | −2.2500 | 1 | $[-2.3827 \times 10^{-6}, -1.3913 \times 10^{-7}]$ | 0.02826 |

**Table 17.** Results of the Wilcoxon rank-sum test on the data in Tables 1, 4, 7, 10 and 13.

| Comparison | Case Study | z-Value | h | p-Value |
|---|---|---|---|---|
| | **Benchmark Datasets** | | | |
| | Case Study 1 | −6.645692 | 1 | $3.017967 \times 10^{-11}$ |
| | Case Study 2 | −6.527324 | 1 | $6.695519 \times 10^{-11}$ |
| | Case Study 3 | −6.616030 | 1 | $3.689726 \times 10^{-11}$ |
| ImCSA versus CSA | **Real Datasets of PV Panels** | | | |
| | Case Study 1 | −6.646061 | 1 | $3.010407 \times 10^{-11}$ |
| | Case Study 2 | −6.645692 | 1 | $3.017967 \times 10^{-11}$ |

## 6. Conclusions

This paper proposed a novel improved variant of CSA called ImCSA for solving the PV models parameter estimation problem based on experimental I-V data of real PV cells and modules. As an enhanced version of CSA, the proposed ImCSA combined three strategies with original CSA to improve its performance. First, a strategy named QOBL scheme was employed in the population initialization step of CSA to accelerate its convergence and enhance its solution accuracy. Second, a dynamic adaptation strategy was developed and introduced for the step size without Lévy flight step in original CSA, which makes the step size with zero parameter initialization adaptively change according to the individual nest's fitness value over the course of the iteration and the current iteration number. This strategy is useful for optimization with a faster rate. Third, a dynamic adjustment mechanism for the fraction probability or discovery rate ($P_a$) was proposed to achieve better tradeoff between the exploration and exploitation to increase searching ability. In this paper, the PV models parameter estimation problem was firstly converted into an optimization problem, and an OF was formulated to quantify the overall difference between the simulated and experimental current data. And then, a new improved CSA, named as ImCSA was proposed and applied for solving the problem of estimating the parameters of PV models based on experimental I-V data. Finally, the performance of proposed ImCSA was comprehensively verified on the parameter estimation of different PV models, i.e., SDM, DDM and PMM of various PV cell/modules.

Experimental comparison results from both benchmark datasets and real datasets with CSA and some other parameter estimation methods available literature, such as TLABC, CIABC, MSSO, IJAYA, SATLBO, GOTLBO, EHA-NMS, CARO, IABC, MABC, ABC, BBO-M, MPCOA, $R_{cr}$-IJADE, ABSO, HS, SA, PS, CPSO, and GA implied that the proposed ImCSA remarkably enhanced the performance of the original CSA and can obtain similar or better results. And they also showed that our proposed ImCSA was capable of finding the best values of parameters for the PV models in such effective way for giving the best possible approximation to the experimental I-V data of real PV cells and modules. Therefore, the proposed ImCSA can be recommended as a promising option to accurately and reliably estimate PV models parameters.

In future work, we hope the applicability of the proposed ImCSA will be expanded to the FACTS devices allocation problem, power economic dispatch problem and some other real-world optimization problems. Moreover, parameter estimation of PV models under partial shading [46] condition needs to be investigated in further research.

**Author Contributions:** All the authors have contributed in the article. Tong Kang conceived and designed the simulations under the supervision and with the help of Jiangang Yao. Tong Kang performed the experiments, analyzed the data and wrote the paper. Jiangang Yao, Min Jin, Shengjie Yang and ThanhLong Duong reviewed the manuscript and provided some valuable suggestions.

**Acknowledgments:** This work was supported by the National Natural Science Foundation of China (Grant Nos. 51277059, 61374172 and 61773157) and the Outstanding Youth Project of Hunan Provincial Education Department (Grant No. 16B143).

**Conflicts of Interest:** The authors declare no conflict of interest.

## References

1. Askarzadeh, A.; Rezazadeh, A. Parameter identification for solar cell models using harmony search-based algorithms. *Sol. Energy* **2012**, *86*, 3241–3249. [CrossRef]
2. Oliva, D.; Cuevas, E.; Pajares, G. Parameter identification of solar cells using artificial bee colony optimization. *Energy* **2014**, *72*, 93–102. [CrossRef]
3. Askarzadeh, A.; Rezazadeh, A. Extraction of maximum power point in solar cells using bird mating optimizer-based parameters identification approach. *Sol. Energy* **2013**, *90*, 123–133. [CrossRef]
4. Yuan, X.; He, Y.; Liu, L. Parameter extraction of solar cell models using chaotic asexual reproduction optimization. *Neural Comput. Appl.* **2015**, *26*, 1227–1239. [CrossRef]

5. Jordehi, A.R. Parameter estimation of solar photovoltaic (PV) cells: A review. *Renew. Sustain. Energy Rev.* **2016**, *61*, 354–371. [CrossRef]
6. *Snapshot of Global Photovoltaic Markets*, 5th ed. IEA-PVPS T1-31. 2017. Available online: http://www.iea-pvps.org/fileadmin/dam/public/report/statistics/IEA-PVPS_-_A_Snapshot_of_Global_PV_-_1992-2016__1_.pdf (accessed on 6 March 2018).
7. Reinders, A.; van Sark, W.; Verlinden, P. Introduction. In *Photovoltaic Solar Energy: From Fundamentals to Applications*; Reinders, A., Verlinden, P., van Sark, W., Freundlich, A., Eds.; John Wiley & Sons, Ltd.: Chichester, UK, 2017; pp. 1–12.
8. Jordehi, A.R. Time varying acceleration coefficients particle swarm optimisation (TVACPSO): A new optimisation algorithm for estimating parameters of PV cells and modules. *Energy Convers. Manag.* **2016**, *129*, 262–274. [CrossRef]
9. Humada, A.M.; Hojabri, M.; Mekhilef, S.; Hamada, H.M. Solar cell parameters extraction based on single and double-diode models: A review. *Renew. Sustain. Energy Rev.* **2016**, *56*, 494–509. [CrossRef]
10. Chen, X.; Yu, K.; Du, W.; Zhao, W.; Liu, G. Parameters identification of solar cell models using generalized oppositional teaching learning based optimization. *Energy* **2016**, *99*, 170–180. [CrossRef]
11. Oliva, D.; Ewees, A.A.; Aziz, M.A.E.; Hassanien, A.E.; Cisneros, M.P. A chaotic improved artificial bee colony for parameter estimation of photovoltaic cells. *Energies* **2017**, *10*, 865. [CrossRef]
12. Jain, A.; Kapoor, A. Exact analytical solutions of the parameters of real solar cells using Lambert W-function. *Sol. Energy Mater. Sol. Cells* **2004**, *81*, 269–277. [CrossRef]
13. Lun, S.X.; Du, C.J.; Guo, T.T.; Wang, S.; Sang, J.S.; Li, J.P. A new explicit I-V model of a solar cell based on Taylor's series expansion. *Sol. Energy* **2013**, *94*, 221–232. [CrossRef]
14. Easwarakhanthan, T.; Bottin, J.; Bouhouch, I.; Boutrit, C. Nonlinear minimization algorithm for determining the solar cell parameters with microcomputers. *Int. J. Sol. Energy* **1986**, *4*, 1–12. [CrossRef]
15. Chatterjee, A.; Keyhani, A.; Kapoor, D. Identification of photovoltaic source models. *IEEE Trans. Energy Convers.* **2011**, *26*, 883–889. [CrossRef]
16. Shongwe, S.; Hanif, M. Gauss-Seidel iteration based parameter estimation for a single diode model of a PV module. In Proceedings of the 2015 IEEE Electrical Power and Energy Conference (EPEC), London, ON, Canada, 26–28 October 2015; pp. 278–284.
17. Tossa, A.K.; Soro, Y.M.; Azoumah, Y.; Yamegueu, D. A new approach to estimate the performance and energy productivity of photovoltaic modules in real operating conditions. *Sol. Energy* **2014**, *110*, 543–560. [CrossRef]
18. Laudani, A.; Riganti Fulginei, F.; Salvini, A. High performing extraction procedure for the one-diode model of a photovoltaic panel from experimental I–V curves by using reduced forms. *Sol. Energy* **2014**, *103*, 316–326. [CrossRef]
19. Jervase, J.A.; Bourdoucen, H.; Al-Lawati, A. Solar cell parameter extraction using genetic algorithms. *Meas. Sci. Technol.* **2001**, *12*, 1922–1925. [CrossRef]
20. Wei, H.; Cong, J.; Lingyun, X.; Deyun, S. Extracting solar cell model parameters based on chaos particle swarm alogorithm. In Proceedings of the 2011 International Conference on Electric Information and Control ngineering (ICEICE), Wuhan, China, 15–17 April 2011; pp. 398–402.
21. Alhajri, M.F.; El-Naggar, K.M.; Alrashidi, M.R.; Al-Othman, A.K. Optimal extraction of solar cell parameters using pattern search. *Renew. Energy* **2012**, *44*, 238–245. [CrossRef]
22. AlRashidi, M.R.; AlHajri, M.F.; El-Naggar, K.M.; Al-Othman, A.K. A new estimation approach for determining the I–V characteristics of solar cells. *Sol. Energy* **2011**, *85*, 1543–1550. [CrossRef]
23. El-Naggar, K.M.; AlRashidi, M.R.; AlHajri, M.F.; Al-Othman, A.K. Simulated Annealing algorithm for photovoltaic parameters identification. *Sol. Energy* **2012**, *86*, 266–274. [CrossRef]
24. Askarzadeh, A.; Rezazadeh, A. Artificial bee swarm optimization algorithm for parameters identification of solar cell models. *Appl. Energy* **2013**, *102*, 943–949. [CrossRef]
25. Gong, W.; Cai, Z. Parameter extraction of solar cell models using repaired adaptive differential evolution. *Sol. Energy* **2013**, *94*, 209–220. [CrossRef]
26. Yuan, X.; Xiang, Y.; He, Y. Parameter extraction of solar cell models using mutative-scale parallel chaos optimization algorithm. *Sol. Energy* **2014**, *108*, 238–251. [CrossRef]
27. Niu, Q.; Zhang, L.; Li, K. A biogeography-based optimization algorithm with mutation strategies for model parameter estimation of solar and fuel cells. *Energy Convers. Manag.* **2014**, *86*, 1173–1185. [CrossRef]

28. Jamadi, M.; Mehdi, F.M. Very accurate parameter estimation of single- and double-diode solar cell models using a modified artificial bee colony algorithm. *Int. J. Energy Environ. Eng.* **2015**, *7*, 13–25. [CrossRef]

29. Wang, R.; Zhan, Y.; Zhou, H. Application of artificial bee colony in model parameter identification of solar cells. *Energies* **2015**, *8*, 7563–7581. [CrossRef]

30. Chen, Z.; Wu, L.; Lin, P.; Wu, Y.; Cheng, S. Parameters identification of photovoltaic models using hybrid adaptive Nelder-Mead simplex algorithm based on eagle strategy. *Appl. Energy* **2016**, *182*, 47–57. [CrossRef]

31. Yu, K.; Chen, X.; Wang, X.; Wang, Z. Parameters identification of photovoltaic models using self-adaptive teaching-learning-based optimization. *Energy Convers. Manag.* **2017**, *145*, 233–246. [CrossRef]

32. Yu, K.; Liang, J.J.; Qu, B.Y.; Chen, X.; Wang, H. Parameters identification of photovoltaic models using an improved JAYA optimization algorithm. *Energy Convers. Manag.* **2017**, *150*, 742–753. [CrossRef]

33. Lin, P.; Cheng, S.; Yeh, W.; Chen, Z.; Wu, L. Parameters extraction of solar cell models using a modified simplified swarm optimization algorithm. *Sol. Energy* **2017**, *144*, 594–603. [CrossRef]

34. Chen, X.; Xu, B.; Mei, C.; Ding, Y.; Li, K. Teaching–learning–based artificial bee colony for solar photovoltaic parameter estimation. *Appl. Energy* **2018**, *212*, 1578–1588. [CrossRef]

35. Wolpert, D.H.; Macready, W.G. No free lunch theorems for optimization. *IEEE Trans. Evol. Comput.* **1997**, *1*, 67–82. [CrossRef]

36. Yang, X.S.; Deb, S. Cuckoo search via Lévy flights. In Proceedings of the 2009 World Congress on Nature & Biologically Inspired Computing, 2009 (NaBIC 2009), Coimbatore, India, 9–11 December 2009; pp. 210–214.

37. Ljouad, T.; Amine, A.; Rziza, M. A hybrid mobile object tracker based on the modified cuckoo search algorithm and the kalman filter. *Pattern Recogn.* **2014**, *47*, 3597–3613. [CrossRef]

38. Fister, I., Jr.; Fister, D.; Fister, I. A comprehensive review of cuckoo search: Variants and hybrids. *Int. J. Math. Model. Numer. Optim.* **2013**, *4*, 387–409.

39. Kang, T.; Yao, J.; Duong, T.L.; Yang, S.; Zhu, X. A hybrid approach for power system security enhancement via optimal installation of flexible AC transmission system (FACTS) devices. *Energies* **2017**, *10*, 1305. [CrossRef]

40. Tizhoosh, H.R. Opposition-based learning: A new scheme for machine intelligence. In Proceedings of the 2005 International Conference on Computational Intelligence for Modelling, Control and Automation, and International Conference on Intelligent Agents, Web Technologies and Internet Commerce (CIMCA-IAWTIC'05), Vienna, Austria, 28–30 November 2005; pp. 695–701.

41. Rahnamayan, S.; Tizhoosh, H.R.; Salama, M.M.A. Opposition-based differential evolution. *IEEE Trans. Evol. Comput.* **2008**, *12*, 64–79. [CrossRef]

42. Xu, Q.; Wang, L.; Wang, N.; Hei, X.; Zhao, L. A review of opposition-based learning from 2005 to 2012. *Eng. Appl. Artif. Intell.* **2014**, *29*, 1–12. [CrossRef]

43. Rahnamayan, S.; Tizhoosh, H.R.; Salama, M.M.A. Quasi-oppositional differential evolution. In Proceedings of the 2007 IEEE Congress on Evolutionary Computation (CEC 2007), Singapore, 25–28 September 2007; pp. 2229–2236.

44. Ergezer, M.; Simon, D. Mathematical and experimental analyses of oppositional algorithms. *IEEE Trans. Cybern.* **2014**, *44*, 2178–2189. [CrossRef] [PubMed]

45. Tong, N.T.; Pora, W. A parameter extraction technique exploiting intrinsic properties of solar cells. *Appl. Energy* **2016**, *176*, 104–115. [CrossRef]

46. Pannebakker, B.B.; de Waal, A.C.; van Sark, W.G.J.H.M. Photovoltaics in the shade: One bypass diode per solar cell revisited. *Prog. Photovolt. Res. Appl.* **2017**, *25*, 836–849. [CrossRef]

*Article*

# Variations of PV Panel Performance Installed over a Vegetated Roof and a Conventional Black Roof

Mohammed J. Alshayeb * and Jae D. Chang

School of Architecture & Design, The University of Kansas, Lawrence, KS 66045, USA; jdchang@ku.edu
* Correspondence: alshayebm@hotmail.com; Tel.: +1-785-864-4281

Received: 7 March 2018; Accepted: 26 April 2018; Published: 1 May 2018

**Abstract:** The total worldwide photovoltaic (PV) capacity has been growing from about 1 GW at the beginning of the twenty-first century to over 300 GW in 2016 and is expected to reach 740 GW by 2022. PV panel efficiency is reported by PV manufacturers based on laboratory testing under Standard Testing Condition with a specific temperature of 25 °C and solar irradiation of 1000 W/m$^2$. This research investigated the thermal interactions between the building roof surface and PV panels by examining the differences in PV panel temperature and energy output for those installed over a green roof (PV-Green) and those installed over a black roof (PV-Black). A year-long experimental study was conducted over the roof of an educational building with roof mounted PV panels with a system capacity of 4.3 kW to measure PV underside surface temperature (PV-UST), ambient air temperature between PV panel and building roof (PV-AT), and PV energy production (PV-EP). The results show that during the summer the PV-Green consistently recorded lower PV-UST and PV-AT temperatures and more PV-EP than PV-Black. The average hourly PV-EP difference was about 0.045 kWh while the maximum PV-EP difference was about 0.075 kWh, which represents roughly a 3.3% and 5.3% increase in PV-EP. For the entire study period, EP-Green produced 19.4 kWh more energy, which represents 1.4% more than EP-Black.

**Keywords:** PV energy performance; PV thermal performance; thermal interaction; conventional roof membrane; vegetated/green roof; Renewable Energy

## 1. Introduction

The global building sector accounts for more than one-fifth of total worldwide energy consumption and is expected to increase by 38% from 2010 to 2040 [1]. Buildings in the United States consume 41% of primary energy in the country and 7% of total primary energy worldwide [2,3]. The challenge is that three-fourths of the world's energy infrastructure heavily depends on fossil fuels. The challenge with fossil fuels as a source of energy is that they are considered a major producer of greenhouse gases and humans deplete them faster than they are generated. Although the worldwide fuel sources used to generate electricity have changed over the last decades, coal and natural gas accounted for more than 60% of the overall worldwide electricity production in 2010 [4].

Solar energy is a promising energy source that has received greater public attention in the last decade. The total worldwide installed photovoltaic (PV) capacity was about 1 GW in 2000 and it surpassed 138 GW of installed power in 2013 [1]. In 2014, around 40 GW of PV power was added to reach a global total capacity of 177 GW [5]. The total PV capacity exceeded 300 GW in 2016 with an addition of over 74 GW to the global capacity, which grew faster than any other fuel source [6,7]. The total global PV capacity is expected to reach 740 GW by 2022 [6].

Building roofs in general are ideal spaces for solar technology because they are usually larger in size, contiguous, and have minimal shade. Also, when electricity is generated on the roof and consumed in the building, the average losses of 7% from transmission and distribution lines can be avoided [8]. According to the National Renewable Energy Laboratory (NREL), rooftop mounted systems accounted for 74% of the installed PV generation capacity in the United States in 2008 [9]. According to the U.S. Energy Information Administration (EIA), the electricity generation from solar power in the building sector was around 5000 GWh in 2010 and reached around 15,000 GWh in 2015 and is expected to reach 100,000 GWh in 2040 [10].

The growth in the PV market is due to several factors; such as government incentives, environmental concerns, and increase in PV efficiency and decrease in cost [11–13]. The conversion efficiency of crystalline silicon, which represents over 85% of the market share, is 20% to 27%. The conversion efficiency of a PV module is tested in laboratories under a controlled environment using a specific procedure called Standard Test Conditions (STC). Standard Test Conditions create uniform test conditions which make it possible to conduct uniform comparisons of photovoltaic modules by different manufacturers. Ratings under the STC involve a constant temperature of 25 °C, a constant solar irradiance of 1000 W/m$^2$, and a constant sunlight spectrum of AM (air mass) 1.5 G (global). However, the conversion efficiency of field installed PV modules is lower than that measured under the STC due to several factors, such as: electrical circuit resistance, dirt and dust accumulation, shade, a range of operating temperatures, solar irradiances, and sunlight spectra. The operation temperature of PV modules has an impact on the conversion efficiency. Therefore, PV manufacturers publish temperature coefficients relating losses in efficiency for each degree the temperature fluctuates from the base of 25 °C.

There are several roofing types that have a range of thermal performance. Conventional roofing materials during summer months can reach temperatures of 80 °C while a green roof, also known as vegetated roof, stays below 50 °C [14]. Green roof temperatures depend on the roof's composition, moisture content, geographic location, and solar exposure [15]. Most green roof surfaces stay cooler than conventional rooftops under summertime conditions because of plant shading and evapotranspiration. Conventional roofing materials have higher surface temperatures during summertime because the material's solar reflectance ranges between 5% and 25%, which means 75–95% of the Sun's energy is absorbed [16]. An experimental study conducted in Chicago, IL during the summer compared the roof surface and ambient air temperatures of a green roof to a conventional roof and it found that the ambient air temperature over the conventional roof measured about 45.5 °C while the green roof was 41.6 °C. The green roof surface temperature ranged from 32.7 °C to 48.3 °C while the black roof was 76 °C [17].

PV panels installed over building roofs have a thermal interaction with the roof surface. This study investigated the thermal interactions between PV panels and roof surface for PV panels installed over a green roof and PV panels installed over a conventional roof. The PV panels over both roof types were the same type, with the same height, tilt, racking system, and inverter. The main goal is to quantify the differences in PV operation temperatures and energy out-put as an impact of different roofing materials. A few studies have examined the impact of roofing choice on the performance of PV panels. Several of these studies, such as the studies of [18–20] compared the performance of PV panels over green roofs with PV panels over black roofs while the studies of [21,22] compared the performance of PV panels over green roofs with PV panels over gravel roofs. These works report that PV panels installed over a green roof have output increase ranging from 0.5% to 4.8% in reference to PV panels over conventional roofing materials; however, none of these works had compared full scale identical installation over a long period of time. Both studies of [18,20] conducted a long term study with full scale installations but in ref. [18] study the PV panels were not identical in terms of number and type and in ref. [20] study the distances from the roof surface to the PV panels were not identical. The experiment measurements for ref. [22] study were conducted for a couple of months and for [19] the study was conducted only for several hours. Also, the PV panel installations in [21,22] studies were at laboratory scale. This study improves on previous literature by quantifying the relationship between roof surface temperatures and PV electrical output by comparing the performance of PV-Green and PV-Black through a 12-month experiment with the same PV panel type, tilt, inverter, racking system, and heights. Investigating the performance of PV-Green and PV-Black through an experimental study will help to fill the lack of quantitative data that identifies the impact of roofing materials on the performance of PV panels.

## 2. Method

### 2.1. Experimental Bed Setting

A field experimental study was conducted to investigate the thermal interaction between building roofing materials and PV panels. This research examined the performance of PV panels over a green roof (PV-Green) to the performance of PV panels over a black roof (PV-Black). The roof of the Center for Design Research (CDR) at the University of Kansas in Lawrence, Kansas, USA was used to conduct the study. The CDR has roof mounted Yingi Solar 235 W PV panels. The total system capacity is 4.23 kW. Nine panels are over a green roof and nine others are over a black roof as shown in Figure 1. The PV modules' electrical features are presented in Table 1. The tilt angle for all the panels was fixed at 10° facing south. The PV panel dimensions are 1.65 m by 0.99 m and mounted in landscape orientation. The height of the panels' bottom frame, with respect to the roofing system, is 0.2 m. The distance between each row of PVs is 0.4 m.

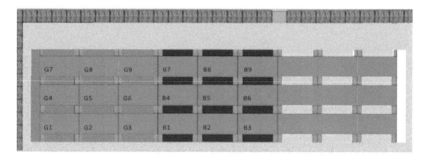

**Figure 1.** PV panel configurations schematic.

**Table 1.** PV module electrical features.

| Characteristics | Unit | Value |
|---|---|---|
| Maximum Power Output at STC | (W) | 235 |
| Module Efficiency | (%) | 14.4 |
| Nominal Operating Cell Temperature (NOCT) | (°C) | 46 ± 2 |
| Temperature coefficient of Voc | (V/°C) | −0.0037 |
| Temperature coefficient of Isc | (I/°C) | +0.0006 |
| Temperature coefficient of Pmpp | (W/°C) | −0.0045 |

The green roof portion is a tray-based green system. Concerns were raised regarding the survival of the vegetated roof under the PV panels. The native sedum plants that were selected need a minimal amount of solar radiation and irrigation. There were no major issues with the plants growing under the PV panels. The black roof membrane is a bitumen membrane. The roof configuration is shown in Figure 2. Enphase M250 Microinverters were installed under each PV panel to monitor the energy production of each individual solar panel. Enphase Energy Inc. is an energy technology company headquartered in Petaluma, CA, USA. The Enphase Microinverter specifications are presented in Table 2.

**Figure 2.** Experimental bed configurations.

**Table 2.** Enphase microinverter specifications.

| Characteristics | Unit | Value |
|---|---|---|
| Peak Output Power | (W) | 250 |
| Maximum continuous output power | (W) | 240 |
| Peak inverter efficiency | (%) | 96.5 |
| Static maximum power point tracking (MPPT) efficiency | (%) | 99.4 |
| Ambient temperature range | (°C) | −40 to +65 |

## 2.2. Data Aquisition

An Onset HOBO U30 weather station was installed over the CDR roof to record data on wind speed and direction, solar radiation, ambient temperature, and relative humidity. In addition, Onset HOBO U12 data loggers were mounted under each panel to measure ambient air temperature, roof surface temperature, relative humidity, and PV panel underside surface temperature. The data acquisition technical specifications are listed in Table 3. The PV underside surface temperature (UST) sensors were attached at the center of each PV panel. The ambient temperature (AT) sensors were placed at the center of each PV panel between the roof surface and the PV panel. The roof surface temperature (RST) sensors were also placed at the center of each PV panel and not exposed to direct solar irradiation. The green roof soil temperature and moisture content were monitored using Onset Hobo soil temperature sensors and EC-5 smart sensors. The sensor locations and types are shown in Figure 3. All monitored data were collected every five minutes; however, the data were averaged hourly to minimize the effects caused by sudden changes in wind speed or passing clouds.

**Table 3.** Data acquisition technical specifications.

| Sensor Type | Accuracy | Operating Temperature |
|---|---|---|
| Onset Hobo Data Logger—U30 | ±8 s/month | −40 °C to 60 °C |
| Onset Hobo Data Logger—U12 | ±0.35 °C | −20 °C to 70 °C |
| Onset Hobo Temp Smart Sensors | ±0.2 °C | −40 °C to 100 °C |
| Air/Water/Soil Temperature Sensor | ±0.25 °C | −40 °C to 100 °C |
| Onset Hobo Pyrometer | ±10 W/m$^2$ | −40 °C to 75 °C |
| Onset Hobo Wind Speed/Gust | ±1.1 m/s | −40 °C to 75 °C |
| Onset Hobo Wind Direction | ±5 degrees | −40 °C to 70 °C |

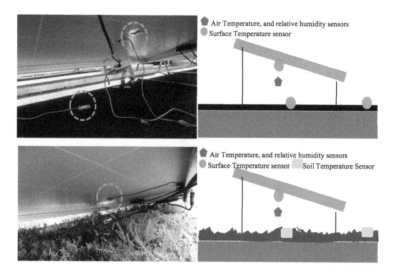

**Figure 3.** Sensor locations (**left**) and schematic of sensor types and locations (**right**).

*2.3. Calibration Tests*

Calibration tests were performed for two months (June and July) to verify that the experimental bed had similar thermal performance before applying any treatment. The ambient temperature (AT), underside surface temperature (UST), and PV panel power output of each panel were measured prior to any treatments. For the pre-treatment test period, the sensors were placed at the same locations that are shown in Figure 3. The ambient temperatures (AT) under the PV panels had small differences during the peak of hot days but were otherwise similar. Since temperature fluctuation happens only when high temperatures occur, the average ambient temperature of days with high temperatures were selected to quantify the fluctuation difference. For the majority of the pre-treatment test period, all the monitoring points recorded almost the same ambient temperature with maximum differences of less than 1 °C. During days with high ambient temperature, the differences in ambient temperature happened during the peak of the day with an average difference of less than 2 °C. The maximum average temperature was 37.6 °C and the minimum average temperature was 35.9 °C, which means that all the monitoring points fit between these two values. The difference between the maximum and minimum temperature is about 1.7 °C. The peak differences occurred during the hottest day of the pre-treatment test period. This test verified that the experimental bed performed almost similarly before any treatments were applied.

## 3. Results

*3.1. Overall Performance Analysis*

The performance of PV-Green was compared with the performance of PV-Black to measure the hourly energy production (EP), ambient temperature (AT), and underside surface temperature (UST) differences for an entire year. The hours of the year that EP-Green produced more power than EP-Black are shown in Figure 4. The average hourly energy production difference was about 45 Wh while the maximum energy production difference was about 75 Wh, which represents roughly a 3.3% and 5.3% increase in energy production. The fluctuations in power output happened more during hot days and the maximum differences occurred during the day's peak temperature. EP-Green produced more energy for a few hours during the cold days than EP-Black with an average difference of about 30 Wh, which represents about a 2.4% increase in energy production.

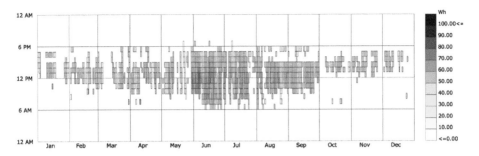

**Figure 4.** Hourly energy production differences of EP-Green vs. EP-Black.

The hours of the year that AT-Black recorded higher ambient temperatures than AT-Green are shown in Figure 5. The average hourly ambient temperature difference was about 2.5 °C whereas the maximum ambient temperature difference was about 5 °C. Similar to the energy production, the fluctuations in ambient temperature happened more during hot days, and the maximum differences occurred during the peak temperature of the day.

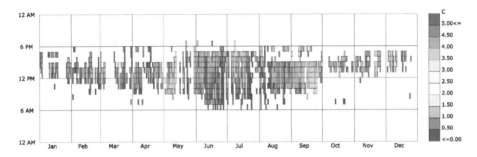

**Figure 5.** Hourly ambient temperature differences of AT-Green vs. AT-Black.

The hours of the year that UST-Black recorded higher underside surface temperature than UST-Green are show in Figure 6. The average hourly underside surface temperature difference was about 3 °C whereas the maximum underside surface temperature was approximately 6 °C. The underside surface temperature differences are consistent with the ambient temperature differences. The maximum differences happened during the peak temperature of the day. As shown in the figure, for a few hours during the cold days, UST-Black's underside surface temperature was higher than UST-Green. The average difference during cold days was about 1 °C.

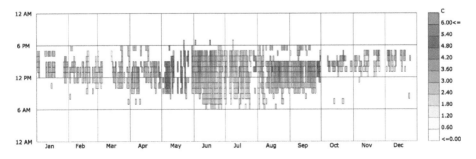

**Figure 6.** Hourly underside surface temperature differences of UST-Green vs. UST-Black.

The previous figures show the correlation between the differences in air and underside surface temperature and PV energy production. However, the differences in the temperatures do not always impact the energy production differences as explained in the following detailed performance analysis. There are hours of the year that recorded temperature differences between the two roofing types and the differences in energy production were minimal. It is also important to mention that a couple of days in early May and late August have slightly higher temperature differences, but fewer differences in energy production. During these specific periods, some work took place over the building roof that cast shade over some of the PV panels for an extended period of time. In addition, the percentage of energy production differences during cold periods, like December, seemed to be high because the actual energy production was low and small differences recorded a high percentage. For about 6% of the year, EP-Black was slightly higher than EP-Green by an average of 15 Wh. These occurrences were scattered throughout the day but did not occur during the day's peak times. The average ambient temperature difference was 0.3 °C and the average underside surface temperature difference was 0.8 °C.

### 3.2. Detailed Performance Analysis

Detailed analyses for several days represent the peak and the average differences throughout the year in energy production and thermal performance. Previous figures show only the energy production, ambient temperature, and underside surface temperature differences. Therefore, this section studies in detail the trends in energy production and temperature change over time. The time period selected for the detailed performance analysis was based on the times that showed the peak, average, and lowest performance differences.

Figure 7 shows the energy production profiles for two days, 4–5 June. On the first day, the maximum energy production by EP-Green at peak hours was 1868 Wh, and EP-Black was 1801 Wh. On the second day, the maximum energy production by EP11-Green at peak hours was 1777 Wh, and EP17-Black was 1707 Wh. The energy production profiles show maximum temperature differences of roughly 67 Wh and 70 Wh at the peak production for the first day and the second day, respectively. This translates to differences of 3.7% and 4.1% between EP-Green and EP-Black in the peak energy production values. The solar irradiation and energy production profiles show similar patterns. The peak solar irradiation values were 972 W/m$^2$ on the first day and 976 W/m$^2$ on the second day.

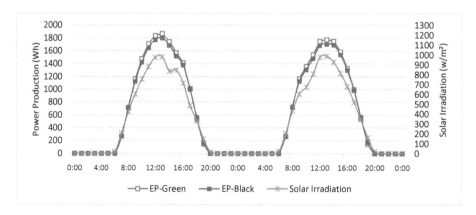

**Figure 7.** Energy production performance of EP-Green vs. EP-Black (4–5 June).

The ambient temperature and the underside surface temperature comparisons between the AT Green and the AT-Black and between UST-Green and the UST-Black are shown in Figure 8. On the first day, the ambient temperature profiles at the peak show a maximum temperature difference of

2.4 °C. The AT-Green reached a peak temperature of 30.5 °C whereas the AT-Black reached a peak temperature of 32.9 °C. The ambient temperature from the weather station (Station AT) was 27.2 °C. The AT-Green and the AT-Black at the peak were 3.3 °C and 5.7 °C hotter than Station AT, respectively. The underside surface temperature profile shows maximum temperature differences of 3.8 °C at the peak temperature. The UST-Green and the UST-Black reached a peak temperature of 47.1 °C and 50.9 °C. On the second day, the ambient temperature profiles at the peak show a maximum temperature difference of 2.6 °C. The AT-Green reached a peak temperature of 31.7 °C, whereas the AT-Black reached a peak temperature of 34.4 °C. The ambient temperature from the weather station was 27.6 °C. The AT-Green and the AT-Black at the peak were 4.1 °C and 6.8 °C hotter than Station AT, respectively. The underside surface temperature profile shows maximum temperature differences of 4.2 °C at the peak temperature. The UST-Green and the UST-Black reached a peak temperature of 50.1 °C and 54.3 °C.

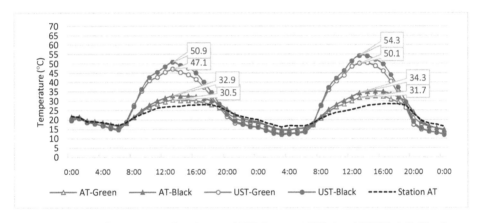

**Figure 8.** Thermal performance of AT-Green and UST-Green vs. AT-Black and UST-Black (4–5 June).

The energy production profiles for 15–16 June are shown in Figure 9. On the first day, the maximum energy production by EP-Green at peak hours was 1695 Wh, and EP-Black was 1641 Wh. On the second day, the maximum energy production by EP-Green at peak hours was 1733 Wh, and EP-Black was 1666 Wh. The energy production profiles show maximum temperature differences of about 54 Wh and 67 Wh at the peak production for the first day and the second day, respectively. This translates to differences of 3.3% and 4% between EP-Green and EP-Black in peak energy production values. The solar irradiation and energy production profiles show similar patterns and the peak solar irradiation values were 964 W/m$^2$ on the first day and 959 W/m$^2$ on the second day.

The ambient temperature and underside surface temperature comparisons between the AT-Green and AT-Black and between UST-Green and the UST-Black are shown in Figure 10. On the first day, the ambient temperature profiles at the peak show a maximum temperature difference of 4 °C. The AT-Green reached a peak temperature of 42.5 °C while the AT-Black reached a peak temperature of 46.5 °C. The ambient temperature from the weather station was 37.2 °C. The AT-Green and the AT-Black at the peak were 5.3 °C and 9.3 °C hotter than Station AT, respectively. The underside surface temperature profile shows maximum temperature differences of 4.3 °C at the peak temperature. The UST-Green and UST-Black reached a peak temperature of 62.1 °C and 66.4 °C, respectively. On the second day, the ambient temperature profiles at the peak show a maximum temperature difference of 2.6 °C. The AT-Green reached a peak temperature of 41.5 °C whereas the AT-Black reached a peak temperature of 44 °C and the ambient temperature from the weather station was 34.7 °C. The AT-Green and AT-Black at the peak were 6.8 °C and 9.3 °C hotter than Station AT. The underside surface temperature profile showed a maximum temperature difference of 3.3 °C at the peak temperature.

The UST-Green and the UST-Black reached a peak temperature of 63.3 °C and 66.6 °C, as evidenced in the data.

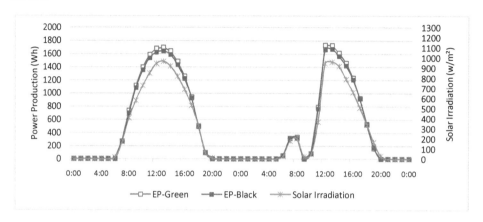

**Figure 9.** Energy production performance of EP-Green vs. EP-Black (15–16 June).

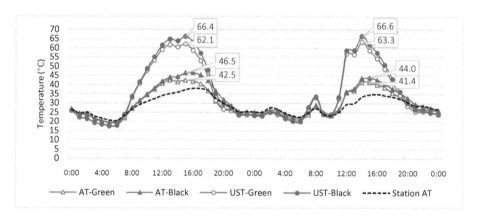

**Figure 10.** Thermal performance of AT-Green and UST-Green vs. AT-Black and UST-Black.

The energy production profiles for 19–20 July are shown in Figure 11. On the first day, the maximum energy production by EP-Green at peak hours was 1715 Wh, and EP-Black was 1698 Wh. On the second day, the maximum energy production by EP-Green at peak hours was 1667 Wh, while EP-Black was 1627 Wh. The energy production profiles show maximum temperature differences of about 47 Wh and 40 Wh at the peak production for the first day and the second day, respectively. This translates to differences of 2.8% and 2.5% between EP-Green and EP-Black in the peak energy production values. The solar irradiation and energy production profiles show similar patterns and the peak solar irradiation values were 922 W/m$^2$ on the first day and 908 W/m$^2$ on the second day.

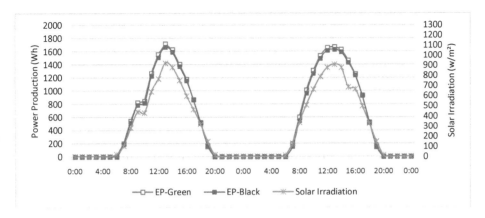

**Figure 11.** Energy production performance of EP-Green vs. EP-Black (19–20 July).

The ambient temperature and the underside surface temperature comparisons between AT-Green and AT-Black and between UST-Green and UST-Black are shown in Figure 12. On the first day, the ambient temperature profiles at the peak show a maximum temperature difference of 0.5 °C. The AT-Green reached a peak temperature of 39.9 °C while the AT-Black reached a peak temperature of 40.4 °C and the ambient temperature from the weather station was 33.4 °C. The AT-Green and AT-Black at the peak were 6.5 °C and 7 °C hotter than Station AT, respectively. The underside surface temperature profile showed maximum temperature differences of 1.9 °C at the peak temperature. The UST-Green and the UST-Black reached a peak temperature of 56.3 °C and 58.2 °C, respectively. On the second day, the ambient temperature profiles at the peak show a maximum temperature difference of 0.5 °C. The AT-Green reached a peak temperature of 40.3 °C whereas the AT-Black reached a peak temperature of 40.8 °C. The ambient temperature from the weather station was 34.9 °C. The AT11-Green and the AT-Black at the peak were 5.4 °C and 5.9 °C hotter than Station AT, respectively. The underside surface temperature profile showed maximum temperature differences of 2 °C at the peak temperature. In addition, the UST-Green and UST-Black reached a peak temperature of 54.2 °C and 56.2 °C.

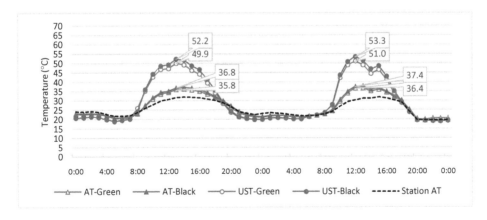

**Figure 12.** Thermal performance of AT-Green and UST-Green versus AT-Black and UST-Black.

The energy production profiles for 3–4 April are shown in Figure 13. On the first day, the maximum energy production by EP-Green at peak hours was 1795 Wh, and EP-Black was 1769 Wh. On the second day, the maximum energy production by EP-Green at the peak hours was 1748 Wh, and EP-Black was

1735 Wh. The energy production profiles show maximum temperature differences of about 26 Wh and 13 Wh at the peak production for the first day and the second day, respectively. This translates to differences of 1.5% and 0.8% between EP-Green and EP-Black in the peak energy production values. The solar irradiation and energy production profiles show similar patterns and the peak solar irradiation values were 832 W/m$^2$ on the first day and 838 W/m$^2$ on the second day.

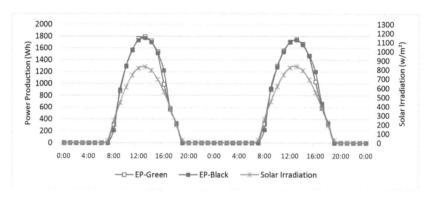

**Figure 13.** Energy production performance of EP-Green versus EP-Black (3–4 April).

The ambient temperature and the underside surface temperature comparisons between the AT-Green and AT-Black and between UST-Green and UST-Black are shown in Figure 14. On the first day, the ambient temperature profiles at the peak show a maximum temperature difference of 1.4 °C. The AT-Green reached a peak temperature of 26.8 °C, whereas the AT-Black reached a peak temperature of 28.2 °C. The ambient temperature from the weather station was 26.7 °C. The AT-Green and AT-Black at the peak were 0.1 °C and 1.5 °C hotter than Station AT, respectively. The underside surface temperature profile shows maximum temperature differences of 2.2 °C at the peak temperature. The UST-Green and the UST-Black reached a peak temperature of 36.6 °C and 38.8 °C. On the second day, the ambient temperature profiles at the peak show a maximum temperature difference of 2.1 °C. The AT-Green reached a peak temperature of 23 °C whereas the AT-Black reached a peak temperature of 25.1 °C. The ambient temperature from the weather station was 20.7 °C. The AT-Green and the AT-Black at the peak were 2.3 °C and 4.4 °C hotter than Station AT, respectively. The underside surface temperature profile showed maximum temperature differences of 2.8 °C at the peak temperature. The UST-Green and the UST-Black reached a peak temperature of 39.4 °C and 42.2 °C, respectively.

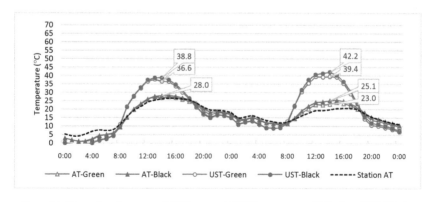

**Figure 14.** Thermal performance of AT-Green and UST-Green versus AT-Black and UST-Black.

## 4. Conclusions and Discussion

Building roof surface types have an impact on the performance of PV panels. The variations of roof surface temperatures and ambient air above the roof impact the PV panel underside surface temperature which also impacts energy production. Since the green roof recorded lower surface temperatures than the conventional roof, PV panels over the green roof produced more energy than the PV panels over the conventional roof. The differences in energy production between the two systems are correlated with ambient air temperature. During hot days of the year, the magnitude of energy production differences is greater as shown in Table 4.

**Table 4.** Monthly energy production and differences for EP-Green and EP-Black.

| Month | EP-Green (kWh) | EP-Black (kWh) | Difference (kWh) | Difference (%) | Ambient Temperature (°C) | | |
|---|---|---|---|---|---|---|---|
| | | | | | Min | Avg | Max |
| January | 48.18 | 47.67 | 0.51 | 1.1% | −14.8 | 3.1 | 20.0 |
| February | 102.86 | 102.47 | 0.39 | 0.4% | −7.4 | 8.9 | 24.5 |
| March | 84.36 | 83.93 | 0.42 | 0.5% | −0.3 | 14.5 | 25.5 |
| April | 141.82 | 141.86 | −0.04 | 0.0% | 7.8 | 18.2 | 26.7 |
| May | 141.59 | 139.77 | 1.82 | 1.3% | 9.4 | 22.1 | 30.9 |
| June | 245.52 | 239.79 | 5.73 | 2.4% | 14.6 | 30.2 | 38.2 |
| July | 198.87 | 194.87 | 3.99 | 2.0% | 17.6 | 30.1 | 36.9 |
| August | 165.85 | 163.13 | 2.72 | 1.7% | 19.2 | 28.8 | 36.7 |
| Sepetember | 131.14 | 130.53 | 0.61 | 0.5% | 15.0 | 27.2 | 33.6 |
| Octorber | 94.28 | 92.86 | 1.41 | 1.5% | 2.3 | 21.0 | 30.0 |
| November | 48.56 | 47.65 | 0.91 | 1.9% | −0.8 | 17.2 | 25.0 |
| December | 36.32 | 35.36 | 0.96 | 2.7% | −2.0 | 8.6 | 18.6 |
| Annual Total | 1439.34 | 1419.90 | 19.43 | 1.4% | - | - | - |

The performance differences varied across the analysis period as shown in Table 4. The larger differences occurred during the peak temperature of the day. During the cooler climate conditions, there were slight temperature differences across the three roofing types. For temperatures below 0 °C, the average AT-Green and average UST-Green were less than 1 °C warmer than AT-Black and UST-Black. For temperatures above 20 °C, the average AT-Black ranged from 0.6 °C to 2.3 °C hotter than the average AT-Green. The average ambient temperature peaks were 42.3 °C for AT-Black and 40 °C for AT-Green when the ambient temperatures from the weather station were above 38 °C. The average UST-Black ranged from 1.1 °C to 2.3 °C hotter than the average AT-Green for temperatures above 20 °C. The average underside surface temperature peaks were 52.6 °C for UST-White and 50.3 °C for UST-Green.

For the entire analysis period, EP-Green produced 19.4 kWh, which represents 1.4%, more electricity than EP-Black. In June, the EP-Green produced 5.7 kWh more electricity than EP-Black, which represents 2.4% more kWh output as shown in Table 4. In January, the EP-Green produced 0.51 kWh more electricity than EP-Black, which represent 1.1% more kWh output. The difference in December was 0.96 kWh, which represents 2.7%. Even though the actual production difference was less than 1 kWh, the percentage was higher than other months with greater production differences.

For this study, the highest ambient air temperatures were during June and July and the greatest variations in energy productions were also during these months. The impact of roofing types on the performance of PV panels is expected to be higher for sites with more days of ambient air temperatures above 25 °C. The height of PV panels from the roof surface was constant in the implementation of this study. The distance between the PV panels and the roof surface can also impact the thermal interaction. For a future study, it is recommended to study the performance of PV panels over a green roof and a conventional roof with different distances from the roof surface.

**Author Contributions:** Jae D. Chang and Mohammed J. Alshayeb worked together to form the project idea, research methodology, and execution plan. Mohammed J. Alshayeb was responsible for the daily activities and for preparing the experimental bed, monitoring, collecting and analyzing data. Jae D. Chang supervised the project, ensured the quality of the work, and accuracy of the data.

**Acknowledgments:** This project was supported in part by Enphase Energy Inc. (Petaluma, CA, USA) through their donation of Enphase Microinvertors which were used to monitor the performance of each PV panel. The Center for Design Research provided the researchers access to their PV panels.

**Conflicts of Interest:** The authors declare no conflict of interest.

## References

1. Department of Energy (DOE). *International Energy Outlook 2013*; Report Number: DOE/EIA-0484; US Energy Information Administration (EIA): Washington, DC, USA, 2013.
2. Franklin, C.; Chang, J. Energy Consumption Monitors: Building Occupant Understanding and Behavior. In Proceedings of the ARCC Conference Repository, University of North Carolina, Charlotte, NC, USA, 27–30 March 2013.
3. Architecture2030. *The 2030 Challenge*; Architecture2030: Santa Fe, NM, USA, 2016.
4. Department of Energy (DOE). *International Energy Outlook 2014*; Report Number: DOE/EIA-0484; US Energy Information Administration (EIA): Washington, DC, USA, 2014.
5. Renewable Energy Policy Network for the 21st Century. *Renewables 2015 Global Status Report: Renewables 2015 Global Status Report*; REN 21: Paris, France, 2015; ISBN 978-3-9815934-6-4.
6. International Energy Agency (IEA). *Renewables 2017 Analysis and Forecasts to 2022*; Organisation for Economic Co-operation and Development (OECD)/IEA: Paris, France, 2017.
7. International Energy Agency (IEA). *Renewables Energy Medium-Term Market Report 2022*; OECD/IEA: Paris, France, 2016.
8. Nagengast, A.; Hendrickson, C.; Matthews, S.H. Variations in photovoltaic performance due to climate and low-slope roof choice. *Energy Build.* **2013**, *64*, 493–502. [CrossRef]
9. Scherba, A.; Sailor, D.J.; Rosenstiel, T.N.; Wamser, C.C. Modeling impacts of roof reflectivity, integrated photovoltaic panels and green roof systems on sensible heat flux into the urban environment. *Build. Environ.* **2011**, *46*, 2542–2551. [CrossRef]
10. US Energy Information Administration. *Annual Energy Outlook 2016: With Projections to 2040*; US Energy Information Administration, Office of Energy Analysis, US Department of Energy: Washington, DC, USA, 2016.
11. Zheng, C.; Kammen, D.M. An innovation-focused roadmap for a sustainable global photovoltaic industry. *Energy Policy* **2014**, *67*, 159–169. [CrossRef]
12. Zhang, F.; Gallagher, K.S. Innovation and technology transfer through global value chains: Evidence from China's PV industry. *Energy Policy* **2016**, *94*, 191–203. [CrossRef]
13. Huang, P.; Negro, S.O.; Hekkert, M.P.; Bi, K. How China became a leader in solar PV: An innovation system analysis. *Renew. Sustain. Energy Rev.* **2016**, *64*, 777–789. [CrossRef]
14. Gartland, L. *Heat Islands: Understanding and Mitigating Heat in Urban Areas*; Routledge: Abingdon, UK, 2010.
15. Wong, E.; Akbari, H.; Bell, R.; Cole, D. *Reducing Urban Heat Islands: Compendium of Strategies*; Environmental Protection Agency: Washington, DC, USA, 2011.
16. Ferguson, B.; Fisher, K.; Golden, J.; Hair, L.; Haselbach, L.; Hitchcock, D.; Kaloush, K.; Pomerantz, M.; Tran, N.; Waye, D. *Reducing Urban Heat Islands: Compendium of Strategies-Cool Pavements*; Environmental Protection Agency: Washington, DC, USA, 2008.
17. Scholz-Barth, K.; Tanner, S. *Green Roofs: Federal Energy Management (FEMP) Federal Technology Alert*; National Renewable Energy Lab: Golden, CO, USA, 2004.
18. Köhler, M.; Wiartalla, W.; Feige, R. Interaction between PV-systems and extensive green roofs. In Proceedings of the 5th North American Green Roof Conference: Greening Rooftops for Sustainable Communities, Boston, MA, USA, 29 April–1 May 2007.
19. Hui, S.C.; Chan, S.C. Integration of green roof and solar photovoltaic systems. In Proceedings of the Joint Symposium 2011: Integrated Building Design in the New Era of Sustainability, Hong Kong, China, 22 November 2011.
20. Nagengast, A.L. Energy Performance Impacts from Competing Low-Slope Roofing Choices and Photovoltaic Technologies. Ph.D. Thesis, Carnegie Mellon University, Pittsburgh, PA, USA, 2013.

21. Perez, M.J.; Wight, N.; Fthenakis, V.; Ho, C. Green-roof integrated PV canopies—An empirical study and teaching tool for low income students in the South Bronx. *ASES* **2012**, *4*, 6.

22. Chemisana, D.; Lamnatou, C. Photovoltaic-green roofs: An experimental evaluation of system performance. *Appl. Energy* **2014**, *119*, 246–256. [CrossRef]

Article

# Photovoltaics (PV) System Energy Forecast on the Basis of the Local Weather Forecast: Problems, Uncertainties and Solutions

Kristijan Brecl * and Marko Topič

Laboratory of Photovoltaics and Optoelectronics—LPVO, Faculty of Electrical Engineering, University of Ljubljana, Tržaška 25, SI-1000 Ljubljana, Slovenia; marko.topic@fe.uni-lj.si
* Correspondence: kristijan.brecl@fe.uni-lj.si; Tel.: +386-1-4768-848

Received: 29 March 2018; Accepted: 26 April 2018; Published: 4 May 2018

**Abstract:** When integrating a photovoltaic system into a smart zero-energy or energy-plus building, or just to lower the electricity bill by rising the share of the self-consumption in a private house, it is very important to have a photovoltaic power energy forecast for the next day(s). While the commercially available forecasting services might not meet the household prosumers interests due to the price or complexity we have developed a forecasting methodology that is based on the common weather forecast. Since the forecasted meteorological data does not include the solar irradiance information, but only the weather condition, the uncertainty of the results is relatively high. However, in the presented approach, irradiance is calculated from discrete weather conditions and with correlation of forecasted meteorological data, an RMS error of 65%, and a $R^2$ correlation factor of 0.85 is feasible.

**Keywords:** PV systems; forecast; energy; simulation

## 1. Introduction

Photovoltaics (PV) nowadays has become a very mature and developed technology. Beside the technology advantages, photovoltaics also became one of the cheapest electricity sources. Here, we are not speaking about big PV plants, but also about small household PV systems. Of course the PV electricity price in small household systems in competitive only if the whole consumer electricity price, together with all taxes, is considered. To maximize the rate of return, prosumers have to consume as much PV energy as possible by themselves or should be included in a net-metering/net-billing scheme. The system is economically feasible in all countries where the PV electricity met the household grid parity.

After the rapid drop or even a complete abolishment of the feed-in-tariffs for PV across Europe some years ago, the interest in large PV system plants has dropped. On the other hand, due to the low PV system prices per kW, as also due to new supporting schemes, like net-metering in some countries [1], small household PV systems are becoming more and more attractive. The net-metering scheme is in a way very customer-friendly since the prosumers usually do not have to take care about the self-consumption because the used and the produced energy is balanced and charged on monthly or annual basis. A net-metering prosumer has just to balance the total energy consumption and production.

If a prosumer is not included in any incentive scheme, then he or she has an interest to consume the PV energy directly. In this case the PV energy production forecast becomes very important. The way that we are living today in the developed world is changing. While on one hand, more and more people want to be energy independent or just to lower their electricity bill, on the other hand, new electricity appliances, like smart household appliances, smart heating systems, and last but not least, electric cars, are penetrating into households. Especially, electric cars will affect the household electrical energy and power demand from grid tremendously. However, to keep the electric system reliably like it was in the past and to simultaneously introduce renewable energy sources and new energy intensive consumers, we need to develop a reliable forecasting system that will help to balance the consumption and production.

There are already several PV power and energy forecasting systems or services [2–6], but they are in general too complicated or too expensive for a household prosumer. Additionally, since the electric consumers are not fully automated in a smart-house system [7,8], the user has to manually decide when to use which appliance as a load. i.e., if the prosumer has the energy forecast for the next day he can decide whether to use the washing machine today or tomorrow and at what time; whether to heat up the water tank for the central heating with the electricity form the grid today or to keep the minimum requirements and wait until tomorrow for the sun; and, whether to charge the electric car or the battery bank above the required capacity or not.

In this paper we would like to present a way to predict the power and energy production with the use of the local meteorological forecast and public available models. Since the reliability and quality of the data is low, we will describe all of the problems and solutions when using these data. Additionally, all of the results will be presented together with their uncertainties.

## 2. Energy Forecasting

There are two different ways of PV system energy forecasting: annual and daily. The first solution is used when a new PV system is planned [9]. The calculations are based on the average annual irradiation data on the location. The irradiation data can be obtained from the actual measured values at a local meteorological station [10], from measured satellite data [11], or modelled [12–14]. Measured data are usually averaged over a certain period of time, i.e., over the last 10 years to get the data as reliable as possible. Also, the modelled data are based on measured data from selected meteorological stations and then inter- or extrapolated to the desired location. In addition, to the irradiation data, we have to know or determine the local horizon to estimate the shading losses of the planed PV system. Simple PV system energy forecasting calculations use just these data and estimate all other losses (mismatch losses, temperature losses, Joule losses in cabling, inverter losses) with a single empirically defined parameter [15]. More comprehensive calculations also take into account the information of the mean temperature at the location, calculate the mismatch losses with regard to the selected PV module, and include a simulation of cabling and inverter losses. The result of this energy forecast simulation is usually the expected PV system final energy yield ($Y_F$), which is given in kWh/kW, or in some cases, also the expected performance ratio ($PR$). The uncertainty of the calculations is relatively low and can lie below 10% when good quality data is used.

The daily energy forecasting methodology is more complicated since it predicts the actual energy output of a PV system for the next day. The result is usually given on an hourly basis. This energy forecast became more and more important in the recent years when the share of the renewable energy sources became noticeable in the whole electricity portfolio of the local energy distributor. For an energy distributor, it is crucial to know the behaviour of very dynamic energy sources to maintain the electricity supply reliable and stable. But, beside the big energy distributors, the PV energy forecasting is important also to the investor, especially if the produced energy is self-consumed. A good and reliable PV energy forecasting will become more and more important in the future with smart houses, smart appliances, and last but not least, with electric cars. An energy balance in households with self-consumption will become very important since the interest of the prosumer is to consume as much self-produced energy as possible.

Currently, there are already several companies that offer a complete PV monitoring system, together with the energy forecasting service [4,5]. The forecasting usually depends on the satellite meteorological data and weather forecasting simulation software. However, for a prosumer with a small PV system of several kW, this service is too expensive or is too complicated.

To our knowledge, there is no model on the market that relies only on the local weather forecast without the knowledge of the forecasted irradiance value. The existing short-term forecasting models start with the forecasted irradiance value [16–19]. According to [18], forecasting global solar irradiance is the same problem as forecasting the PV power output. The modelling of the clearness index is usually done by an auto regressive function [12,18] or neural networks [20]. In our approach the PV, energy forecast is purely based on the weather forecast of the local meteorological agency [10]. Since we wanted to make a model that is capable to predict the PV energy output on the current weather forecast without the knowledge of the weather history, we could not use any auto regression or self-learning techniques.

## 3. Modelling

Using the weather forecast data of the local meteorological station, the solar irradiation information is hidden in the weather type information like cloudy, partially sunny, sunny … Beside the weather type information, also temperature information and forecast of precipitation are available. While the accuracy of the temperature forecast is usually acceptable, the main problem presents the irradiation forecast, since no irradiance value is given.

The modelling methodology in the paper is divided in three sections with regard to modelled parameter: irradiation modelling form the actual weather forecast, irradiation transposition to plane-of-array, and PV system energy output modelling.

The methodology is based on the daily weather forecast and data of the last four years (2014–2017) by the Slovenian meteorological agency—ARSO [10]. Each forecast was issued in the early hours of the day and it is valid for the day of issuing and two consecutive days.

### 3.1. Weather Conditions—Irradiation Modelling

The forecast weather data is given in three-hour period for the issuing date and two consecutive days. The data includes five different weather conditions: clear sky, mostly clear, partly cloudy, prevailingly cloudy, and overcast. Additionally, there is information of the precipitation in three steps: light, moderate and heavy rain or snow-fall. Beside the weather conditions, also the ambient temperature and wind speed data is available. Wind data was not considered in the paper, since the local wind conditions usually deviate largely from the forecasted regionally valid value.

The weather data we divide into five main classes (*wc*) (marked from 0 to 4 in the paper) plus one additional class for a really dark weather conditions with heavy rainfalls (marked as −1). To each class, we define a clearness index ($K_t$). The clearness index was derived from the measured irradiance value of the weather report at the same meteorological station and calculated clear sky irradiance ($G_{cs}$). The value is an average clearness index at the selected weather condition, since this approach gave better results than the regression analysis over all the values. We used two clear sky models (Inechien [21] and Heliosat-1 [22]) however, the Heliosat-1 clear sky model gave better results and is presented in this study. Separately, we derived the average clearness index also for overcast days with a heavy rainfall. The data is collected and presented in Table 1.

**Table 1.** Weather data and average clearness indexes for the selected weather condition.

| Weather Condition | Weather Class (*wc*) | $K_t$ |
|---|---|---|
| Clear | 4 | 1.02 |
| Mostly clear | 3 | 0.97 |
| Partly cloudy | 2 | 0.83 |
| Prevailingly cloudy | 1 | 0.60 |
| Overcast | 0 | 0.37 |
| *Overcast + heavy rain* | −1 | 0.20 |

Over the derived data we made a 5th order polynomial interpolation that was used in the further calculations of the clearness index (Figure 1). The clearness indexes were derived for the location

Ljubljana, but could be valid for any location in the central Europe where the weather is forecasted in five classes.

$$K_t = 0.0014 \cdot wc^5 - 0.0108 \cdot wc^4 + 0.0128 \cdot wc^3 + 0.0252 \cdot wc^2 + 0.1836 \cdot wc + 0.3804 \tag{1}$$

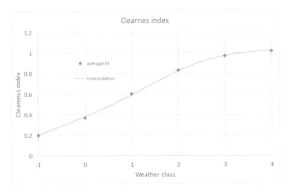

**Figure 1.** Polynomial interpolation of the clearness index over weather classes.

During the input data analysis, we could observe some correlation between the accuracy of the defined clearness index and the constancy of the forecasted weather, as also large changes between the forecasts of the prevailing, current and successive hour. The actual value of the $wc$ was therefore in the first round taken from the weather forecast for the selected hour and in the second round calculated from the previous forecasts ($k_{0,i}$, $k_{-1,i}$, $k_{-2,i}$) for the observed $i$-th hour and the forecast of the preceding ($k_{0-1h,i}$) and successive ($k_{0+1h,i}$) hour. The actual expected weather condition was defined as the weighted arithmetic mean of the five weighted forecasted values (Equation (2)). The weights were defined by the optimisation procedure with the RMS error as a criterion. The resulting weights are presented in Table 2.

$$wc_i = \frac{W_{k_{0,i}} * k_{0,i} + W_{k_{-1,i}} * k_{-1,i} + W_{k_{-2,i}} * k_{-2,i} + W_{k_{0-1h,i}} * k_{0-1h,i} + W_{k_{0+1h,i}} * k_{0+1h,i}}{W_{k_{0,i}} + W_{k_{-1,i}} + W_{k_{-2,i}} + W_{k_{0-1h,i}} + W_{k_{0+1h,i}}} \tag{2}$$

**Table 2.** Weights for the corresponding forecasted weather class.

| Weather Condition | $W_{wc}$ Filtered Data $-1 \leq (wc_{simulated} - wc_{actual}) \leq 1$ | | | | |
|---|---|---|---|---|---|
| | $W_{k_{0,i}}$ | $W_{k_{-1,i}}$ | $W_{k_{-2,i}}$ | $W_{k_{0-1h,i}}$ | $W_{k_{0+1h,i}}$ |
| Clear | 20 | 0 | 1 | 0 | 2 |
| Mostly clear | 16 | 0 | 0 | 1 | 0 |
| Partly cloudy | 10 | 2 | 0 | 1 | 0 |
| Prevailingly cloudy | 19 | 3 | 1 | 3 | 1 |
| Overcast | 12 | 20 | 1 | 7 | 4 |

Table 2 shows that the weights of the latest forecast $W_{k0,\,i}$ are dominant in all cases, except in the case of an overcast weather condition. The influence of the precipitation was taken into account in a way that the light rainfall changes the $wc$ for one third, moderate rainfall for two thirds, and heavy rain for the whole weather class. The solar irradiance for the observed location was calculated by multiplying the clearness index with the clear sky irradiance value.

Since only five weather classes are given in the weather forecast, the reliability of the irradiance is not at the highest level. The uncertainties are presented by the Root-Mean-Square (RMS) error of the relative deviation of the measured value [10] from the simulated one, and the R-squared ($R^2$) factor, which is similar to the correlation factor:

$$R^2 = \left( \frac{n\sum xy - \sum x \sum y}{\sqrt{[n\sum x^2 - (\sum x)^2][n\sum y^2 - (\sum y)^2]}} \right)^2 \tag{3}$$

Figure 2 shows the measured vs. simulated hourly irradiation for Ljubljana, Slovenia for the last three and a half years. The RMS error of the relative difference between the measured and the simulated value is 44%. The $R^2$ factor is 0.79.

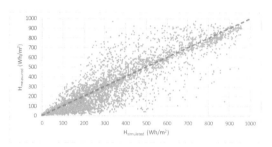

**Figure 2.** Measured vs. simulated solar hourly irradiation of the horizontal surface (red line is to guide your eyes and presents an ideal case).

To improve the reliability of the data and to remove all of the outliers, we decided to work only with data where the actual weather condition deviates from the forecasted one for not more than one weather class. In the four years around 65% of the data match the criteria. The results are presented in Figure 3. The RMS error is much lower (34%) and the $R^2$ factor reaches 0.91. The main problem to simulate the irradiance is at overcast conditions where the solar irradiance is relatively low. Table 3 presents the RMS error and the $R^2$ factor for all weather conditions separately.

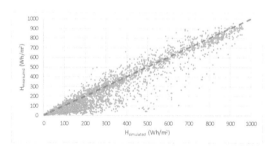

**Figure 3.** Measured vs. simulated hourly irradiation at $-1 \leq (wc_{forecast} - wc_{measured}) \leq 1$(red line is to guide your eyes).

**Table 3.** Simulation errors with regard to the weather class.

| Weather Class | $R^2$ | RMS |
|---|---|---|
| Clear | 0.99 | 9.0% |
| Mostly clear | 0.95 | 13.7% |
| Partly cloudy | 0.80 | 29.4% |
| Prevailingly cloudy | 0.64 | 39.7% |
| Overcast | 0.52 | 47.5% |
| All weather conditions | 0.91 | 34.1% |

When only the latest forecasted weather class value is used, the RMS error over all weather conditions changes very slightly to a value of 33.7%. However, the use of averaged weighted forecasted values is advisable when the whole input data (not limited to $-1 \leq (wc_{forecast} - wc_{measured}) \leq 1$) is used. In that case, the RMS error drops from 55.9 to 46.7%.

### 3.2. Transposition of Irradiation to Plane-of-Array

To calculate the plane-of-array irradiation, we have to know the diffuse share of the irradiation. Several transposition models were tested [23–26], but finally our previously developed solar irradiation model SISIM [27] with the modified Louche [23] global to diffuse model was used to calculate the diffuse part of the light. Plane-of-array irradiation was calculated by the most used and cited Perez et al. [28] model. According to our previous work, the error of the horizontal to plane-of array irradiation model is around 10% for irradiance values over 200 W/m$^2$ and 40% for low irradiance conditions when the transposition to south 30° is made.

In our case, we transform the irradiation to an orientation of 25° east and an inclination of 30°, which is the position of our PV system that is used as a reference and verification object for our study (Figure 4). The R$^2$ factor between the simulated and measured values is 0.98, while the RMS error is 33%.

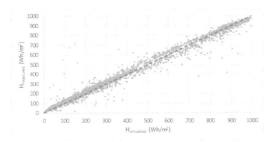

**Figure 4.** Plane-of-array irradiation model (red line is to guide your eyes).

### 3.3. PV System Output Modelling

There are several ways how to model the power output of the PV system. The simplest way is to calculate with the basic equation [29]:

$$P_{sim,i} = P_{rated} * \frac{G_{sim,i}}{G_{STC}} * (1 + \gamma * (T_{module,i} - 25\,°C)) * PLF, \qquad (4)$$

where $P_{rated}$ is the rated PV system power, $G_{sim,i}$ the simulated/measured irradiance, $G_{STC} = 1000$ W/m$^2$, $T_{module,i}$ module temperature and $\gamma$ power temperature coefficient of the module, and $PLF$ is the power loss factor. The drawback of the equation is the absence of the system losses and the influence of the degradation mechanisms. However, all these influences can be included in the equation as a $PLF$, which is due to the degradation time dependent.

The other was is to use analytical or heuristic models to calculate the PV system performance. In general, these models need some additional information, like $I$-$V$ curve of a module or at least the measured weather data at the observed location, which are usually not available for an owner of a household size PV system. While the analytical models, that need knowledge of many measured $I$-$V$ curves at different combinations of irradiance and temperature, heuristic models rely on measured power output and weather data. The benefit of the heuristic model is that the parameters can be frequently and automatically defined from the PV system data, which means that the models automatically include the influence of the degradation mechanisms, seasonal or other temporary changes in the energy output. Anyhow, measurements of the PV system power, solar irradiance, and ambient or module temperature are needed. Several heuristic model are already available [30–33].

They differ from each other in how many coefficients they need and the type of interpolation (linear, bilinear, non-linear). We decided to use a relatively simple but efficient non-linear model with three coefficients [30].

$$P_{sim}'(G_{poa}) = a * G_{poa} + b * G_{poa}^2 + c * G_{poa} * \ln \frac{G_{poa}}{G_{STC}}, \tag{5}$$

where $G_{poa}$ is plane-of-array irradiance and a, b, c fitting parameters that were derived from the measurements of our PV system [33]. Since the model parameters are usually extracted from temperature compensated values, the temperature compensation is added to the model:

$$P_{sim} = P_{sim}' * (1 + \gamma * (T_{module,i} - 25 \,^{\circ}\text{C})). \tag{6}$$

To improve the model even further, especially at lower irradiance values at larger incidence angles, we can add incidence angle dependence factor [34]:

$$F(\theta_s) = 1 - \frac{e^{(-\frac{\cos\theta_s}{a_r})} - e^{(-\frac{1}{a_r})}}{1 - e^{(-\frac{1}{a_r})}}, \tag{7}$$

where $\theta_s$ is the solar incidence angle and $a_r$ the angular loss factor, typically 0.2.

In Figure 5, the measured vs. simulated PV system output is presented. The RMS error of the heuristic and linear model is 12% and 14%, respectively.

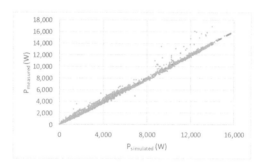

**Figure 5.** Measured vs. simulated photovoltaics (PV) system output power (red line is to guide your eyes).

## 4. Results

The PV system forecast methodology that is presented in the previous section is applied to a rooftop PV system for verification. The PV system is mounted on the roof of Faculty of Electrical Engineering in Ljubljana (46.067° N, 14.517° E), Slovenia. The orientation of PV modules is 25° east with an inclination angle of 30°. There are no obstacles that could shade the PV modules and influence the PV system energy production. A photo of the PV system is presented in Figure 6.

**Figure 6.** LPVO PV system on the roof of Faculty of Electrical Engineering in Ljubljana.

The PV system was installed at the end of 2010, and since June 2014 the weather forecast data from ARSO [10] are logged.

### 4.1. PV System Energy Forecast Validation

As already mentioned, we simulate the PV system energy output on the forecasts from the local meteorological station and use the measured data of irradiance, temperature, and PV system energy production for verification from our PV system test site. To present all of the steps together with the uncertainties of the results we used the weather data that deviate from the actual measured conditions for not more than one weather class.

The solar irradiation on a horizontal plane with the RMS error of 34.1% and $R^2$ of 0.91 was already presented in Figure 3. Transposition of the simulated solar irradiation to the plane-of-array adds another 15% of error to the results (Figure 7). The RMS error of the plane-of-array irradiation is 44.4%, while the correlation factor $R^2$ is 0.89.

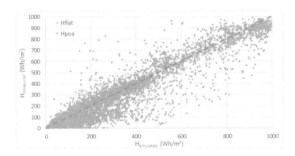

**Figure 7.** Transposition of the solar irradiation to plane-of-array (red line is to guide your eyes).

From the plane-of-array irradiation values, we calculated the predicted energy output of the modules. The module temperature was calculated from the solar irradiance and ambient temperature values by the Equation [35]:

$$T_{module} = T_{ambient} * (T_{NOCT} - 20\,°C) \frac{G}{800\,\frac{W}{m^2}} \qquad (8)$$

where $T_{NOCT}$ is the module temperature at nominal operating conditions (*NOCT*: $G = 800$ W/m$^2$, $T_{ambient} = 20\,°C$). $T_{NOCT}$ of the modules is 44 °C. The predicted PV system energy output using Equations (4) and (5) vs. the measured PV system energy output is presented in Figure 8.

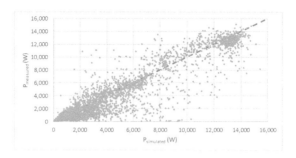

**Figure 8.** Simulated vs. measured power output of the PV system (red line is to guide your eyes).

The RMS error is 64.6% and the $R^2$ factor 0.85. The errors for each weather class are presented in Table 4.

**Table 4.** Simulation errors with regard to the weather class.

| Weather Class | $R^2$ | RMS |
|---|---|---|
| Clear | 0.90 | 22.6% |
| Mostly clear | 0.82 | 24.6% |
| Prevailingly cloudy | 0.72 | 41.3% |
| Mostly overcast | 0.50 | 61.0% |
| Overcast | 0.38 | 91.8% |
| All weather conditions | 0.85 | 64.6% |

*4.2. PV Energy Output Forecast for Three Successive Days*

While in the previous subsection, we were using the filtered forecast data to present the best possible results, we are here using the unfiltered forecast data like they are available to the end user. In Figure 9 and Table 5, the forecasted energy output and uncertainties are presented for the three successive days. The correlation factor $R^2$ drops from the first to the third forecasted day, while the RMS error of the third day is lower, which could be due to a better weather forecast correlation.

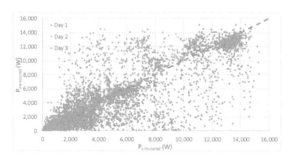

**Figure 9.** Simulation of PV system power output for three consecutive days (red line is to guide your eyes).

**Table 5.** Simulation errors for three consecutive days.

| Forecast | $R^2$ | RMS |
|---|---|---|
| First day | 0.85 | 75.4% |
| Second day | 0.73 | 87.7% |
| Third day | 0.72 | 69.8% |

## 5. Discussion

It is not so easy to get a reliable power energy output forecast of PV system of the forthcoming days if the high quality data is not available. The most important parameter that influences the reliability of the forecasted results is irradiance. If the value of the solar irradiance is not directly modelled and forecasted, then we have to take into account large deviations in the results. When the irradiance is calculated from discrete weather conditions, we have to be aware that just the discretisation introduces an error of around 20%. This value we could observe when we compare the measured irradiance data with the irradiance data simulated from the observed discrete weather conditions that exclude the forecasting uncertainty. The forecasting uncertainty is the highest in our research. In the observed four years only 65% of the forecasted weather conditions differ from the actual one for maximum one class.

When including the forecast uncertainty, the plane-of-array transposition, and PV system output simulation uncertainty, we end at an RMS error of more than 80%. However, the $R^2$ factor that shows the correlation between the measured and simulated data is still over 0.80. The main deviations we could observe at mostly sunny and partly cloudy day, where scattered clouds can have no influence on the energy production although we predicted a lower production. On the other hand, we could observe forecasted and observed clear sky weather conditions, but a low irradiance level due to fog in the winter months. But, this is related to the weather forecast data quality. Our weather forecast source does not have the information about the fog.

To improve the reliability of the PV energy output forecast from the weather condition data we have to constantly calculate the averaged expected $wc$ (Equation (2)) and detect the weather stable and non-stable periods. During our work, we could improve the reliability for some forecasted unstable weather periods with the use of the $wc$ information of previous days and preceding forecasts, but the solutions were not applicable to the whole set of available data, since no correlation of the behaviour between these unstable weather periods could be found.

## 6. Conclusions

We have presented a PV energy forecast solution from public available weather forecast data. Since the forecast data is usually given in discrete weather conditions, the simulation of the needed hourly irradiance data is a subject of lower reliability. However, following the results and the correlation between the simulation and measurements, the methodology is useful for PV energy prediction in households where prosumers do not have the weather parameter data and do not want to pay for a PV system monitoring and forecast service.

When only discrete weather conditions are available, we have no information about the solar irradiance, which is crucial to predict the PV system power output. The solar irradiance level has to be simulated out of the selected clearness index and the calculated clear sky irradiance. Only this step due to discrete weather conditions introduces an RMS error of around 35%. To that value, we have to add the transposition RMS error of at least of 30% and power output prediction RMS error of 20%. Finally, the RMS of the resulted PV system energy output over four years is 85%. With the correlation of forecasted data, we could achieve an RMS error of 65%. In favor of this simple approach speaks also the $R^2$ factor, which is around 0.85 for the forecasted energy power output.

However, we have to be aware that the results include also the weather forecast uncertainty. In our set of data, we could observe that only 65% of the forecasted $wc$ deviates from the observed one for one class, and only 43% of the forecasts met the actual condition. Additionally, we get a 20% error due to the weather classes instead of using an irradiance model. The uncertainties of the used transposition and PV output models are 33% and 12%, respectively.

Our simulations showed that due to high uncertainties in the weather data, also the simplest linear equation for power output could be used.

**Author Contributions:** K.B. analyzed the data and develops the methodology in cooperation with M.T. K.B wrote the manuscript and M.T. reviewed and improved the manuscript.

**Acknowledgments:** The authors acknowledge the financial support of the Slovenian Research Agency (Research Programme P2-0197).

**Conflicts of Interest:** The authors declare no conflict of interest.

## References

1.  GmbH, Eclareon RES Legal Europe. Available online: http://www.res-legal.eu/search-by-country/ukraine/single/ (accessed on 5 March 2018).
2.  Nnergix, Data Mining & Forecasting Services for Energy Markets. Available online: https://www.nnergix.com/ (accessed on 5 March 2018).
3.  SteadySun–Solar Power Forecasting Services. Available online: http://steady-sun.com/ (accessed on 5 March 2018).
4.  Solar Radiation and PV Power Forecast. Available online: https://solargis.com/products/forecast/overview/ (accessed on 5 March 2018).
5.  Solar Power Forecasting–Meteocontrol. Available online: https://www.meteocontrol.com/en/services/solar-power-forecasting/ (accessed on 5 March 2018).
6.  EuroWind–Making Renewables Predictable: Forecasts and Actual Data. Available online: http://www.eurowind.info/en/services/forecasts-and-actual-data/ (accessed on 5 March 2018).
7.  Zipperer, A.; Aloise-Young, P.A. Electric Energy Management in the Smart Home: Perspectives on Enabling Technologies and Consumer Behavior: Preprint. 2013; p. 12. Available online: https://www.nrel.gov/docs/fy13osti/57586.pdf (accessed on 5 March 2018).
8.  Tischer, H.; Verbic, G. Towards a smart home energy management system—A dynamic programming approach. In Proceedings of the 2011 IEEE PES Innovative Smart Grid Technologies, Perth, Australia, 13–16 November 2011; pp. 1–7.
9.  PVSYST. Available online: http://www.pvsyst.com/en/ (accessed on 5 March 2018).
10. Slovenian Environment Agency. Available online: http://www.arso.gov.si/en/ (accessed on 5 March 2018).
11. Hersbach, H.; Peubey, C.; Simmons, A.; Berrisford, P.; Poli, P.; Dee, D. ERA-20CM: A twentieth century atmospheric model ensemble. *Q. J. R. Meteorol. Soc.* **2015**, *141*. [CrossRef]
12. Meteonorm Meteonorm: Irradiation Data for Every Place on Earth. Available online: http://www.meteonorm.com/en/ (accessed on 5 March 2018).
13. JRC's Directorate C. Energy, Transport and Climate–PVGIS–European Commission. Available online: http://re.jrc.ec.europa.eu/pvgis/ (accessed on 5 March 2018).
14. Simón-Martín, M.; Alonso-Tristán, C.; Montserrat, D.-M. Diffuse solar irradiance estimation on building's façades: Review, classification and benchmarking of 30 models under all sky conditions. *Renew. Sustain. Energy Rev.* **2017**, *77*, 783–802. [CrossRef]
15. Marion, B.; Adelstein, J.; Boyle, K.; Hayden, H.; Hammond, B.; Fletcher, T.; Canada, B.; Narang, D.; Kimber, A.; Mitchell, L.; et al. Performance parameters for grid-connected PV systems. In Proceedings of the 2005 Conference Record of the Thirty-First IEEE Photovoltaic Specialists Conference, Lake Buena Vista, FL, USA, 3–7 January 2005; pp. 1601–1606.
16. Malvoni, M.; De Giorgi, M.G.; Congedo, P.M. Forecasting of PV Power Generation using weather input data-preprocessing techniques. *Energy Procedia* **2017**, *126*, 651–658. [CrossRef]
17. Da Silva, F.J.J.G.; Takashi, O.; Hideaki, O.; Takumi, T.; Kazuhiko, O. Regional forecasts of photovoltaic power generation according to different data availability scenarios: A study of four methods. *Prog. Photovolt. Res. Appl.* **2014**, *23*, 1203–1218. [CrossRef]
18. Bacher, P.; Madsen, H.; Nielsen, H.A. Online short-term solar power forecasting. *Sol. Energy* **2009**, *83*, 1772–1783. [CrossRef]
19. Paulescu, M. (Ed.) *Weather Modeling and Forecasting of PV Systems Operation*; Green Energy and Technology; Springer: London, UK, 2012; ISBN 978-1-4471-4648-3.
20. Sfetsos, A.; Coonick, A.H. Univariate and multivariate forecasting of hourly solar radiation with artificial intelligence techniques. *Sol. Energy* **2000**, *68*, 169–178. [CrossRef]

21. Ineichen, P.; Perez, R. A new airmass independent formulation for the Linke turbidity coefficient. *Sol. Energy* **2002**, *73*, 151–157. [CrossRef]
22. Gueymard, C.A. Clear-sky irradiance predictions for solar resource mapping and large-scale applications: Improved validation methodology and detailed performance analysis of 18 broadband radiative models. *Sol. Energy* **2012**, *86*, 2145–2169. [CrossRef]
23. Louche, A.; Notton, G.; Poggi, P.; Simonnot, G. Correlations for direct normal and global horizontal irradiation on a French Mediterranean site. *Sol. Energy* **1991**, *46*, 261–266. [CrossRef]
24. Maxwell, E.L. *A Qusi-Phisycal Model for Converting Hourly Global Horizontal to Direct Normal Insolation 1987*; Solar Energy Research Institute: Golden, CO, USA, 1987.
25. Reindl, D.T.; Beckman, W.A.; Duffle, J.A. Diffuse Fraction Correlations. *Sol. Energy* **1990**, *45*, 1–7. [CrossRef]
26. Batlles, F.J.; Rubio, M.A.; Tovar, J.; Olmo, F.J.; Alados-Arboledas, L. Empirical modeling of hourly direct irradiance by means of hourly global irradiance. *Energy* **2000**, *25*, 675–688. [CrossRef]
27. Brecl, K.; Topič, M. Development of a Stochastic Hourly Solar Irradiation Model. *Int. J. Photoenergy* **2014**, *2014*, 1–7. [CrossRef]
28. Perez, R.; Ineichen, P.; Seals, R.; Michalsky, J.; Stewart, R. Modeling daylight availability and irradiance components from direct and global irradiance. *Sol. Energy* **1990**, *44*, 271–289. [CrossRef]
29. Kurnik, J.; Jankovec, M.; Brecl, K.; Topic, M. Outdoor testing of PV module temperature and performance under different mounting and operational conditions. *Sol. Energy Mater. Sol. Cells* **2011**, *95*, 373–376. [CrossRef]
30. Ding, K.; Ye, Z.; Reindl, T. Comparison of Parameterisation Models for the Estimation of the Maximum Power Output of PV Modules. *Energy Procedia* **2012**, *25*, 101–107. [CrossRef]
31. Huld, T.; Gottschalg, R.; Beyer, H.G.; Topič, M. Mapping the performance of PV modules, effects of module type and data averaging. *Sol. Energy* **2010**, *84*, 324–338. [CrossRef]
32. Kirn, B.; Brecl, K.; Topic, M. A new PV module performance model based on separation of diffuse and direct light. *Sol. Energy* **2015**, *113*, 212–220. [CrossRef]
33. Kirn, B.; Topic, M. Diffuse and direct light solar spectra modeling in PV module performance rating. *Sol. Energy* **2017**, *150*, 310–316. [CrossRef]
34. Martin, N.; Ruiz, J.M. Calculation of the PV modules angular losses under field conditions by means of an analytical model. *Sol. Energy Mater. Sol. Cells* **2001**, *70*, 25–38. [CrossRef]
35. Models to Predict the Operating Temperature of Different Photovoltaic Modules in Outdoor Conditions. Mora Segado, 2014, Progress in Photovoltaics: Research and Applications–Wiley Online Library. Available online: http://onlinelibrary.wiley.com/doi/10.1002/pip.2549/full (accessed on 5 March 2018).

*Article*

# Quantitative Prediction of Power Loss for Damaged Photovoltaic Modules Using Electroluminescence

**Timo Kropp \*, Markus Schubert and Jürgen H. Werner**

Institute for Photovoltaics and Research Center SCoPE, University of Stuttgart, 70569 Stuttgart, Germany; Markus.Schubert@ipv.uni-stuttgart.de (M.S.); Juergen.Werner@ipv.uni-stuttgart.de (J.H.W.)
\* Correspondence: timo.kropp@ipv.uni-stuttgart.com, Tel.: +49-711-685-67246

Received: 13 April 2018; Accepted: 4 May 2018; Published: 7 May 2018

**Abstract:** Electroluminescence (EL) is a powerful tool for the *qualitative* mapping of the electronic properties of solar modules, where electronic and electrical defects are easily detected. However, a direct *quantitative* prediction of electrical module performance purely based on electroluminescence images has yet to be accomplished. Our novel approach, called "EL power prediction of modules" (ELMO) as presented here, used just two electroluminescence images to predict the electrical loss of mechanically damaged modules when compared to their original (data sheet) power. First, using this method, two EL images taken at different excitation currents were converted into locally resolved (relative) series resistance images. From the known, total applied voltage to the module, we were then able to calculate absolute series resistance values and the real distribution of voltages and currents. Then, we reconstructed the complete current/voltage curve of the damaged module. We experimentally validated and confirmed the simulation model via the characterization of a commercially available photovoltaic module containing 60 multicrystalline silicon cells, which were mechanically damaged by hail. Deviation between the directly measured and predicted current/voltage curve was less than 4.3% at the maximum power point. For multiple modules of the same type, the level of error dropped below 1% by calibrating the simulation. We approximated the ideality factor from a module with a known current/voltage curve and then expand the application to modules of the same type. In addition to yielding *series* resistance mapping, our new ELMO method was also capable of yielding *parallel* resistance mapping. We analyzed the electrical properties of a commercially available module, containing 72 monocrystalline high-efficiency back contact solar cells, which suffered from potential induced degradation. For this module, we predicted electrical performance with an accuracy of better than 1% at the maximum power point.

**Keywords:** silicon; photovoltaics; modules; electroluminescence; defects; cracks

---

## 1. Introduction

Electroluminescence (EL) imaging is a powerful tool to delineate the local and overall electrical and electronic properties of photovoltaic (PV) modules [1–6]. Mapping of EL allows for the characterization of not only single solar cells, but also modules and even complete module strings in large area photovoltaic systems. Thus, this method is capable of detecting electronic or electrical effects on the length scale, from micrometers to tens of meters, depending on the spatial resolution of the camera (and on the diffusion length of carriers). Most commonly, EL images provide *qualitative* information about the *pure existence* of defective parts of cells or modules in a photovoltaic system [1,2]. Sometimes, from the shape of luminescence patterns it is also possible to conclude *on the type* of defects [6]. Moreover, if EL measurements are combined with other characterization techniques, it is sometimes also possible to *quantitatively* predict the electrical performance of the cell, module, or module string [5]. However, so far it has not been possible to make quantitative predictions on the performance of cells, modules, or module strings, *just from EL measurements alone*.

This contribution presents a novel method for the *quantitative* prediction of the electrical properties (i.e., the current/voltage curve) of all cells in a photovoltaic module, just from EL measurements. The method is particularly appropriate for PV modules, which can be damaged by mechanical impact, such as hail, wind, snow, and earthquakes, and/or for modules which suffer from potential induced degradation (PID). In both cases, either the series resistances of the cell fragments or the shunt resistance of a total cell have changed when compared to the original, undamaged state. For these cases, we are able to convert electroluminescence images into either a series resistance or a parallel resistance map. With these maps, together with data from the original data sheet of the undamaged module, we are then able to predict the complete current/voltage curve of the damaged modules. Therefore, we are also able to quantitatively predict the electrical power loss of the damaged modules. The current/voltage curves, which we predict by means of our novel ELMO method are in excellent agreement with the directly measured curves.

## 2. Modelling Principle

### 2.1. Basic Principle of the ELMO Method

The electroluminescent signal in solar cells or modules stems from the recombination of electrons and holes, which, due to applied voltage at the junction, are in non-equilibrium. The luminescence signal therefore locally originates from the diode itself as well as from the bulk of the material within a radial distance of the order of a diffusion length. Luminescent intensity depends on the current across the junction. The junction current, in turn, depends on:

- The junction voltage (which is only part of the externally applied voltage due to series resistances);
- The ideality factor and saturation current density of the diode; and
- The shunt currents that circumvent the diode.

Our ELMO method makes use of two independent principle ideas for modelling mechanically damaged modules as well as PID affected modules.

#### 2.1.1. Series Resistance Mapping of Mechanically Damaged Modules

The mechanical damage of modules, in a first order approximation, does not change the quality of the junction (ideality factor, saturation current density, shunt resistance), but only the series resistances due to broken contact fingers and bus bars, for example. Therefore, it is possible to generate a series resistance map just by comparing two EL images. The first image is taken at a relatively low current (with all the external voltage dropping across the junction). As introduced by Potthoff et al. [3], in this case, the highest luminescence intensity in each cell measures the voltage share (operating voltage) of each cell to the total applied voltage. Potthoff et al. used this method to calculate the overall module series resistance from a second EL image with higher current injection (with parts of the voltage also dropping at the local series resistances as well as connecting resistances between the cells). Here, we extended the approach of Potthoff et al.

We determined the maximal possible luminescence of each cell at high current injection, as predicted from the low current EL image without series resistance dependence.

Consequently, each local series resistance was directly related to the reduced luminescence compared to the maximal luminescence in each cell. In addition, in the high current EL image, the series resistance at the location of maximal luminescence was no longer negligible.

Therefore, we determined the resistance at the location of maximal luminescence and thereby quantified all local series resistance of the cell in respect to the local resistance at the location of maximal luminescence.

For this purpose, based on data sheet information, we calculated the mean series resistance of the originally defect-free cell. We assumed that in each cell at least one partial area persisted with a good connection to the bus bars, which was appropriately described by the respective portion of the mean series resistance of the defect free cell. Therefore, the highest EL signal of each cell in the high current EL image represented the lowest series resistance in this particular cell. These lowest series resistances were in a constant ratio to the mean resistance of the defect-free cell. Therefore, in each cell we quantified its local series resistances relative to the lowest series resistance (i.e., the mean resistance of the defect-free cell) by evaluating its local luminescence intensities relative to the maximal luminescence of the particular cell.

### 2.1.2. Parallel Resistance Mapping of PID Modules

The PID of modules, within the most simplified approximation, does neither change the junction (ideality factor, saturation current density) nor the series resistances, but the parallel resistances of the constituting cells. As a consequence, if there is still one defect free cell left in the module, we are able to calculate all parallel resistances with respect to this defect free cell, which exhibits the highest mean luminescence signal in the low current EL image. This reference cell calibrates the EL intensities of all other cells, at a relatively low current with all the external voltage dropping across the junction. Therefore, the reduction in luminescence of the PID affected cells originates from leakage/shunt currents through the parallel resistances. As a consequence, the low current EL image directly allows for the quantification of the parallel resistance of each cell and maps the parallel resistances of a PID module.

### 2.2. Mathematical Description of ELMO

### 2.2.1. Series Resistance Mapping from EL Images

The simplest and most common approach for describing the electrical properties of solar cells and even total PV modules is the evaluation of the one-diode model. Figure 1a shows the equivalent circuit of the one-diode model, with the photo current source providing the photo generated current density $J_{ph}$, series resistance $r_{i,s}$, the diode parameters with the ideality factor $n_{id}$ and saturation current density $J_0$, as well as the parallel resistance $r_{i,p}$, which reproduces the current($I$)/voltage($V$) curve of each solar cell with cell index i in a module with a total number $N$ of solar cells. Further simplified for a module in the dark ($J_{ph} = 0$), Figure 1b shows the simplified equivalent circuit in the case of an idealized parallel resistance $r_{i,p} = \infty$. The junction voltage $V_i = V_{i,c} - J_{i,c}r_{i,s}$ generates the luminescence intensity $\Phi_i$.

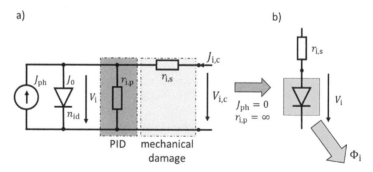

**Figure 1.** Equivalent circuit of each solar cell i in a PV module (**a**) and simplified circuit for an idealized solar cell ($r_{i,p} = \infty$) in the dark ($J_{ph} = 0$) (**b**). Mechanically damaged and cracked solar cells influence the junction voltage $V_i$ by means of a changed series resistance $r_{i,s}$. In contrast, PID can be modeled as a change in parallel resistance of cell i and influences the junction voltage $V_i$ by reducing the current through the junction. The luminescence intensity $\Phi_i$ results from the junction voltage $V_i$.

Specific defects such as mechanically damaged and cracked solar cells only change the local series resistance $r_{i,s}$, i.e., the equivalent connecting resistance of partially separated segments. The diode parameters ($J_0$ and $n_{id}$) are assumed to be unchanged and still the same as for the originally defect free solar cells. In contrast, defects like PID, in the simplest approximation, only change the parallel resistance $r_{i,p}$ of a cell.

These assumptions allow us to calculate the change in series resistance $r_{i,s}$ or parallel resistance $r_{i,p}$ directly from EL images, which essentially map the local junction voltages $V_i'$. Each luminescence intensity captured in each pixel of an EL image or even larger segments of approximately the same intensity are described with equivalent circuits, as shown in Figure 1a,b.

Figure 2 shows the parallel connected spatial distribution of equivalent circuits for evaluating the local luminescence intensities $\Phi_i'(x, y)$ inside a solar cell i with cell voltage $V_{i,c}$. The cell current density $J_c = J_{i,c}$ is identical for all series connected cells inside the module.

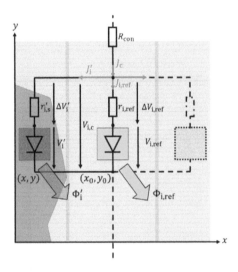

**Figure 2.** Equivalent circuit for referencing the local luminescence $\Phi_i'(x, y)$ to the reference luminescence $\Phi_{i,ref}(x_0, y_0)$, the highest luminescence intensity of cell i. The resistance $R_{con}$ cumulates the contact resistance as well as bus bar resistance between the individual cells. The local junction voltage $V_i'$ generates the local luminescence $\Phi_i'(x, y)$ as part of to the cell voltage $V_{i,c}$ taking the voltage drop $\Delta V_i'$ at each local series resistance $r_{i,s}'(x, y)$ into account. The reference luminescence $\Phi_{i,ref}(x_0, y_0)$ is generated by the local reference junction voltage $V_{i,ref}(x_0, y_0)$.

Since the local luminescence intensities,

$$\Phi_i' = C \, exp\left(\frac{V_i'}{n_{id} V_{th}}\right) \text{ and } \Phi_{i,ref} = C \, exp\left(\frac{V_{i,ref}}{n_{id} V_{th}}\right) \tag{1}$$

captured in an EL image depend exponentially on the local junction voltage $V_i'(x, y)$ of cell i, an increase in local series resistance $r_{i,s}'(x, y)$ decreases the local junction voltage $V_i'(x, y)$ and therefore also decreases the local luminescence intensity $\Phi_i'(x, y)$.

The index i represents the cell number inside the module up to the total number N of cells. Each locally assigned value is indicated by an upper quote throughout this paper. Depending on the spatial resolution of one cell in the EL image, each local value is also mapped by x and y inside each cell i.

Potthoff et al. proposed that in each cell, even with cracks and inactive areas, there is always one spot $(x_0, y_0)$ with good connection to the bus bars [3]. As a consequence, the series resistance $r_{i,ref}(x_0, y_0)$ at this location is the lowest series resistance. Therefore, this part of the cell shows the highest luminescence intensity $\Phi_{i,ref}(x_0, y_0)$ in the EL image. If this EL image is generated with low current density $J_c$ (approximately $J_c < 10\%$ of short circuit density $J_{sc}$ [3]), the voltage drops are negligible across the local reference series resistance $r_{i,\text{ref}}(x_0, y_0)$ as well as connecting resistances $R_{\text{con}}$ between the cells.

Based on this simplification, Potthoff et al. calculated from the low current EL image the contribution of each cell to the overall module voltage $V_{\text{mod}}$ used for generating the EL signal. Their evaluation of the low current EL image quantitatively connected the local luminescence $\Phi'_i(x, y)$, measured in arbitrary units, to the local junction voltages $V'_i(x, y)$ of each cell i in the module with the calibration factor:

$$C = \sqrt[N]{\frac{\prod_{i=1}^{N} \Phi_{i,\text{ref}}}{\exp\left(\frac{V_{\text{mod}}}{V_{\text{th}}}\right)}}, \tag{2}$$

calculated with the reference luminescence intensity $\Phi_{i,\text{ref}}(x_0, y_0)$ of each cell i up to the total number $N$ of cells, the module voltage $V_{\text{mod}}$ for generating the EL signal and the thermal voltage $V_{\text{th}}$ [3].

At this point, we extend Potthoff et al.'s approach. For EL images from higher current injections ($J_c \gtrsim 30\%\ J_{\text{sc}}$), the local series resistance, even at the brightest spot, is no longer negligible. Though, if the series resistance $r_{i,\text{ref}}(x_0, y_0)$ at the brightest spot is well known, all other series resistances $r_{i,s}(x, y)$ can be calculated in relation to the reference series resistance $r_{i,\text{ref}}(x_0, y_0)$ at the brightest spot. Here, we assume that this reference resistance $r_{i,\text{ref}}(x_0, y_0)$ remains unchanged by the structural defect and is therefore directly proportional to the mean series resistance $\bar{r}_{i,s,\text{ds}}$ of the cell in the originally produced defect free module (data sheet module).

Our approach calculates the mean series resistance $\bar{r}_{I,s,\text{ds}}$ of undamaged and all identical cells from the data sheet of the undamaged PV module. The mean series resistance $\bar{r}_{I,s,\text{ds}} = \bar{r}_{s,\text{ds}} = R_{\text{mod},s,\text{ds}}/(A_c N)$ is derived from a lumped one-diode model with the total module resistance $R_{\text{mod},s,\text{ds}}$, the cell area $A_c$ and the number of cells $N$ that represents the data sheet information of the module.

The reference series resistance $r_{i,\text{ref}}(x_0, y_0)$ at the brightest spot always represents the minimal series resistance of the cell. This minimal resistance $r_{i,\text{ref}}(x_0, y_0) = d_{rs}\ \bar{r}_{i,s,\text{ds}}$, is calculated taking the statistical deviation $d_{rs}$ from of the mean resistance $\bar{r}_{i,s,\text{ds}}$ of the defect free cell into account. For the similar spatial fluctuation of the series resistance of defect free cells, the statistical deviation $d_{rs}$ is assumed to be constant and identical for all cells.

Based on the equivalent circuit of Figure 2, we evaluate the local luminescence intensities $\Phi'_i(x, y)$ in each cell i. The influence of parallel resistance is neglected. Therefore, the total local current density $J'_i(x, y)$ is equal to the total current through the local diode and thereby proportional to the local luminescence $\Phi'_i(x, y)$.

Using the exponential dependence between the local current density $J'_i(x, y)$ and the local voltage $V'_i(x, y)$, the local luminescence intensities $\Phi'_i(x, y)$ as well as the reference luminescence $\Phi_{i,\text{ref}}(x, y)$ are proportional to their local current densities,

$$J'_i = J_0\left(\frac{\Phi'_i}{C}\right) \text{ and } J_{i,\text{ref}} = J_0\left(\frac{\Phi_{i,\text{ref}}}{C}\right) \tag{3}$$

at each local region $(x, y)$ of a cell i and the reference region $(x_0, y_0)$ of the same cell. Therefore, the voltage difference,

$$V_{i,\text{ref}} - V'_i = V_{i,c} - \Delta V_{i,\text{ref}} - (V_{i,c} - \Delta V'_i) = \Delta V'_i - \Delta V_{i,\text{ref}} = n_{\text{id}} V_{\text{th}}\left[\ln\left(\frac{\Phi_{i,\text{ref}}}{C}\right) - \ln\left(\frac{\Phi'_i}{C}\right)\right] = n_{\text{id}}\ V_{\text{th}}\ln\left(\frac{\Phi_{i,\text{ref}}}{\Phi'_i}\right) \tag{4}$$

is defined with respect to the local luminescence intensities $\Phi'_i(x, y)$. Combining Equations (3) and (4) results in:

$$n_{\text{id}} V_{\text{th}}\ \ln\left(\frac{\Phi_{i,\text{ref}}}{\Phi'_i}\right) = J'_i\ r'_{i,s} - J_{i,\text{ref}}\ r_{i,\text{ref}} \tag{5}$$

for each region of interest $(x, y)$ with the local series resistance $r'_{i,s}(x, y)$ in relation to the reference region $(x_0, y_0)$ with the local series resistance $r_{i,ref}(x_0, y_0)$. As a result, the local series resistance,

$$r'_{i,s} = \frac{n_{id} V_{th} \, ln\left(\frac{\Phi_{i,ref}}{\Phi'_i}\right) + J_0\left(\frac{\Phi_{i,ref}}{C}\right) d_{rs} \bar{r}_{s,ds}}{J_0\left(\frac{\Phi'_i}{C}\right)} \tag{6}$$

for each individual cell i is calculated combining Equations (3) and (5). The thermal voltage $V_{th}$, the ideality factor $n_{id}$, as well as the saturation current density $J_0$ are assumed to be identical and constant for all cells and unrelated to the structural defect. Thereby, all local luminescence intensities $\Phi'_i(x, y)$ from the high current EL image (compare Figure 1a) are transposed into the series resistance mapping $r'_{i,s}(x, y)$ further discussed in Section 2.3.

Since the statistical deviation $d_{rs}$ of series resistances is at first unknown and itself depends on the calculated series resistance map $r'_{i,s}(x, y)$, we use the bisection method to calculate the factor $d_{rs}$ from Equation (6) iteratively.

We start from a guessed initial statistical deviation $(d_{rs} \gg 1)$ and calculate the module series resistance map $r'_{i,s}(x, y)$ from the high current EL image. Then, we derive the mean series resistance $\bar{r}_{k,s}$ of one defect free cell (i = k) from the sum of all local series resistances $r'_{k,s}(x, y)$ divided by the number of pixels per cell. Using the bisection method, we adjust the statistical deviation $d_{rs}$ iteratively until the mean series resistance $\bar{r}_{k,s}$ derived from series resistance map $r'_{k,s}(x, y)$ equals the series resistance $\bar{r}_{s,ds}$ of one defect-free cell derived from the data sheet. This calibration of the series map with respect to the data sheet allows for the transposition of the EL image into a series resistance input for the simulation model, shown in Section 2.3.

However, the resistance map $r'_{i,s}$ as well as the mean series resistance $\bar{r}_{s,ds}$ derived from the data sheet, depend on the assumed ideality factor $n_{id}$. Therefore, all further simulated results need to be evaluated in relation to an ideality factor expectation interval. As a further approximation, the ideality factors of all cells are assumed to be identical. Nevertheless, these approximations already lead to excellent results when comparing simulated and measured $I/V$ curves.

### 2.2.2. Parallel Resistance Mapping from EL Images

Figure 3 shows the equivalent circuit for evaluating the mean luminescence $\bar{\Phi}_i$ of each cell i in relation to the maximal mean luminescence $\bar{\Phi}_{ref}$ of all cells. Here, the same principle approach as shown in Section 2.2.1 evaluates the total parallel resistance $r_{i,p}$ of each cell i. However, this parallel resistance approximation (PRA) only holds true if the evaluated defect type is limited to change in parallel resistance and is not additionally influencing the local series resistance of the cell.

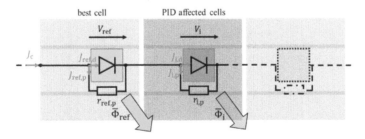

**Figure 3.** Equivalent circuit for referencing the mean luminescence $\bar{\Phi}_i$ of individual cells i (PID affected cells) to the cell with the maximal mean luminescence $\bar{\Phi}_{ref}$ (best cell). The cell current density $J_c$ is identical for all cells connected in series and for each cell i divided into the current $J_{i,p}$ through the parallel resistance and the current density $J_{i,d}$ through the local diode.

The connecting resistance $R_{con}$ between the cells and the local series resistance $r'_{i,s}$ is neglected for the cell current density $J_c < 10\%\ J_{sc}$, used for generating the low current EL image. The cell current $J_c = J_{i,c}$ being the same for every in series connected cell, is divided into parallel resistance current density $J_{i,p}$ and local diode current density $J_{i,d}$. The diode current density $J_{i,d}$ of a cell i is proportional to mean luminescence $\overline{\Phi}_i$, defined analogously by Equation (3). Solving for the parallel resistance current density,

$$J_{ref,d} = J_c - J_{ref,p} \tag{7}$$

at the reference cell with the maximal mean luminescence $\overline{\Phi}_{ref}$ and using Equations (1) and (3) results in the relation,

$$J_0 \frac{\overline{\Phi}_{ref}}{C_{PRA}} = J_c - \frac{n_{id} V_{th}}{r_{ref,p}} ln\left(\frac{\overline{\Phi}_{ref}}{C_{PRA}}\right). \tag{8}$$

In this case, taking into account the reference parallel resistance $r_{ref,p}$, the calibration factor,

$$C_{PRA} = \frac{J_0\ r_{ref,p} \overline{\Phi}_{ref}}{n_{id} V_{th}} W_0 \left\{ \frac{J_0\ r_{ref,p}}{n_{id} V_{th}} \exp\left(\frac{J_0\ r_{ref,p}}{n_{id} V_{th}}\right) \right\}^{-1} \tag{9}$$

is calculated by solving Equation (8) with the main branch $W_0$ of the Lambert-W function. Using Equation (8) analogously for the mean luminescence $\overline{\Phi}_i$ of all other cells instead of the reference cell, the parallel resistance,

$$r_{i,p} = \frac{n_{id} V_{th}\ ln\left(\frac{\overline{\Phi}_i}{C_{PRA}}\right)}{J_c - J_0 \frac{\overline{\Phi}_i}{C_{PRA}}} \tag{10}$$

is calculated for each cell i. The cell current $J_c$ is well-known, since it needs to be adjusted and measured to generate the low current EL image. The recombination current density $J_0$ is calculated from the data sheet information of the PV module and only the ideality $n_{id}$ remains a scalable parameter. Again, all further simulated results will be evaluated in relation to an ideality factor expectation interval.

## 2.3. Series Resistance Segmentation and Module Simulation

Figure 4a shows the high current EL image and Figure 4b the low current EL image of a photovoltaic module with 60 multicrystalline silicon solar cells (cell area $A_c = 243\ cm^2$). In both images, darker regions with lower EL intensity indicate cracked cells with fully or partially disconnected cell areas.

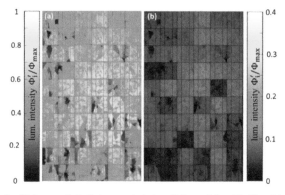

**Figure 4.** False color EL image of a hail-damaged photovoltaic module with 60 multicrystalline silicon solar cells (module A) at (**a**) supplied current of 37% of short circuit current $I_{sc} = 8.3$ A with 100 s exposure time and (**b**) with 7% $I_{sc}$ with 360 s exposure time. The local intensities $\Phi'_i$ in the low current EL image in (**b**) are downscaled proportional to the exposure time ratio. For the false color visualization, both images are normalized to the individual maximal luminescence intensity $\Phi_{max}$ of the high current image in (**a**).

To quantify the impact of those structural defects we convert the luminescence image shown in Figure 4a into a series resistance map. Based on series resistance mapping, our simulation model predicts the overall power loss induced by the change in local series resistance of the individual cells with all cells connected in series.

Starting from the series resistance map, Figure 5 shows the simulation principle for calculating the predicted power loss compared to the data sheet performance. Figure 5a shows the series resistance map $r'_{i,s}(x,y)$ with colored segmentation categories in units of $\Omega$ cm$^2$. The space in between the cells is cut out and excluded for the series resistance segmentation. Figure 5b (red box) qualitatively shows the histogram of categorized series resistances $r'_{i,s}(x,y)$ of one cell i with the relative area $A_{i,seg}/A_c$ of a segment, as a share of the total cell area $A_c$.

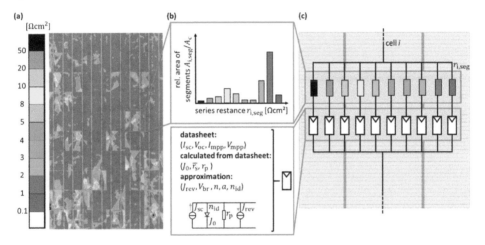

**Figure 5.** False color segmentation of series resistance mapping (**a**). Predefined and separated into ten segments with the relative area $A_{i,seg}/A_c$ as a share of the total cell area $A_c$, all cells are individually modeled, as shown by the area segmentation histogram in (**b**). The model shown in (**c**) contains the series resistance segments $r_{i,seg}$ and the constant diode model parameters extracted from the data sheet (green box). The ideality factor $n_{id}$ is identical for all cells and varied between $1.0 < n_{id} < 1.4$. The reverse characteristics for describing the reverse current density $J_{rev}$ depending on the breakdown voltage $V_{br} = -15$ V as well as the breakdown model parameters $n = 2.3 \times 10^{-3}$ and $a = 1.9$ are numerically modeled, as proposed by Quaschning [7]. From the data sheet, all other parameters are calculated, transposed into the series resistance map and scaled by the relative area of the segment. All cells i are connected in series and simulated with the bypass-diode configuration of the module.

The mean series resistance $r_{i,seg}$ in each segment is used for the simulation model shown in Figure 5c, predefined into ten segments. Each series resistance is weighted by its relative area $A_{i,seg}/A_c$. All constant diode model parameters (green box in Figure 5b) are directly derived from the data sheet or calculated from a one-diode model. Figure 5c shows the resulting simulation model for cell i.

All cells i = 1, 2, ... , N, with a total number N of cells, are connected in series and simulated with the bypass diode configuration of the module (typically 20 or 24 cells anti-parallel connected to a bypass-diode).

Note, fully and partially disconnected cell areas potentially limit the short circuit current density $J_{sc}$ of the cell. In this case, highly damaged cells with a large part of the cell being disconnected, are no longer acting as a current source. These cells operate under reverse bias ($V_{i,c} < 0$) and limit the power output of the module, depending on their interaction with the bypass diodes.

Therefore, the reverse characteristics for describing the reverse current density $J_{rev}$ under reverse bias ($V_{i,c} < 0$) are numerically modelled as originally introduced by Bishop [8] and further applied by

Quaschning [7]. As proposed by Quaschning, for describing multicrystalline solar cells we incorporate the breakdown voltage $V_{br} = -15$ V, as well as the avalanche effect under reverse bias by the non-linear multiplication factors $n = 2.3 \times 10^{-3}$ and $a = 1.9$. Again, we assume identical cells with the same breakdown model unrelated to the defects.

Finally, by evaluating the ideality factor expectation interval (identical ideality factors for all cells), in this case $1.0 < n_{id} < 1.4$ for the ideality factor of multicrystalline silicon solar cells, we find excellent agreement compared to measured results.

To evaluate PID modules, the simulation model is further simplified. Since the change in parallel resistance influences each total cell i, no segmentation is needed. Therefore, the series resistance of each cell i is extracted in a straightforward manner from the data sheet and only the overall parallel resistance $r_{i,p}$ of each cell is varied in the simulation model based on Equation (10).

## 3. Experimental Results

### 3.1. Hail-Damage Modules with Cracked Cells

As a reference measurement, a multicrystalline silicon PV module (module A) was characterized via the measured I/V curves of each individual cell of this module. The characterized module with 60 multicrsystalline cells (data sheet maximum power $P_{mpp} = 230$ W$_p$) was already installed prior to the experiments and showed multiple cracked cells after on-site hail impacts (compare Figure 4a). The bus bars in between the cells were reached by drilling small holes through the backsheet encapsulation. At the interconnections between the cells, the positive and negative contact of each cell, was accessible. A Keithley 2651a source-meter captured the I/V curves of each cell, as well as of the total module, while a halogen-lamp solar simulator irradiated the module with the equivalent solar irradiance $E_{irrad}$ = 600 W/m$^2$ (0.6 suns). All I/V curves were transposed into standard test condition (STC) equivalent electrical characteristics. The EL images were captured by a cooled Si-CCD camera (FLI MicroLine ML8300M, 8.3 Megapixel) and downsampled to a fixed resolution of $100 \times 100$ pixel/cell.

Figure 6a shows the measured irradiance $E_{irrad}$ on the total module and Figure 6b the measured temperature $T_{i,meas}$ during the acquisition of the I/V curves for each cell i. Figure 6c shows the calculated series resistance $R_{i,s}$, numerically derived from a two-diode equivalent circuit which reproduced the I/V curves for each cell i. These parameters are mandatory and used for deriving the STC characteristics of all cells.

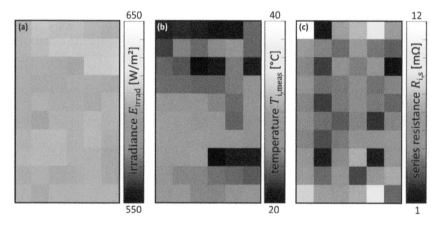

**Figure 6.** Measured irradiance $E_{irrad}$ (**a**) and temperature $T_{i,meas}$ (**b**) during I/V curve acquisition of each cell i of a multicrystalline module with 60 cells (module A). (**c**) Calculated series resistance $R_{i,s}$ for each cell i from a two-diode equivalent circuit reproducing the measured I/V curve. All three values ($E_{irrad}$, $T_{i,meas}$ and $R_{i,s}$) illustrated in the images (a) to (c) were used to calculate the I/V curve at STC.

Based on the temperature coefficient of the total module, the open circuit cell voltages were transposed into temperature values $T_{i,meas}$. The temperature was calibrated by a reference measurement at room temperature immediately before performing all further measurements. Here, we assumed that the cells maintained room temperature during measurement, without any temperature drift. Therefore, depending on the location of the cell inside the module, two tracked open circuit voltages were used to determine each cell temperature $T_{i,meas}$ during *I/V* measurements. Irradiance $E_{irrad}$ was measured via a reference cell directly placed beside the module and showed only weak fluctuations during the measurements.

Since the contacts of all cells were accessible, the cell temperatures $T_{i,meas}$ were derived from the open circuit voltage of one cell in the middle, as well as from one cell at the edge of the module. These two cell voltages acted as temperature sensors inside the module and represented the cell temperatures more accurately than an overall measured module temperature, typically measured by sensors placed at the backside of a module.

Based on the individual *I/V* curves, the maximal power $P_{i,mpp}$ of each cell i was extracted. Figure 7a shows the maximal power $P_{i,mpp}$ of the individual cells extracted from the *I/V* measurements, normalized to the maximal measured cell power $P_{mpp,max}$ in the module. Comparing the measured normalized cell powers of each cell i with the EL image in Figure 4a, the location of strongly damaged cells, indicated by dark regions in the EL images, matched the location of cells with low power output (dark yellow) in Figure 7a. The cells located at the top of the module in Figure 7a showed the strongest deviation, since there were no damaged cells in the EL images. However, this can be explained by the increased series resistance induced by the measurements (only in this case) using the junction box.

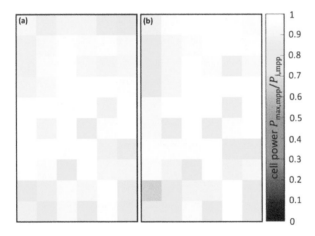

**Figure 7.** Measured maximal power output (**a**) and simulated maximal power output (**b**) for all individual cells. All cell powers were normalized according to the maximal measured or simulated cell power in the module and colored by the same false color scale. The location of defective cells and the normalized cell power was in good agreement comparing simulation and measurement.

Comparing the measured results in Figure 7a with the simulated results in Figure 7b, initially using the ideality factor $n_{id} = 1$ (lowest ideality factor from the expectation interval $1.0 < n_{id} < 1.4$), an excellent agreement between the normalized cell powers was found. Further, the maximal measured power of the total module $P_{mpp,meas} = 209$ W $\pm$ 5% (at STC) was also in good agreement with the simulated power expectation interval 200 W $< P_{mpp,sim} < 209$ W, depending on the chosen ideality factor $1.0 < n_{id} < 1.4$. The highest deviation was found for the ideality factor $n_{id} = 1$, resulting in a maximal deviation of 4.3% compared to the measured power output. Since our *I/V* measurements usually showed an accuracy of ±5% for maximum power (at STC), these simulated results were already in the expected range of accuracy.

In addition to module A with open contacts at the backside of the modules, we further evaluated our method using two untreated modules of the same type (module B and module C). All three modules (A, B and C) were installed on the same site prior to the experiments and showed damage from hail-impact, with varying degrees of power loss. Figure 8 shows the resulting simulated power losses for all three modules using module A as the calibration module for the simulation.

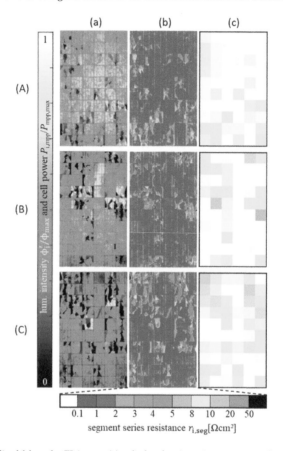

**Figure 8.** Normalized false color EL images (**a**), calculated series resistance maps colored by categorized mean series resistance of segments $r_{i,seg}$ (**b**) and simulated normalized power output of all cells of modules A, B and C (**c**). The simulated and normalized power output of cells identify the most severe defects in each module.

As an additional calibration possibility aimed at a more accurate ideality factor approximation, we assumed that there was always access to one calibration module with a well-known $I/V$ curve and not only data sheet information. Therefore, we used the already characterized $I/V$ curve of module A to find the optimal ideality factor $n_{id}$. The assumed ideality factor for calculating the series resistance map of module A was adjusted until the predicted power loss in the simulation showed a deviation to the measured $I/V$ curve of MPP below 1%. The optimal ideality factor $n_{id} = 1.4$ was then applied to predict the power loss of module B and module C.

Based on this calibration procedure, we found a deviation between the measured and simulated results for MPP power output below 1%. Table 1 summarizes the calibrated simulation results of all three characterized modules (A, B and C) with hail-damage from the same site.

**Table 1.** Calibrated simulation results of three multicrystalline modules (A, B and C) from the same site with hail damage. The calibration module A was used to determine the ideality factor $n_{id} = 1.4$ as the optimal simulation setup ($P_{mpp,meas}(A) \approx P_{mpp,sim,c}(A)$). The calibration of the ideality factor reduced error below 1% between the measured power $P_{mpp,meas}$ and simulated power $P_{mpp,sim,c}$ of modules B and C.

| Module Number | Measured Power $P_{mpp,meas}$ [W] | Measured Power Loss $p$ [%] | Calibrated and Simulated Power $P_{mpp,sim,c}$ [W] |
|---|---|---|---|
| A | 209 W | 9.1% | 209 W |
| B | 198 W | 13.9% | 197 W |
| C | 186 W | 19.1% | 186 W |

### 3.2. PID Affected Module

As proof-of-concept of the parallel resistance approximation, we additionally evaluated EL images of a high-efficiency module (data sheet maximum power $P_{mpp} = 225$ W$_p$) with 72 back-contact monocrystalline solar cells (cell area $A_c = 153$ cm$^2$) partially showing PID.

Figure 9a shows the low current (7% $I_{sc}$) EL image of the PID module and the parallel resistance map of each cell i in Figure 9b. For PID only affecting the parallel resistance of cells, Equations (9) and (10) were used to directly calculate the parallel resistance $R_{i,p} = r_{i,p} A_c$ from the mean luminescence $\overline{\Phi}_i$ of each cell i with the cell area $A_c$. The result is shown in Figure 9b and was used to simulate the expected power output. Again, all other simulation parameters ($J_0$, $n_{id}$ and $r_{i,s}$) were kept constant, assumed to be unaffected by the defect and were extracted from the data sheet.

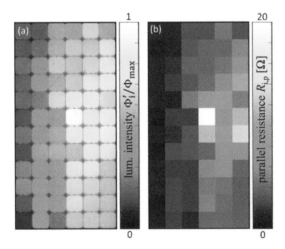

**Figure 9.** False color EL image of a PID affected monocrystalline silicon photovoltaic module with back-contact solar cells at (**a**) supplied current of 7% $I_{sc}$ with 300 s exposure time and (**b**) calculated parallel resistance map based on Equations (9) and (10).

By comparing the simulated results with the measured $I/V$ curve, we found excellent agreement regarding the MPP power output with an error below 1% (measured power $P_{mpp,meas} = 216$ W and simulated power 217 W $< P_{mpp,sim} < 218$ W) using an ideality factor expectation interval $1.0 < n_{id} < 1.2$.

## 4. Discussion

Regarding mechanically damaged modules with cracked solar cells, our modelling principle showed a clear physical relation between defect type and electrical characteristics based on common one-diode-modelling and already leads to very accurate power loss prediction. However, the limits of applicability with respect to the cells' series resistances, as well as combination with shunts, need to be further evaluated.

In practice, the power loss of PID affected modules is evident, but the origin and specific types of PID are still under discussion. In the literature, there are several approaches for describing the origin of PID [2,9–12]. Typically, the resulting power loss is either linked to processes increasing carrier recombination due to degradation of the front side passivation layer, which increases the saturation current $J_0$, or PID is modelled by simply decreasing the parallel resistance for shunt-type PID.

However, by solving Equation (8) analogously for the individual recombination currents $J_{i,0}$ using the mean luminescence $\overline{\Phi}_i$ of the individual cells, we found no clear correlation between measurements and simulations. Hence, we neglected the modelling approach based on a change in recombination current, and we described PID as a change in parallel resistance only, by using Equations (9) and (10).

## 5. Conclusions

This work presented a proof-of-concept of the novel electroluminescence characterization method ELMO based on series resistance mapping and modelling of photovoltaic modules with cracked solar cells. The series resistance map was successfully extracted from two luminescence images with low ($<10\%$ $I_{sc}$) and high ($>30\%$ $I_{sc}$) current injection. Using the data sheet information of the photovoltaic module, the simulation model was capable of predicting the expected power output with an error of less than 4.3%. By calibrating the ideality factor of the simulation using the $I/V$ curve of the reference module, the error in the power loss prediction for the other modules of the same type was reduced to below 1%. By modelling PID as a change in parallel resistance, we approximated the power loss with an error of below 1% without any further calibration. The simultaneous application of parallel resistance approximation and series resistance mapping to predict power loss has yet to be evaluated. Furthermore, the impact of image quality on the simulation results needs to be analyzed.

**Author Contributions:** T.K. developed the theory and simulation model, conducted the experiments and evaluated the data. M.S. supervised the project and contributed with fruitful discussions and valuable suggestions. J.H.W. critically revised the theory and the paper. All authors contributed in writing the manuscript.

**Acknowledgments:** The authors would like to thank L. Stoicescu for helpful discussions. We gratefully acknowledge funding by the German Federal Ministry for Economic Affairs and Energy (BMWi) under contract No. 0324069A.

**Conflicts of Interest:** The authors declare no conflict of interest.

## References

1. Kajari-Schröder, S.; Kunze, I.; Eitner, U.; Köntges, M. Spatial and orientational distribution of cracks in crystalline photovoltaic modules generated by mechanical load tests. *Sol. Energy Mater. Sol. Cells* **2011**, *95*, 3054–3059. [CrossRef]
2. Hara, K.; Jonai, S.; Masuda, A. Potential-induced degradation in photovoltaic modules based on n-type single crystalline Si solar cells. *Sol. Energy Mater. Sol. Cells* **2015**, *140*, 361–365. [CrossRef]
3. Potthoff, T.; Bothe, K.; Eitner, U.; Hinken, D.; Köntges, M. Detection of the voltage distribution in photovoltaic modules by electroluminescence imaging. *Prog. Photovolt. Res. Appl.* **2010**, *18*, 100–106. [CrossRef]
4. Fruehauf, F.; Turek, M. Quantification of Electroluminescence Measurements on Modules. *Energy Procedia* **2015**, *77*, 63–68. [CrossRef]
5. Bauer, J.; Frühauf, F.; Breitenstein, O. Quantitative local current-voltage analysis and calculation of performance parameters of single solar cells in modules. *Sol. Energy Mater. Sol. Cells* **2017**, *159*, 8–19. [CrossRef]

6.  Köntges, M.; Kunze, I.; Kajari-Schröder, S.; Breitenmoser, X.; Bjorneklett, B. The risk of power loss in crystalline silicon based photovoltaic modules due to micro-cracks. *Sol. Energy Mater. Sol. Cells* **2011**, *99*, 1131–1137. [CrossRef]
7.  Quaschning, V.; Hanitsch, R. Numerical simulation of current-voltage characteristics of photovoltaic systems with shaded solar cells. *Sol. Energy* **1996**, *56*, 513–520. [CrossRef]
8.  Bishop, J.W. Computer simulation of the effects of electrical mismatches in photovoltaic cell interconnection circuits. *Sol. Cells* **1998**, *25*, 73–89. [CrossRef]
9.  Oh, J.; Bowden, S.; TamizhMani, G. Potential-Induced Degradation (PID): Incomplete Recovery of Shunt Resistance and Quantum Efficiency Losses. *IEEE J. Photovolt.* **2015**, *5*, 1540–1548. [CrossRef]
10.  Lausch, D.; Naumann, V.; Breitenstein, O.; Bauer, J.; Graff, A.; Bagdahn, J.; Hagendorf, C. Potential-Induced Degradation (PID): Introduction of a Novel Test Approach and Explanation of Increased Depletion Region Recombination. *IEEE J. Photovolt.* **2014**, *4*, 834–840. [CrossRef]
11.  Naumann, V.; Geppert, T.; Großer, S.; Wichmann, D.; Krokoszinski, H.; Werner, M.; Hagendorf, C. Potential-induced degradation at interdigitated back contact solar cells. *Energy Procedia* **2014**, *55*, 498–503. [CrossRef]
12.  Naumann, V.; Lausch, D.; Hähnel, A.; Bauer, J.; Breitenstein, O.; Graff, A.; Werner, M.; Swatek, S.; Großer, S.; Bagdahn, J.; Hagendorf, C. Explanation of potential-induced degradation of the shunting type by Na decoration of stacking faults in Si solar cells. *Sol. Energy Mater. Sol. Cells* **2015**, *120*, 383–389. [CrossRef]

 *energies*

*Article*

# Visualization of Operational Performance of Grid-Connected PV Systems in Selected European Countries

Bala Bhavya Kausika *, Panagiotis Moraitis and Wilfried G. J. H. M. van Sark

Copernicus Institute, Utrecht University, Heidelberglaan 2, 3584 CS Utrecht, The Netherlands;
P.Moraitis@uu.nl (P.M.); w.g.j.h.m.vansark@uu.nl (W.G.J.H.M.v.S.)
* Correspondence: b.b.kausika@uu.nl

Received: 20 April 2018; Accepted: 21 May 2018; Published: 23 May 2018

**Abstract:** This paper presents the results of the analyses of operational performance of small-sized residential PV systems, connected to the grid, in The Netherlands and some other European countries over three consecutive years. Web scraping techniques were employed to collect detailed yield data at high time resolution (5–15 min) from a large number (31,844) of systems with 741 MWp of total capacity, delivering data continuously for at least one year. Annual system yield data was compared from small and medium-sized installations. Cartography and spatial analysis techniques in a geographic information system (GIS) were used to visualize yield and performance ratio, which greatly facilitates the assessment of performance for geographically scattered systems. Variations in yield and performance ratios over the years were observed with higher values in 2015 due to higher irradiation values. The potential of specific yield and performance maps lies in the updating of monitoring databases, quality control of data, and availability of irradiation data. The automatic generation of performance maps could be a trend in future mapping.

**Keywords:** performance ratio; annual yield; GIS; PV system; spatial analyses

## 1. Introduction

Recent years have seen a constant growth in solar photovoltaic technology (PV). Several countries have utilized this potential to create a competitive market in view of a green energy future, which led to an increase in small and medium-sized residential solar PV installations [1,2]. These small-sized installations (with capacities less than 10 kWp) are scattered and operate under diverse conditions without adequate monitoring equipment. Studies show that most of these systems perform adequately, but due to a lack of systematic data collection, performance validation was mainly focused on specific geographic areas with a limited amount of systems [3,4].

A "Photovoltaic Geographical Information System" (PVGIS) system was designed to provide performance assessments to an accuracy that is suitable for small installations and for estimating the potential solar energy over large regions at any location in Europe [5]. Although this large-scale GIS (Geographic Information System) database of solar radiation and ambient temperature has been created to estimate energy output from crystalline silicon PV systems and solar water heating systems, it does not provide continuous monitoring or performance evaluation for small-sized, grid-connected PV systems.

Currently available monitoring technology in the market is capable of providing owners with sophisticated web tools to monitor their production and system performance at any point during the day, besides measuring energy production. With the advent of such hardware and smart-metering technology, high-resolution monitoring data is publicly available, which is uploaded daily on web platforms, however, in some cases only owners can view this data.

With the huge amount of data that is available due to the monitoring equipment, abnormalities can be compared with additional data (remote sensing imagery) for identification of reasons for underperformance or fault detections [6]. Monitoring small grid-connected PV systems to minimize financial losses has also been explored [7] along with the need for long-term monitoring for reliability and increased PV performance [4]. In [1,8], the authors show the importance of using a graphical supported analysis of monitoring and operation of PV installations for fault detections. In our earlier work [9,10], we show how technical aspects and geographical location of PV systems affect PV performance. In this paper, geographic information systems (GIS) are employed to analyze, visualize, and map PV monitoring data from five countries, namely, The Netherlands, France, Germany, Belgium, and Italy. We also present and discuss methods for visualization and detection of underperforming or overperforming systems for further analysis, performance ratio analysis of systems, and spatio-temporal mapping of performance differences.

## 2. Method

Data used for the analyses was collected using online services provided by Solar Log [11] and SMA [12], which also ensured data legitimacy. Solar Log has users over a hundred countries and is one of the major key players in monitoring applications, though it has lost a lot of its clients after 2015. SMA is one of the specialists in photovoltaic inverter system technology. The code used in this research was developed to extract online data and was designed using Python programming [13]. The objective of the web scraping code was to mimic human navigation through web pages of SMA and Solar Log, and to locate and save information that was available to the user [9]. This means that the monitoring information pages of different PV systems was retrieved and saved accordingly. This information was later organized into datasheets. In this way, high temporal resolution yield data (5 min) and other system metadata like orientation, tilt, type of module, etc. were obtained. Recently, privacy constraints have been put on the data, and these data are not available publicly anymore. In total, data from about 31,844 systems were collected for the years 2012–2016 from 5 different countries in Europe, namely, The Netherlands, France, Germany, Belgium, and Italy.

In order to calculate the performance ratio (PR) of all the systems, system yield and reference yield are required. System yield is obtained from the data collected by web scraping and reference yield is calculated using the Olmo model [14]. This model requires irradiation data [15]. Hourly global horizontal irradiation data obtained from the 31 ground-based stations of the Royal Netherlands Meteorological Institute (KNMI) were used to compute the reference yield for The Netherlands. These stations cover the entire country. For each installation in the database, irradiation data was collected by linking it to the closest ground station, in order to minimize the uncertainties in the irradiation data. Note that no system was further away than ~30 km from the nearest KNMI station. The tilt and orientation for every system has also been obtained from web-scraped data of PV systems. The Olmo model was then used to calculate the total irradiation in the plane of array on an hourly basis. This study does not take into account effects such as shading as the aim of the paper is to visualize performance rather than detect reasons behind over- or underperformance [16]. The PR was calculated using Equation (1), where $Y_f$ is the final system yield and $Y_r$ is the reference yield [15]. Since high-resolution, up-to-date annual irradiation data was not available for the rest of the countries, PR was calculated only for The Netherlands.

$$PR = \frac{Y_f}{Y_r} \qquad (1)$$

Geographic Information System (GIS) is a "powerful set of tools for collecting, storing, retrieving at will, transforming, and displaying spatial data from the real world" [17]. Based on the principles of geography, cartography, etc., GIS is used for integration of different data types. It is a very powerful tool when it comes to analyses of spatial information, layering or organizing layers of information into visualizations using maps and 3D scenes [18]. There are several GIS software packages available in the market today, but ArcGIS [19] is a leading licensed tool for performing powerful geo-analyses, which will be used in this paper as an example tool.

Visualizations of the performance ratio, the locations of the installations, and yield and performance maps were created using the ArcGIS platform. An inverse distance weighted (IDW) interpolation technique was used to create the performance ratio maps and specific yield maps for different years of data collection. Although data from around 31,800 systems was available (2012–2016), only those systems that recorded data continuously for three consecutive years (between 2014 and 2016) were used to compare the differences in yield generation. This provides an understanding of how system performance varies spatially (over geographic areas) and helps in identifying outliers in the data. In addition to being able to visualize the results, looking into the diffusion of distributed systems within a country or area allows for the computation of geo-statistics pertaining to the region which are useful for policy implementation.

## 3. Results and Discussion

From 2011 to 2016, data from more than 31,800 systems was collected and analyzed. However, only 7894 of them were consistently delivering valid data for more than 350 days per year for at least three consecutive years (2014–2016). The total capacity of these systems was about 102 MWp with 56% of them having a lower capacity than 10 kWp and only 1.1% being larger than 50 kWp (see Figure 1). The mean value was 12 kWp. The spatial distribution of all the installations with system size information is illustrated in Figure 2. The variation in average size and composition of the systems in each country is a direct reflection of the country's policies on PV subsidy schemes.

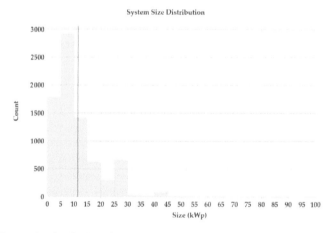

**Figure 1.** System size distribution of systems with capacity of less than 100 kWp for five countries. The red line illustrates the mean value of 12 kWp.

A high concentration of small-scale domestic installations is observed in Germany, Belgium, and The Netherlands with 64% of the systems in Germany. The Netherlands and Belgium have most of the systems' total capacity under 5 kWp. While in Germany only 7.2% of the installations fall in this category, 45% of the PV of the systems installed are still below 10 kWp. Though the monitoring procedure might have started at a later time, most of the systems from the sample were installed between 2008 and 2014.

Figure 2 shows the location of each PV system, categorized by system capacity. Data collected from the monitoring systems and organized in a database was imported into GIS to create this map. Clearly, large numbers of systems are concentrated in Germany, Belgium, and The Netherlands. Some of the systems had faulty location information and hence were not included in the map. Most of the systems are also concentrated in the North where irradiation is lower, rather than in the South where there is higher irradiation.

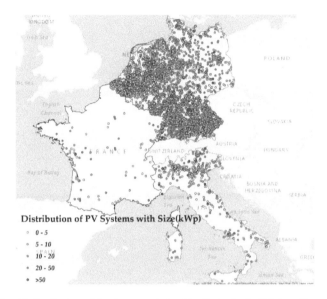

**Figure 2.** Spatial distribution of the data sample for The Netherlands, Belgium, France, Germany, and Italy.

*3.1. Yield Analysis and Performance Ratio*

The available data was found to be varying through different time periods as new installations were added every year. Also, not all the systems recorded data for all the years. Therefore, only those systems that had been consistently delivering data for the three consecutive years (2014–2016) have been considered for analysis. Furthermore, since the interest is in monitoring small-scale installations, annual yield analysis of systems below 20 kWp for the years 2014–2016 has been conducted for The Netherlands, France, Germany, Belgium, and Italy. These countries were found to have the highest amount of data records from the data collection.

The mean value and the standard deviation of the performance of systems of each sample is shown in Figure 3: here we show the annual specific yield, i.e., generated amount of energy divided by system capacity (kWh/kWp). These are known to be affected by a number of environmental and operational factors [20]. Moreover, a wider spread of yearly yield values can be expected from countries covering larger areas as a result of the variation of irradiation levels at different latitudes. The distribution of annual system yield for The Netherlands, Belgium, Germany, and Italy is shown in Figure 4. France has only 95 installations between 2014 and 2016 out of which 76 systems are below 20 kWp capacity, while Germany has nearly 3900 systems, and Belgium 1700. Therefore, France has higher mean yields and only four outliers due to sample size. Between 2014 and 2016, the annual yield increased in 2015. However, the decrease or increase in yields falls within standard deviations, but at the same time relates directly to the decrease or increase in solar irradiation on a country level.

Performance ratio (PR) analysis was conducted for The Netherlands which revealed a mean PR value of 79% for the year 2016 and 80% for 2014 and 2015. The PR values were calculated with an average daily PR value over a year. These values are close to the results of an earlier study performed in Germany [21]. The sample size for this estimation was about 600 installations. The number of PV installations in The Netherlands significantly increased from 2009, but their performance dropped in 2016 in comparison to 2014 (Figure 5). Systems installed in 2013 performed well in 2014 and 2015, while in 2016, a lot of outliers were observed. In some cases, the large variation in PR values could also be due to technical errors in data collection.

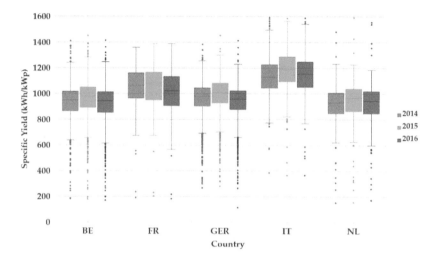

**Figure 3.** Distribution of specific yield by country from 2014 to 2016 for systems less than 20 kWp. Highest yields were recorded in 2015, with Italy (IT) having the highest mean, followed by France (FR), Germany (GER), Belgium (BE), and The Netherlands (NL), respectively.

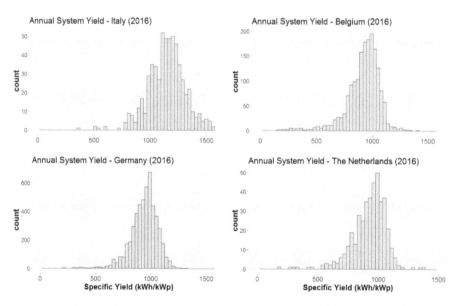

**Figure 4.** Distribution of specific yield for the year 2016 for Italy, Germany, Belgium, and The Netherlands.

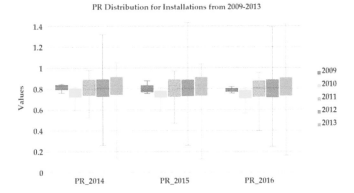

**Figure 5.** Distribution of performance ratio of The Netherlands between 2014 and 2016 for systems that have been installed from 2009 to 2013.

*3.2. Geographical Variation of Specific Yield*

Point data (vector information) collected from the web monitoring services has been converted to images (raster information) by using interpolation techniques. Interpolated data is visualized using color scales stretched using specific bins of annual yield values. From these images/maps, outliers can be quickly discerned to locate PV systems with minimum or maximum yields, thus providing a starting point for further analyses into the reason behind the system's under- or overperformance. The maps can be compared to the country irradiance maps to check for irradiation trends in the particular year, as yield values are related to irradiation values. This provides a quick approximation of the variation of performance over the country.

Figure 6 shows an interpolated map of annual yield of The Netherlands and Italy for three years with dots representing the location of the systems. Inverse Distance Weighted (IDW) interpolation technique was used to generate the maps. Higher yield values have warmer and darker shades (reds) and lower yield values have a blue shade. A variation in yield values is observed within the countries, while it should be noted that these variations can further be optimized using different color scales and data stretching methods. A few examples of this are shown below.

Although variations over the years are not very prominent because of the type of stretch used for data visualization and the data sample (system size up to 20 kWp), it could still be distinguished that 2015 has higher annual yields. A min–max stretch was used to visualize data with the same scale of minimum and maximum values for the three years to maintain consistency.

When using a different data stretching method (see, for example, Figure 7a), extreme deviations in data (bright spots) can be identified around systems with extreme yield values. These extreme values carry a higher weight factor during interpolation causing the spot or bleeding effect. As mentioned earlier, these spots can be separated out as outliers or as inadequately performing systems. Moreover, if high resolution irradiation data is available, the database of collected information could be explored by irradiation zones in addition to spatial diffusion or technical criteria.

An example of different stretching techniques is shown in Figure 7, in which two types of data stretching were used over a color scale. In Figure 7a, a percentage clip method was used, where the values displayed are cut-off percentages of highest and lowest values, while Figure 7b displays values between the actual or set minimum and maximum. A smoothing effect can also be seen in Figure 7b, while it is easier to pick out underperforming or overperforming systems to analyze them further from Figure 7a.

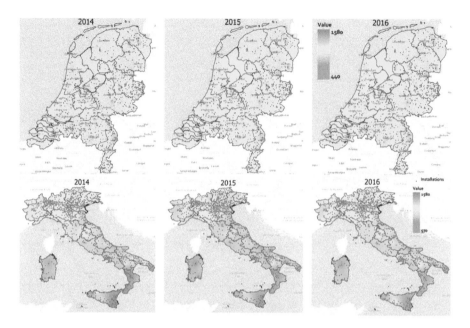

**Figure 6.** Annual specific yield variation from the installations (up to 20 kWp) in The Netherlands (**above**) and Italy (**below**) for the years 2014–2016, visualized using interpolation techniques.

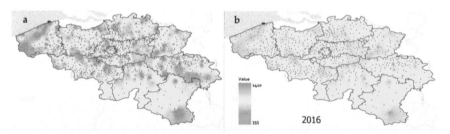

**Figure 7.** Annual specific yield variation from the installations (up to 20 kWp) for Belgium visualized using two different types of data stretching for 2016. Data stretching techniques (**a**) percent clip and (**b**) min-max used for data visualization.

Another example of the power of GIS in visualization is shown in Figure 8, where the mean specific yield for Germany using different thresholds of system sizes is presented. The variation in specific yield of systems up to 20 kWp is much smoother compared to when only systems up to 10 kWp are considered. For example, the systems in the highlighted area (red box in both images) shows underproduction when compared with larger systems (<20 kWp) while, on comparison with smaller systems (<10 kWp), they seem to be performing adequately. Also, it can be seen that a few systems seem to be less efficient in both categories, which means they could actually suffer from a malfunction.

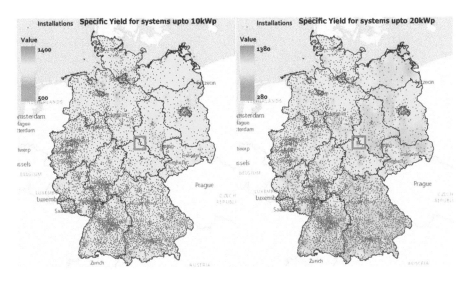

**Figure 8.** Specific yield variation from the installations in the Germany, for different system sizes.

*3.3. Mapping Performance Differences*

Differences in specific yield for three years for The Netherlands are shown in Figure 9. This has been calculated based on the yield maps generated by interpolation. Areas in red show decrease in yield, while areas in green show increase in yield for different years. Yellow regions are regions of no change. The limits for no change were set at −20 to +20 kWh/kWp (~2% of annual specific yield) and anything higher or lower than these values was recorded as increase or decrease in yields. Increase in yields were observed for most of the regions from 2014 to 2015, while from 2015 to 2016, the yields were either constant or decreased. When looking at differences from 2014 to 2016 compared to 2014–2015, lower yield values were observed in the south of The Netherlands. In general, these differences can also be visualized with scatter plots. The advantage of using mapping techniques to visualize difference data lies not only in knowing how large the change is, but also in being able to see where the change is taking place. Maps of yield differences should be used in conjunction with maps of irradiation differences to explain the yield differences, or one can map the performance ratio differences.

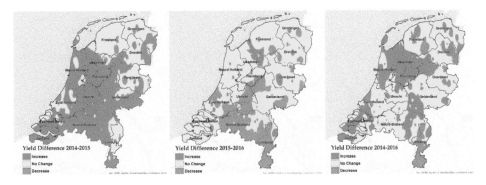

**Figure 9.** Specific yield difference maps for 2014–2016 for The Netherlands.

## 4. Conclusions

In this study, GIS has proven to be an excellent tool for visualization of yields and performance of scattered, small-sized, residential PV systems over wide-spread areas. We were able to successfully demonstrate this for The Netherlands and a few other countries like Italy, Germany, and Belgium. Additionally, geo-processing tools (hot-spot analyses, network analyses) could provide useful information to individuals or policy-makers to make informed decisions. This could be done if information (system metadata) pertaining to all the installed PV systems is available.

This paper further provides an update on performance of residential PV systems scattered in a few European countries. It was found that the year 2015 showed a higher specific yield in kWh/kWp compared to the years 2014–2016. Performance ratio for The Netherlands did not change with respect to earlier years, although there is an increase in extreme values with the increase in number of installations. Access to high-resolution irradiation data for all the countries is necessary to analyze temporal variations in performance ratios of PV systems. Recent reports suggest an increase in performance ratio values [21–23], however, long-term changes over expected lifetime of the systems should be analyzed to show if performance ratio values still are increasing.

Building up and expanding the present PV performance databases to other countries will provide up-to-date performance maps of more countries. In addition, irradiation maps can be combined with yield maps in order to construct maps of performance to understand the relationship between climatic zones across the world and performance of PV systems.

**Author Contributions:** Conceptualization and Methodology: B.B.K.; Acquisition of data and initial analysis: P.M.; Analysis and data interpretation: B.B.K.; Writing original draft: P.M. and B.B.K.; Writing Review and Editing: B.B.K. and W.G.J.H.M.v.S.; Funding acquisition: W.G.J.H.M.v.S.

**Acknowledgments:** This work is part of the International Energy Agency-Photovoltaic Power Systems (IEA-PVPS) Task 13 "Performance and Reliability of Photovoltaic Systems" [24]; we would like to thank all members of this task for their support. This project is partly financially supported by The Netherlands Enterprise Agency (RVO).

**Conflicts of Interest:** The authors declare no conflict of interest.

## References

1. Santiago, I.; Trillo Montero, D.; Luna Rodríguez, J.; Moreno Garcia, M.; Palacios Garcia, E.J. Graphical Diagnosis of Performances in Photovoltaic Systems: A Case Study in Southern Spain. *Energies* **2017**, *10*, 1964. [CrossRef]
2. Mayer, D.; Wald, L.; Poissant, Y.; Pelland, S. *Performance Prediction of Grid-Connected Photovoltaic Systems Using Remote Sensing*; International Energy Agency: Paris, France, 2008.
3. Nordmann, T.; Clavadetscher, L.; van Sark, W.; Green, M. *Analysis of Long-Term Performance of PV Systems*; International Energy Agency: Paris, France, 2014.
4. Jahn, U.; Nasse, W. Operational performance of grid-connected PV systems on buildings in Germany. *Prog. Photovolt. Res. Appl.* **2004**, *12*, 441–448. [CrossRef]
5. Huld, T.; Suri, M.; Dunlop, E. A GIS-based System for Performance Assessment of Solar Energy Systems over Large Geographical Regions. In *Proceeding of Solar 2006 Conference*; American Solar Energy Society: Boulder, CO, USA, 2006.
6. Stegner, C.; Dalsass, M.; Luchscheider, P.; Brabec, C.J. Monitoring and assessment of PV generation based on a combination of smart metering and thermographic measurement. *Sol. Energy* **2018**, *163*, 16–24. [CrossRef]
7. Drews, A.; de Keizer, A.C.; Beyer, H.G.; Lorenz, E.; Betcke, J.; van Sark, W.G.J.H.M.; Heydenreich, W.; Wiemken, E.; Stettler, S.; Toggweiler, P.; et al. Monitoring and remote failure detection of grid-connected PV systems based on satellite observations. *Sol. Energy* **2007**, *81*, 548–564. [CrossRef]
8. Mau, S.; Jahn, U. Performance analysis of grid-connected PV systems. In Proceedings of the European Photovoltaic Solar Energy Conference and Exhibition (EU PVSEC), Dresden, Germany, 4–9 September 2006; pp. 2676–2680.
9. Moraitis, P.; van Sark, W.G.J.H.M. Operational performance of grid-connected PV systems. In Proceedings of the 40th IEEE PVSC, Denver, CO, USA, 8–13 June 2014; pp. 1953–1956.

10. Moraitis, P.; Kausika, B.B.; van Sark, W.G.J.H.M. Visualization of operational performance of grid-connected PV systems in selected European countries. In Proceedings of the 42nd IEEE PVSC, New Orleans, LA, USA, 14–19 June 2015; pp. 1–3.
11. Solar Log GmbH. Available online: https://www.solar-log.com/en/ (accessed on 15 April 2018).
12. SMA Solar Technology AG—Inverter & Photovoltaics Solutions. Available online: https://www.sma.de/en/ (accessed on 15 April 2018).
13. Moraitis, P. Review of the Operational Performance of Grid Connected PV Systems. Master's Thesis, Utrecht University, Utrecht, The Netherlands, 2014.
14. Olmo, F.J.; Vida, J.; Foyo, I.; Castro-Diez, Y.; Alados-Arboledas, L. Prediction of global irradiance on inclined surfaces from horizontal global irradiance. *Energy* **1999**, *24*, 689–704. [CrossRef]
15. International Electrotechnical Commission. *Photovoltaic System Performance Monitoring—Guidelines for Measurement, Data Exchange and Analysis*; International Electrotechnical Commission: Geneva, Switzerland, 1998.
16. Tsafarakis, O.; Sinapis, K.; van Sark, W.G.J.H.M. PV System Performance Evaluation by Clustering Production Data to Normal and Non-Normal Operation. *Energies* **2018**, *11*, 977. [CrossRef]
17. Burrough, P.A.; McDonnell, R.A.; Lloyd, C.D. *Principles of Geographical Information Systems*; OUP Oxford: Oxford, UK, 2015; ISBN 978-0-19-874284-5.
18. What Is GIS? | Geographic Information System Mapping Technology. Available online: https://www.esri.com/en-us/what-is-gis/overview (accessed on 15 April 2018).
19. ArcGIS Platform. Available online: http://www.esri.com/software/arcgis (accessed on 13 June 2016).
20. Fouad, M.M.; Shihata, L.A.; Morgan, E.I. An integrated review of factors influencing the perfomance of photovoltaic panels. *Renew. Sustain. Energy Rev.* **2017**, *80*, 1499–1511. [CrossRef]
21. Reich, N.H.; Mueller, B.; Armbruster, A.; Sark, W.G.; Kiefer, K.; Reise, C. Performance ratio revisited: Is PR > 90% realistic? *Prog. Photovolt. Res. Appl.* **2012**, *20*, 717–726. [CrossRef]
22. Müller, B.; Hardt, L.; Armbruster, A.; Kiefer, K.; Reise, C. Yield predictions for photovoltaic power plants: Empirical validation, recent advances and remaining uncertainties. *Prog. Photovolt. Res. Appl.* **2015**, *24*, 570–583. [CrossRef]
23. Tsafarakis, O.; Moraitis, P.; Kausika, B.B.; van der Velde, H.; Hart't, S.; de Vries, A.; de Rijk, P.; de Jong, M.M.; van Leeuwen, H.-P.; van Sark, W. Three years experience in a Dutch public awareness campaign on photovoltaic system performance. *IET Renew. Power Gener.* **2017**. [CrossRef]
24. IEA-PVPS Task 13—Performance and Reliability of Photovoltaic Systems Website. Available online: http://iea-pvps.org/index.php?id=57 (accessed on 28 March 2018).

 *energies*

Article

# Urban Environment and Solar PV Performance: The Case of the Netherlands

Panagiotis Moraitis *, Bala Bhavya Kausika, Nick Nortier and Wilfried van Sark

Utrecht University, Copernicus Institute, Heidelberglaan 2, 3584 CS Utrecht, The Netherlands;
B.B.Kausika@uu.nl (B.B.K.); n.s.nortier@uu.nl (N.N.); W.G.J.H.M.vanSark@uu.nl (W.v.S.)
* Correspondence: P.Moraitis@uu.nl; Tel: +31-30-253-55385

Received: 16 April 2018; Accepted: 15 May 2018; Published: 23 May 2018

**Abstract:** The modern urban landscape creates numerous challenges for the deployment of solar Photovoltaic (PV) technology. The large structures that dominate the skyline of every city create compactness, which, in turn, limits the available rooftop area and creates unpredicted shading patterns. The majority of research today relies on modern applications such as geographical information system (GIS) software to evaluate urban morphology; however, this approach is computationally intensive and therefore it is usually limited to a small geographical area. In this paper, we approach this issue from another perspective, utilizing the enormous amount of high resolution PV yield data that is available for the Netherlands. Our results not only correlate performance losses with urban compactness indicators, but they also reveal a significant seasonality effect that can reach 15% in some cases.

**Keywords:** performance ratio; GIS; PV module; system; population density; urban compactness

## 1. Introduction

Today, approximately half of the world's population lives and works in cities and it is there that about 75% of global resources are consumed [1,2]. With a continuously increasing urban population, the alarming issues of energy security and climate change are becoming more prominent than ever, requiring careful planning and effective measures. The European political instruments have already addressed this challenge from 2010 with the Energy Performance of Buildings Directive (EPBD) that provides guidelines towards the net zero-energy buildings [3]. The goal is to reach a point where buildings minimize their carbon emissions and the dependence on the electrical grid through an extended integration of renewable energy generators and innovative architectural design.

In this direction, solar Photovoltaics (PV) is one of the most promising forms of renewable energy production. Fast technological improvements, cost reduction and public acceptance are the key factors that accelerate the global demand for solar systems. A strategic advantage of solar generators is the modular nature of the technology, which makes it ideal for onsite energy production and consumption, leading to a critical reduction in transformation and transmission losses.

The complex urban environment, however, is challenging for the deployment of PV technology [4]. While outside of the city boundaries solar energy generation might only be constraint by the stochastic nature of the meteorological conditions, in a metropolitan area, there are a number of threats that need to be overcome. Besides the scarcity of available space and the high cost of land, it is mainly the modern architecture that develops predominantly in terms of height that poses the most important hindering factors. An inherent characteristic of this type of development is the small rooftop that corresponds to a large operational volume for each construction. While this can be tackled with the use of Building Integrated Photovoltaic (BIPV) solutions [5], the dynamic overshadowing and the unevenly distributed solar irradiation caused by urban compactness can significantly affect the solar potential of those areas.

Early attempts to quantify the solar potential of an urban region were mostly based on a hierarchical methodology that would examine restrictions on a variety of different levels to identify physical, geographical and technical potential mainly on a 2D scale [6–8]. The challenge of estimating the available rooftop area was tackled with the use of municipal data or image processing of aerial photographs. Today, however, thanks to vast improvements in computational power and the development of new modeling techniques, a more accurate representation of the urban landscape can be acquired. Software tools such as Geographical Information System (GIS) and Light Detection And Ranging (LiDAR) are able to manipulate large amounts of geo-related data and to perform almost any kind of spatial analysis on both microscale and macroscale levels [9–11]. By introducing accurate 3D models of neighborhoods or even entire cities, researchers could evaluate a large number of significant parameters such as the inclination of the roofs and the total surface of the exterior walls. Therefore, on one hand, studies on the potential impact of solar energy generated could be extended and enhanced with high level of detail [12,13], but also the effect of various urban features such as the height-to-width ratio of street canyons, site coverage, plot ratio, horizontal distribution, and vertical uniformity of buildings could be thoroughly evaluated [14,15].

Regardless of the methodology or the simulation tool, the majority of the scientific research conducted up to date highlights the fact that the contribution of solar energy can be significant in the urban environment and that shading effects caused by adjusting buildings is the main threat that can limit the energy produced. Despite the impressive technological advances, the accurate computation of solar irradiation in a 3D environment is still a challenge and, in most cases, it is performed first by numerical computation and ray tracing simulations software before it is introduced into the GIS [13]. Therefore, these types of studies are limited to a small number of buildings or a very specific part of a city [16] and the main target is to provide guidelines either for urban designers [17] or to evaluate architectural designs [18].

The scope of research is to approach large scale PV performance from a big data perspective. In this way, it will elegantly circumvent the dependence on computational intensive simulation software, and correlate the performance of PV systems with urban compactness indicators by a thorough analysis of a large number of systems, operating in various locations under real life conditions. This approach benefits from the enormous amount of PV yield data produced constantly by monitoring systems, to give an accurate year round image of solar energy performance over extended geographical areas. Our results provide a comprehensive framework suitable to address decision-making for assessment and integration of PV technology in dense built environments.

## 2. Methodology

The research area is the entire country of the Netherlands, which provides a variety of morphologically different locations in close proximity that are also characterized, however, with very similar meteorological and irradiation conditions. According to provisional statistics that were published by the Dutch Central Bureau of Statistics, the Dutch market recorded the largest annual growth so far in 2017, with the addition of 700 MW of new installations reaching a total cumulative capacity of 2.7 GW [19]. The majority of those systems are equipped with an independent monitoring system or they are monitored through specialized software through the inverter. For this research, data were collected from 3271 systems with 5-minute resolution, provided by monitoring vendors or public available online sources covering the years 2014 to 2016.

The first step was to separate the country into three zones, urban, suburban and rural based on population density. In this way, it is possible to acquire a broad overview of the performance output per zone throughout the year, examine differences in system configurations that may occur and evaluate the effects of location and seasonality. The categorization was done at neighborhood level for the whole country within a GIS environment. Once the neighborhoods were classified, the PV systems were then categorized based on their location within the neighborhoods (see Figure 1). However, each rural, urban and suburban location has been developed under different social and economical circumstances and therefore locations with similar population density values might exhibit completely different morphological characteristics. Consequently, a number of indicators has been identified that can better describe the different aspects of urban compactness.

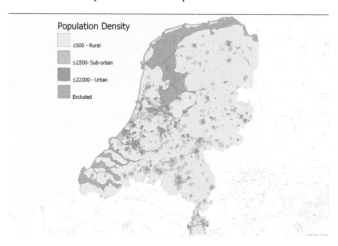

**Figure 1.** Map division in urban, suburban and rural areas in the Netherlands based on population density.

Urban compactness has a horizontal, or a 2D component that quantifies the site coverage from human made constructions and a vertical, or 3D component that describes the shape of the cityscape. The site coverage, which is directly linked with the available open space can be measured by (a) the building density, as the number of buildings per km$^2$, (b) the plot ratio, as the ratio of building surface to the total surface of the area and (c) the Average Building Distance (ABD) as it is measured from the centroids of each building. The most straightforward indicator to describe the 3D aspect of the urban compactness is the (d) the Average Building Height (ABH) in an area. However, a uniform distribution of building heights can have nearly zero shading effects and therefore the (e) Standard Deviation of Building Height (SDBH) is also examined in this research, as high values of SDBH can potentially lead to shading from adjusting buildings. In an ideal situation, a flat roof free of other constructions is the most suitable place to install a PV system as it can give unlimited freedom to tilt and orientation angles while having a direct and unobstructed view to the sun. The (f) Internal Building Height Standard Deviation (IBHSD) shows the distribution of height within a building polygon when it is projected to a GIS map and then it is averaged over the entire neighborhood. This indicator is suitable for quantifying the effect of any object that is placed on a rooftop such as chimneys, air condition and ventilation equipment or even architectural elements that exceed 0.5 × 0.5 m$^2$ in size. In case the average internal height standard deviation (std) is zero, then the roof of the building is completely flat without tilted parts or any extra constructions.

In order to effectively evaluate the performance of a PV system, the energy produced is compared with a corresponding reference system. In the most simple form, the reference system is the actual system size of the PV installation and the final system yield is measured in kWh of energy generated per kW$_p$ of capacity installed. This indicator has the advantage of being straightforward and easy to calculate; however, since the energy generated is dependent on several factors such as the Plane of Array (POA) irradiance and temperature, the system yield is inadequate to properly evaluate the PV system by itself, or in comparison with other systems. For a homogeneous and systematic assessment of the technical quality of PV installations, the comparison is set between the actual energy production of a system, and a reference system that hypothetically operates with the exact same characteristics under the same conditions, but it is free of losses. The most widely used indicator that follows this principle is Performance Ratio ($PR$), which shows the degree of utilization of an entire PV system. It is a dimensionless quantity and it is calculated by dividing the final system yield $Y_f$ by the reference yield $Y_r$ during the same time interval:

$$PR = \frac{Y_f}{Y_r}. \tag{1}$$

The final yield is defined by total measured power output $P_{out}$ (kW) multiplied with the time recording interval ($\tau$), to the rated power output ($P_0$). The reference yield is defined by the total plane of array irradiance (kWh) divided by the reference irradiation under Standard Test Conditions (STC), which is 1000 W/m$^2$ [20]:

$$Y_f = \sum_k \frac{P_{out,k} \times \tau_k}{P_0}, \tag{2}$$

$$Y_r = \sum_k \frac{G_{i,k} \times \tau_k}{G_{i,ref}}. \tag{3}$$

Any difference between 1 and $PR$ aggregates all the possible energy losses including inverter efficiency, wire losses, real power of the PV modules below nominal rating, mismatch, shades, dust, thermal, failures and in larger systems mid-voltage transformer losses that are also influencing the final value. As thermal losses depend on the location but also on the time of the year, the $PR$ of a given system will have a fluctuating $PR$ value for different operational environments but also through the course of the year, or even a day. A temperature corrected $PR_{STC}$ has been suggested that shows to be independent of the time and geographical location of the system. However, it requires more complex calculations and onsite measurements of the temperature and therefore it is not easy to apply in large scale research, where only basic features are monitored.

The irradiation data that are necessary for the calculation of $PR$ were acquired from the 31 meteorological stations of the Royal Meteorological Institute of the Netherlands (KNMI). The stations are able to provide Global Horizontal Irradiation measurements with hourly resolution. However, the majority of the PV installations that were monitored are not positioned on the horizontal level, but, on the contrary, they exhibit a very large variation of tilt and orientation angles. Therefore, the estimation of the solar irradiance on the tilted plane of a PV system requires the application of a transposition model that will convert the Global Horizontal Irradiance (GHI) to that of the tilted plane. A variety of different transposition models have been developed to predict solar irradiation when on site measurements are not available and many studies have focused on the validation of these models. The most successful predictions [21] come from Perez [22] and Hay [23], which come, however, with a high computational cost or high level of input details, which are not available in this case. Therefore, for this study, the Olmo model [24] is used, which depends on the clearness index and the corresponding GHI measurements, avoiding the direct and diffuse components of irradiation.

An other issue that is inherent with large scale data sets provided by private users is that it can be expected to include errors coming either from incorrect data entry, such as miscalculation of systems tilt and orientation angle or caused by malfunctioning monitoring equipment. In order to exclude them and achieve a good degree of confidence, Tukey's method is used to isolate outliers from symmetrical distributions was used [25].

## 3. Results

### 3.1. PV Systems Overview

All the PV systems of the data set were divided in three categories based on the population density of the location that they were installed. This separation resulted in 1166 systems located in urban areas,

981 in suburban areas and 1119 in rural areas. With the sample almost evenly distributed between the three different types of regions, the installation date is the first parameter that was examined. Long-term outdoor exposure and operation of the modules might lead to performance decay [26], which could affect our final results. However, the distribution of the installation year, which is depicted in Figure 2 as a percentage of the total number of systems in each group, confirms that most of the modules on Dutch rooftops are relatively new. As it can be clearly seen, the majority of the systems were installed between 2012 and 2014, a time period which is not long enough for a commercial module to exhibit any form of degradation effect [27].

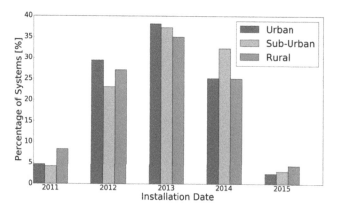

**Figure 2.** Installation date distribution.

The characteristic that is affected the most from the geographical location is the average system size. Scarcity of available space is probably one of the most representative features of urban areas and unavoidably solar installations could not be an exception. While in cities the average rooftop hosts a 3.4 $kW_p$ system, in sub-urban locations, this capacity increases to 4.7 $kW_p$, and, in rural areas, it almost doubles that figure reaching 9.2 $kW_p$. The effects of urban compactness are illustrated in Figure 3 where 90% of installations placed in cities are smaller than 5 $kW_p$ and only 2% has a capacity higher than 15 $kW_p$. The lower population density of rural areas results in higher availability of space, which can be better utilized by allowing room for installation of larger PV generators. Small size systems are still dominant in rural areas as 5 $kW_p$ are sufficient for the average household; however, large systems reach 19% of the total number of installations. This number accounts for the 54.3% of the total installed capacity in rural locations.

A very important factor for the electricity yield of a PV system is the tilt and orientation of the modules as they will determine the total amount of POA irradiance. A question that might arise here is whether urban compactness limits the available options for proper installation of PV modules that will result in less favorable tilt and orientation angles for systems positioned in cities. This is not expected to affect the final results of this research since the main outcome is based on the *PR*, which inherently takes into account the POA irradiance of each system, but it will exhibit at which degree the urban landscape leads to orientation dispersion and consequently to lower yields. Therefore, in order to examine whether systems are properly positioned, the average POA for the years 2014 to 2016 was calculated for every tilt and orientation angle using hourly resolution values from the KNMI meteorological stations (Figure 4).The average annual energy irradiation for this period is maximized at 1045 $kWh/m^2$ for PV generators oriented at 200° (with South being 180°) and having an inclination of 37°. The displacement from the South is due to the clearing of morning clouds in the afternoon, which results in a larger share of radiation in the afternoon, and thus leading to off-South optimum orientations [28]. In order for a PV system to be able to harvest at least 95% of the incoming irradiation the tilt and the orientation angle should fall within the grey shaped area in Figure 4b, which extends from 152° to 252° for the orientation and 10° to 64° for inclination. The grey colored area in Figure 4c represents the range of orientation and inclination angles that are necessary to collect at least 90% of incoming irradiation. Since, in most cases, the exact configuration of each installation is dictated by the roof that the PV is mounted on a large variation is revealed as expected. However, the extent of this variation is a an indicator of how well a system is planned before installation. By considering as good oriented systems the ones that were able to harvest at least 90% of the maximum incoming irradiation (grey colored area in Figure 4c), in the data set, we find that 90%, 88% and 92% of urban, suburban and rural systems, respectively, fall in this category. More specifically by setting the optimum harvested POA to 95% (shaded area in Figure 4b) the same values are 75%, 74% and 77%. These numbers demonstrate that, at least in this case, urban compactness is not leading to extreme misplacement of installations especially in locations with limited availability in rooftop area.

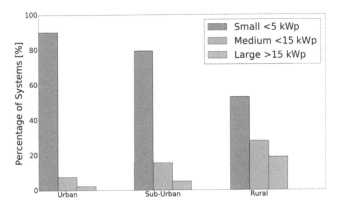

**Figure 3.** System size distribution per location.

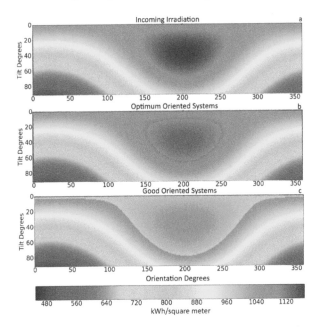

**Figure 4.** (**a**) Plane of Array (POA) irradiation; (**b**) optimum oriented systems; (**c**) well oriented systems.

*3.2. Performance Analysis—Seasonal Variation*

As it was specified in the introduction, 5-minute resolution power data were collected from each site. Each system is characterized by its latitude and longitude values completed with the corresponding tilt and orientation of the panels. In that way, the calculation of both the Annual Yield and the Performance Ratio was feasible. Due to system failures, in very few cases, part of data was either not recorded or not transmitted. To encounter the issue of missing entries, only installations that operated at least 97% percent of the time each year were taken into account for the calculation of annual yield and *PR*. The summary of the performance indicators is presented in Table 1, and it is in compliance with similar values that are reported for the Netherlands in previous research [29,30].

**Table 1.** Average annual yield and *PR* values with the corresponding standard deviations

| Year | Annual Yield kWh/kW$_p$ | PR % |
|------|--------------------------|------|
| 2014 | $919 \pm 78$ | $79 \pm 6$ |
| 2015 | $970 \pm 126$ | $79 \pm 6$ |
| 2016 | $945 \pm 89$ | $80 \pm 7$ |

To determine whether there is a variation in performance between PV generators located in urban, suburban and rural areas, the dataset of 2016 is presented as a boxplot in Figure 5. Comparison of *PR* values can reveal losses caused by wiring, system failures, Alternating Current/Direct Current (AC/DC) conversion and dust, which are mainly system dependent, but it can also unveil a situation were a system is shaded for a long period of time or it is exposed to higher temperatures. To further investigate if the means of the three groups demonstrate significant statistical difference, one way Analysis of Variance (ANOVA) was performed, which is a common practice to identify differences in PV monitoring [31]. The null hypothesis was set so that the three distributions had no difference in the mean values.

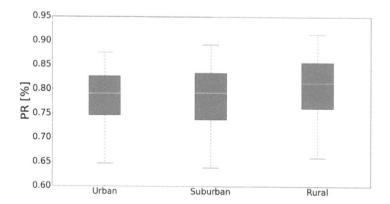

**Figure 5.** Boxplots of annual *PR* distribution for 2016.

The result shows that there is indeed an effect of the location on the performance of the PV installation at the $p < 0.05$ level for the three samples ($F = 18.56$, $p = 1.1 \times 10^{-8}$). However, the ANOVA test states whether there is an overall difference between the groups, but it does not allow us to know which specific groups differed. Therefore, it is necessary to conduct a post hoc test, since it is designed for situations where additional exploration of the differences among means is required to provide specific information on which means are significantly different from each other. Post hoc comparisons using the Tukey Honest Significant Difference (HSD) test indicated that the mean performance of rural systems (mean = 81, std = 6.3) was significantly higher than the mean performance of urban systems (mean = 78.4, std = 5.6), as it was compared to suburban systems (mean = 78.5, std = 6.4). However, urban and suburban systems did not show a significant difference between them, and, therefore, for seasonal variation, only the urban and rural systems will be taken into consideration. Taken together, these results suggest that the location of a PV system really does have an effect on the annual performance.

To further identify the causes of this performance mismatch in the annual mean values, it is necessary to investigate whether the difference is leveled throughout the year, or it is subjected to a seasonal variation. For that purpose, the monthly $PR$ value differences between rural and urban systems will be compared. As it is shown in Figure 6, there is a specific pattern that the percentage difference between the monthly $PR$ values follows consistently for the entire period of monitoring. During winter months (December–February), PV generators that are located in rural areas are outperforming the ones located in urban areas by up to 16% in some cases. The difference gradually declines during spring and summer months where it reaches a minimum of 0.2–3.4%.

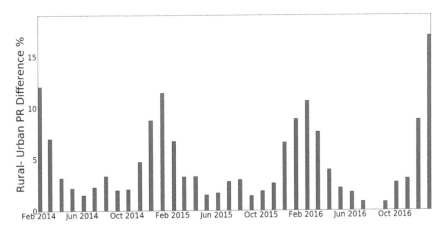

**Figure 6.** Monthly $PR$ difference between PV installations in rural and urban areas.

The seasonality and the consistency of the effect can be attributed only to variations in environmental parameters. Since the highest difference is observed during winter months, additional focus will be set on irradiation conditions, and specifically the trajectory of the sun. For the geographical coordinates of the Netherlands (Amsterdam 52.3702° N, 4.8952° E), the maximum solar elevation is limited to 15° in December and 20° in January, but it can reach 61° in June. In order to examine the effect of the solar elevation on PV performance, $PR$ is correlated to solar elevation degrees (Figure 7) using the PVLIB library of Python [32] and the specific coordinates of every installation. As it is observed, for low elevation angles below 18°, the difference in $PR$ values can reach 15% in favor of rural systems. While low solar paths are in general not favorable for the energy yield of PV installations regardless their specific location, it appears that it has a stronger effect on the ones located in urban areas. A possible explanation is that low solar elevation is more likely to cause shading in the dense environment of a city rather than in a sparsely occupied rural area. To consider the overall effect on the energy production on a yearly basis, it is highlighted that, during late autumn and winter months (November–February), which corresponds to 1/3 of a year, the sun spends 79% of its time in the sky below the threshold of 18°. However, besides this behavior that is observed for a part of the day or the year, a systematic difference that can reach 5% in favor of rural systems is noticed for the rest of the time and therefore a more detailed analysis is conducted over specific urban indicators.

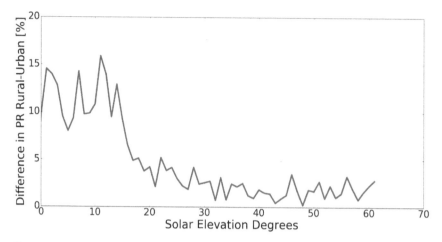

**Figure 7.** Difference in *PR* between PV installations in rural and urban areas versus the solar elevation.

### 3.3. Performance Analysis—Urban Indicators

Thus far, it is demonstrated that the urban environment can have a strong effect on the overall performance of a PV system. The characterization of different locations was based on the population density; however, not all the urban and rural environments exhibit the same morphology neither does the population density necessarily dictate the urban landscape, since large structures such as public or office buildings could be in non-residential areas. Therefore, it is necessary to identify specific indicators that could quantify the effects of urban compactness.

In this case, the six different indicators that were described in Section 2 are compared with the *PR* values of the systems as they all detach the form of the cityscape from the population factor (Figure 8). All the PV installations were grouped according to the overall characteristics of their surrounding location as this was indicated using GIS software. The groups of ABH, Plot Ratio, BHSD and the IBHSD predict the average *PR* with a linear regression model (with slope $\alpha$) and the high R-squared values show a good correlation within the boundaries of the independent variables. The reasons behind the low *R*-squared values of the last two indicators, the Number of Buildings and the ABD, will be further examined.

For the categorization according to the ABH, the first group in Figure 8a consists of systems placed in areas with an average building height of 7 meters, which is approximately the height of the typical Dutch household with a tiled rooftop. As the height increases, the average *PR* exhibits a reduction of 0.583 units for every added meter, which corresponds a decrease of 1.75 units for every added floor (assuming that the average height of every additional floor is 3 m). The plot ratio defines the degree of built environment compactness of an area and again exhibits linear reduction with the *PR* as it can be seen in Figure 8b. A plot ratio of 0.1 indicates that the specific location is sparsely developed in terms of building infrastructure as the total surface of every buildings equals with just the 10% of the total area and the average *PR* in those locations is 78.7%. On the contrary a plot ratio of 1.1 represents a a densely built environment, where the total building surface equals with 110% of the total surface of this area and the average *PR* is limited to 75.7%. Shading from adjusting buildings is probably one of the major causes of low PV performance in metropolitan areas. The ABH indicates the level of vertical development of a neighborhood but it is not the only indicator able to predict the possibility for one construction to shade another. This can also be done by examining the slope of the linear model between std of building heights and average *PR* in Figure 8c, and it is revealed that every extra meter of deviation in vertical development is expected to reduce the average *PR* by 2.45 units. The almost linear relationship between the size of various constructions/equipment installed on rooftops and the average *PR* values can be seen in Figure 8d. The effect in this case can be direct shading, total or partial which in a compact environment it is also possible to affect nearby rooftops, or indirect, as these elements might significantly limit the available space and lead to a non-optional placement of the PV system. According to the figure, even a chimney, shed or the ventilation equipment that does not exceed 1–1.5 m in height is able to reduce by more than 2% the average *PR* compared to a totally flat rooftop.

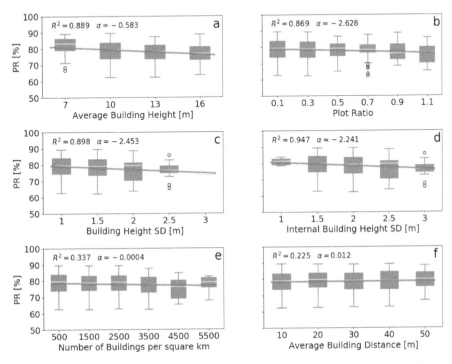

**Figure 8.** Comparison of the average *PR* using linear regression with (**a**) ABH; (**b**) plot ratio; (**c**) BHSD; (**d**) IBHSD; (**e**) number of buildings per km$^2$; (**f**) ABD.

The low *R*-squared value in Figure 8e indicates a weak linear relationship between the density of buildings and the average *PR*. However, by removing the last group which includes areas with 5000–6000 buildings per square kilometer, the *R*-squared value increases from 0.337 to 0.779. A closer look over the general characteristics of the last, high building concentration group reveals that the majority of these systems were installed in residential areas with low ABH (mean = 8.24 m, std = 0.4 m) and relatively flat rooftops (IBHSD mean = 1.45 m, std = 0.14 m), showing that the use of a single indicator can be insufficient to evaluate the effects of urban compactness on the average *PR* of an area. Finally, the average *PR* seems to be unaffected by the ABD between buildings despite the fact that this can be a direct way to measure urban compactness. However, this can also be done by examining a slightly different indicator, the Nearest Neighbor Index (NNI), which is expressed as the ratio of the Average Building Distance to the Expected Mean Distance (EMD). The EMD is the average distance between neighbors in a hypothetical situation where all the buildings of an area were randomly distributed. Every value below 1 indicates clustering, while, for values greater than 1, the trend is towards dispersion (Figure 9).

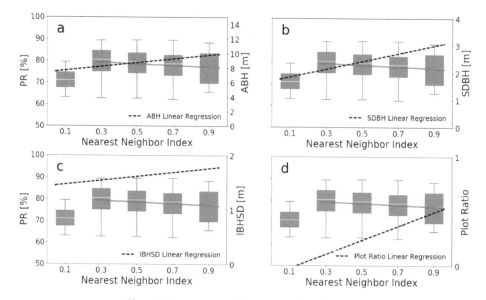

**Figure 9.** *PR* compared with the nearest index indicator.

As it can be seen in Figure 9, the average *PR* is affected only for small NNI values (0–0.2), where urban clustering is high. One would expect that, as the NNI value increases and the horizontal distribution of buildings in an area leans towards scattering, the overall performance of PV systems would increase. However, by considering NNI values above 0.2, an almost negative linear relationship appears ($R^2 = 0.997$, $\alpha = -4.01$) compared to the average *PR*, which is depicted as a red line in Figure 9a–d. This counterintuitive outcome can be explained by taking into consideration the ABH, the SDBH, the IBHSD and the Plot Ratio of those areas. All four indicators exhibit strong, increasing, linear trends (black dashed line in figures 9a–9d) for those specific locations ($\alpha_{ABH} = 3$, $\alpha_{SDBH} = 1.62$, $\alpha_{IBHSD} = 0.41$, $\alpha_{PlotRatio} = 0.73$), proving that even in the case of high clustering, vertical distribution still plays the major role.

*3.4. Errors and Uncertainties*

For this type of research, two types of errors can be identified. The first source of error is related to the data sources themselves as the number of input entries, such as the size of the system or the tilt and orientation, requires hand-digitized information that can be either incorrectly calculated by the owner or incorrectly entered in the database. This type of error usually leads to an abnormal *PR* or annual yield value that can be identified and excluded as an outlier.

The second type of error is somehow inherent to every type of research that focuses on a large scale analysis of PV systems scattered over a large geographical area. While digital data acquired from monitoring equipment are considered to be of high quality, the lack of irradiation measurements on site leads to uncertainty over the performance indicators. Even when measurements are provided by the dense meteorological network of a governmental agency such in this case, it is usually in the form of GHI without separation of the direct and diffuse component. Consequently, the error associated with the transposition model is unavoidable and the related uncertainty depends on the mainly on the overall meteorological conditions and the time of the day, but in some cases it can exceed 10% [21]. Installation and maintenance of independent monitoring equipment are both costly and impractical, and therefore the validity of this type of research will always be relying on the large number of systems for the analysis.

However, by analyzing a large number of systems over a long monitoring period, the validity of the research can be ensured. In Figure 10, the linear relationship of the indicators described in Section 2 is presented in comparison with the final system yield $Y_f$. As in Figure 9, the ABH, BHSD, IBHSD are following a negative relationship with the $Y_f$ even though the linearity is weakened.

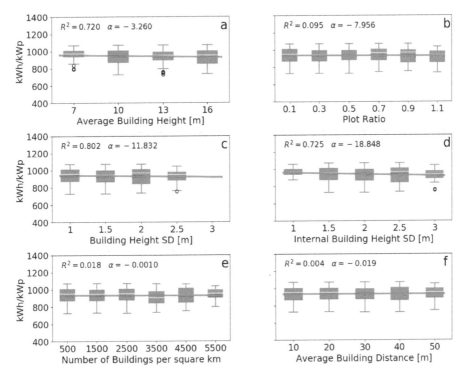

**Figure 10.** Comparison of $Y_f$ using linear regression with (**a**) ABH; (**b**) plot ratio; (**c**) BHSD; (**d**) IBHSD; (**e**) number of buildings per km$^2$; (**f**) ABD.

## 4. Conclusions

In an era where almost everything can be thoroughly monitored, this study proved that the massive amount of PV yield data that are collected can offer a fast and inexpensive way to evaluate the performance potential of large geographical areas. Urban compactness has many forms, and it can be physically separated on a horizontal and vertical level. The results of this analysis show that, for the second case, the indicators that describe the 3D variation over different areas within the Netherlands, a strong linear relationship occurs with relatively high R-squared values ($R^2_{ABH} = 0.889$, $R^2_{BHSD} = 0.898$, $R^2_{IBHSD} = 0.947$). The linearity is also preserved when the average $PR$ is compared to $Y_f$ for these three indicators. Homogeneity in building height and flat roofs are essential to achieve high performance with minimum shading, as every meter of height deviation results in 2.453 units of reduction in the average $PR$ in the first case and 2.241 units in the second. The distance between buildings is important only when the urban environment is highly clustered, but, as buildings become more scattered, the effects of the vertical distribution are driving the performance losses.

It was also proven that the effects of the urban environment are very complex, and, in most cases, the use of a single indicator to assess the overall performance of PV was inadequate. Finally, it revealed that urban effects should not be considered as a constant effect but as an effect that exhibits large seasonal variation that can be over 15% depending on the latitude of the specific location, something that was also mentioned in previous research [33], but it was never quantified up to now according to our knowledge.

The constantly increasing energy demands, which are projected to rise even more due to electrification in mobility, highlight the importance of PV integration in our cities. In this direction, this study could be extended, comparing areas with different urban environments and in combination with software like GIS, can help us develop better ways to construct our buildings and plan our cities.

**Author Contributions:** P.M. and W.v.S. conceived the project, P.M. conceived and designed the simulations, performed the data analysis and wrote the paper, B.B.K. and N.N. performed the GIS analysis.

**Funding:** This project is partly financially supported by the Netherlands Enterprise Agency (RVO).

**Acknowledgments:** This work is part of the International Energy Agency—Photovoltaic Power Systems (IEA-PVPS) Task 13 "Performance and Reliability of Photovoltaic Systems"; we would like to thank all members of this task for their support.

**Conflicts of Interest:** The authors declare no conflicts of interest.

## References

1. Secretariat, U.N. *World Urbanization Prospects*; Technical Report; Department of Economic and Social Affairs: New York, NY, USA, 2012.
2. Performance, I.E. *POLIS, Solar Urban Planning-Manual of Best Practices*; Technical Report; European Commission: Brussels, Belgium, 2010.
3. *Directive 2010/31/ of the European Parliament on the Energy Performance of Buildings*; Technical Report; European Parliament: Brussels, Belgium, 2010.
4. Li, D.; Liu, G.; Liao, S. Solar potential in urban residential buildings. *Sol. Energy* **2015**, *111*, 225–235, doi:10.1016/j.solener.2014.10.045. [CrossRef]
5. Shukla, A.K.; Sudhakar, K.; Baredar, P. Recent advancement in BIPV product technologies: A review. *Energy Build.* **2017**, *140*, 188–195, doi:10.1016/j.enbuild.2017.02.015. [CrossRef]
6. Izquierdo, S.; Rodrigues, M.; Fueyo, N. A method for estimating the geographical distribution of the available roof surface area for large-scale photovoltaic energy-potential evaluations. *Sol. Energy* **2008**, *82*, 929–939, doi:10.1016/j.solener.2008.03.007. [CrossRef]
7. Izquierdo, S.; Montañés, C.; Dopazo, C.; Fueyo, N. Roof-top solar energy potential under performance-based building energy codes: The case of Spain. *Sol. Energy* **2011**, *85*, 208–213, doi:10.1016/j.solener.2010.11.003. [CrossRef]
8. Wiginton, L.; Nguyen, H.; Pearce, J. Quantifying rooftop solar photovoltaic potential for regional renewable energy policy. *Comput. Environ. Urban Syst.* **2010**, *34*, 345–357, doi:10.1016/j.compenvurbsys.2010.01.001. [CrossRef]
9. Hofierka, J.; Zlocha, M. A New 3D Solar Radiation Model for 3D City Models. *Trans. GIS* **2012**, *16*, 681–690, doi:10.1111/j.1467-9671.2012.01337.x. [CrossRef]
10. Jakubiec, J.A.; Reinhart, C.F. A method for predicting city-wide electricity gains from photovoltaic panels based on LiDAR and GIS data combined with hourly Daysim simulations. *Sol. Energy* **2013**, *93*, 127–143, doi:10.1016/j.solener.2013.03.022. [CrossRef]
11. Catita, C.; Redweik, P.; Pereira, J.; Brito, M. Extending solar potential analysis in buildings to vertical facades. *Comput. Geosci.* **2014**, *66*, 1–12, doi:10.1016/j.cageo.2014.01.002. [CrossRef]
12. Verso, A.; Martin, A.; Amador, J.; Dominguez, J. GIS-based method to evaluate the photovoltaic potential in the urban environments: The particular case of Miraflores de la Sierra. *Sol. Energy* **2015**, *117*, 236–245, doi:10.1016/j.solener.2015.04.018. [CrossRef]
13. Freitas, S.; Catita, C.; Redweik, P.; Brito, M. Modelling solar potential in the urban environment: State-of-the-art review. *Renew. Sustain. Energy Rev.* **2015**, *41*, 915–931, doi:10.1016/j.rser.2014.08.060. [CrossRef]
14. Sarralde, J.J.; Quinn, D.J.; Wiesmann, D.; Steemers, K. Solar energy and urban morphology: Scenarios for increasing the renewable energy potential of neighbourhoods in London. *Renew. Energy* **2015**, *73*, 10–17, doi:10.1016/j.renene.2014.06.028. [CrossRef]

15. Redweik, P.; Catita, C.; Brito, M. Solar energy potential on roofs and facades in an urban landscape. *Sol. Energy* **2013**, *97*, 332–341, doi:10.1016/j.solener.2013.08.036. [CrossRef]

16. Mohajeri, N.; Upadhyay, G.; Gudmundsson, A.; Assouline, D.; Kämpf, J.; Scartezzini, J.L. Effects of urban compactness on solar energy potential. *Renew. Energy* **2016**, *93*, 469–482, doi:10.1016/j.renene.2016.02.053. [CrossRef]

17. Kanters, J.; Wall, M. A planning process map for solar buildings in urban environments. *Renew. Sustain. Energy Rev.* **2016**, *57*, 173–185, doi:10.1016/j.rser.2015.12.073. [CrossRef]

18. Lobaccaro, G.; Frontini, F. Solar Energy in Urban Environment: How Urban Densification Affects Existing Buildings. *Energy Procedia* **2014**, *48*, 1559–1569, doi:10.1016/j.egypro.2014.02.176. [CrossRef]

19. CBS. CBS Statistics. 2017. Available online: https://www.cbs.nl/en-gb/our-services/methods/surveys/korte-onderzoeksbeschrijvingen/renewable-energy (accessed on 23 May 2018).

20. Reich, N.H.; Mueller, B.; Armbruster, A.; van Sark, W.G.J.H.M.; Kiefer, K.; Reise, C. Performance ratio revisited: Is PR greater than 90% realistic? *Prog. Photovolt. Res. Appl.* **2012**, *20*, 717–726, doi:10.1002/pip.1219. [CrossRef]

21. Yang, D. Solar radiation on inclined surfaces: Corrections and benchmarks. *Sol. Energy* **2016**, *136*, 288–302, doi:10.1016/j.solener.2016.06.062. [CrossRef]

22. Perez, R.; Seals, R.; Ineichen, P.; Stewart, R.; Menicucci, D. A new simplified version of the perez diffuse irradiance model for tilted surfaces. *Sol. Energy* **1987**, *39*, 221–231, doi:10.1016/s0038-092x(87)80031-2. [CrossRef]

23. Hay, J.E. Calculating solar radiation for inclined surfaces: Practical approaches. *Renew. Energy* **1993**, *3*, 373–380, doi:10.1016/0960-1481(93)90104-o. [CrossRef]

24. Olmo, F.; Vida, J.; Foyo, I.; Castro-Diez, Y.; Alados-Arboledas, L. Prediction of global irradiance on inclined surfaces from horizontal global irradiance. *Energy* **1999**, *24*, 689–704, doi:10.1016/s0360-5442(99)00025-0. [CrossRef]

25. Tukey, J.W. *Exploratory Data Analysis*; Pearson: London, UK, 1977.

26. Polverini, D.; Field, M.; Dunlop, E.; Zaaiman, W. Polycrystalline silicon PV modules performance and degradation over 20 years. *Prog. Photovolt. Res. Appl.* **2012**, doi:10.1002/pip.2197. [CrossRef]

27. Ishii, T.; Masuda, A. Annual degradation rates of recent crystalline silicon photovoltaic modules. *Prog. Photovolt. Res. Appl.* **2017**, *25*, 953–967, doi:10.1002/pip.2903. [CrossRef]

28. Litjens, G.B.; Worrell, E.; van Sark, W.G. Influence of demand patterns on the optimal orientation of photovoltaic systems. *Sol. Energy* **2017**, *155*, 1002–1014. [CrossRef]

29. Moraitis, P.; Kausika, B.B.; van Sark, W.G.J.H.M. Visualization of operational performance of grid-connected PV systems in selected European countries. In Proceedings of the 2015 IEEE 42nd Photovoltaic Specialist Conference (PVSC), Hamburg, Germany, 14 December 2015. [CrossRef]

30. Tsafarakis, O.; Moraitis, P.; Kausika, B.B.; van der Velde, H.; 't Hart, S.; de Vries, A.; de Rijk, P.; de Jong, M.M.; van Leeuwen, H.P.; van Sark, W. Three years experience in a Dutch public awareness campaign on photovoltaic system performance. *IET Renew. Power Gener.* **2017**, *11*, 1229–1233, doi:10.1049/iet-rpg.2016.1037. [CrossRef]

31. Leloux, J.; Narvarte, L.; Trebosc, D. Review of the performance of residential PV systems in France. *Renew. Sustain. Energy Rev.* **2012**, *16*, 1369–1376, doi:10.1016/j.rser.2011.10.018. [CrossRef]

32. Stein, J.S. The photovoltaic performance modeling collaborative (PVPMC). In Proceeding of the Photovoltaic Specialists Conference, Austin, TX, USA, 3–8 June 2012.

33. Kanters, J.; Wall, M.; Dubois, M.C. Typical Values for Active Solar Energy in Urban Planning. *Energy Procedia* **2014**, *48*, 1607–1616, doi:10.1016/j.egypro.2014.02.181. [CrossRef]

*Article*

# Evaluation of the Reactive Power Support Capability and Associated Technical Costs of Photovoltaic Farms' Operation

**Luís F. N. Lourenço [1,*,†], Renato M. Monaro [1], Maurício B. C. Salles [1], José R. Cardoso [1] and Loïc Quéval [2]**

[1] Laboratory of Advanced Electric Grids (LGRID), Escola Politécnica, University of São Paulo, São Paulo 05508-010, Brazil; monaro@usp.br (R.M.M.); mausalles@usp.br (M.B.C.S.); jose.cardoso@usp.br (J.R.C.)

[2] Group of Electrical Engineering—Paris (GeePs), UMR CNRS 8507, Centrale Supélec, Univ. Paris-Sud, Université Paris-Saclay, Sorbonne Université, 3 & 11 rue Joliot-Curie, 91192 Plateau de Moulon Gif-sur-Yvette CEDEX, France; loic.queval@geeps.centralesupelec.fr

* Correspondence: lfnlourenco@usp.br; Tel.: +55-11-3091-5533

† Current address: Av. Prof. Luciano Gualberto, 158-Butantã, São Paulo 05508-900, Brazil.

Received: 15 April 2018; Accepted: 8 June 2018; Published: 14 June 2018

**Abstract:** The share of photovoltaic (PV) farms is increasing in the energy mix as power systems move away from conventional carbon-emitting sources. PV farms are equipped with an expensive power converter, which is, most of the time, used well bellow its rated capacity. This has led to proposals to use it to provide reactive power support to the grid. In this framework, this work presents a step-by-step methodology to obtain the reactive power support capability map and the associated technical costs of single- and two-stage PV farms during daytime operation. Results show that the use of two-stage PV farms can expand the reactive power support capability for low irradiance values in comparison to single-stage ones. Besides, despite losses being higher for two-stage PV farms, the technical cost in providing reactive power support is similar for both systems. Based on the obtained maps, it is demonstrated how the profits of a PV farm can be evaluated for the current ancillary services policy in Brazil. The proposed method is of interest to PV farm owners and grid operators to estimate the cost of providing reactive power support and to evaluate the economic feasibility in offering this ancillary service.

**Keywords:** solar farm; photovoltaics; reactive power support; STATCOM; technical costs

---

## 1. Introduction

Utility-scale photovoltaic (PV) farms are expected to reach an installed capacity of 290 GW by 2019 [1]. This motivates the evolution of grid codes to regulate their connection to the electric grid. In particular, the reactive power support from PV farms is being discussed in many countries [2], and updates are expected from early grid codes that prevented them from providing reactive power support [3,4]. For example, the German grid code [5] now requires reactive power support from PV farms. In parallel, reactive power markets are emerging [6–9] with the prospect of expanding the portfolio of products offered by PV farm owners. These trends would benefit consumers too because significant savings in grid operation costs are expected with flexible reactive power support from distributed renewable sources [10,11].

One of the key components of the PV farm is the power electronics converter that converts the power generated in direct current (DC) by the PV panels into the alternating current (AC) electric grid. This converter, usually a voltage source converter (VSC), is an expensive asset, but most of the time, it is used well bellow its rated power: during nighttime, the converter is idle as there is no irradiance and therefore no active power generation; while during daytime, the converter power follows the solar irradiance cycle and reaches the rated power only just a few minutes per day. The remaining converter capacity could then be used to provide reactive power support to the electric grid.

The use of the PV farm converter to provide reactive power support was called "PV-STATCOM" by [12], where the concept was proposed for nighttime operation when the full converter capacity is available. In [13], it was shown that the use of a PV-STATCOM during nighttime can increase the active power transmission limits without the installation of flexible AC transmission system (FACTS) devices or new expensive transmission lines. Following work from the same authors [14,15] extended the PV-STATCOM concept to daytime operation in scenarios where priority was given to active power generation and only the remaining converter capacity could be used for reactive power support: Ref. [14] used the PV system to provide the necessary reactive power for the steady-state and transient operation of a heat pump and Ref. [15] has shown that daytime active power transmission capacity could also be increased by using a PV-STATCOM.

Reactive power support by VSC-based PV farms is possible thanks to the inherent ability of VSCs to control active and reactive power independently. Several works have discussed PV farms' reactive power control schemes with the constraint of keeping the VSC within its capability limits. The direct power control with space vector modulation (DPC-SVM) of single-stage PV farms with three-phase VSC was discussed by [16,17], the control of two-stage single phase PV farms by [18], the voltage-oriented control (VOC) for three phase single-stage PV farms by [17,19] and the control hardware and strategies for two-stage three-phase PV farms by [17,20], and Ref. [17] also presented a model-based predictive controller (MPC). However, none of these works have dealt with the system losses, nor the technical costs of providing reactive power support. Note that distribution network losses due to reactive power support of PV systems were discussed by [21], where global system losses were evaluated through an active-reactive optimal power flow (A-R-OPF).

In the context of wind energy, a technical-economic discussion of wind farms (WF) providing reactive power support to the grid and a modeling of WF capability based on wind variations were presented by [22].

The work in Ref. [23] presented a method to maximize the reactive power reserves while minimizing system losses, and Ref. [24] presented a method to evaluate the losses focusing on the WF and also an operation strategy to minimize power losses when there is a reactive power dispatch. From an overall grid point of view [11,25–27] presented a reactive power dispatch based on A-R-OPF showing that combined reactive support from WF and energy storage systems could significantly reduce the grid losses. Therefore, as previous works did not deal with the losses from the point of view of the PV farm providing reactive power support, the authors proposed earlier a methodology to evaluate the technical cost of operating a PV farm as a STATCOM for nighttime operation [28,29]. A sequel to the work extended the analysis for daytime operation [30], but focused exclusively on single-stage PV farms and lacked a clear step-by-step methodology.

The objective of this paper is to establish a step-by-step methodology to obtain the reactive power support capability map and the associated technical costs of single- and two-stage PV farms for daytime operation. The proposed method relies exclusively on simulations and manufacturer data instead of experimental data. It is assumed throughout this study that priority is given to active power generation, and therefore, only the remaining converter capacity can be used for reactive power support. This work should benefit PV farm owners to plan their bids in future reactive power markets accordingly and to system operators to evaluate if the price paid for this ancillary service is adequate.

This work is structured as follows: Section 2 presents the single- and two-stage PV farm systems used in this work; Section 3 presents the proposed methodology; Section 4 presents the system losses without reactive power support; Section 5 presents the reactive power support capability and the system losses with reactive power support and derives the technical cost of providing the reactive power support.

## 2. PV Farm Models

This section presents the single- and two-stage PV farms models used throughout this work.

### 2.1. Single-Stage PV Farm Model

A single-stage PV farm is a topology where only one DC/AC converter is used to interface the PV panel array to the electric grid [31–33]. The DC/AC converter is typically a VSC. Single-stage PV farms are usually more efficient than two-stage ones, but the PV panels' output voltage is not fully decoupled from the grid voltage.

In this work, we adopt the single-stage PV farm shown in Figure 1, which is based on [33]. The DC/AC converter is a two-level, three-phase VSC. The DC/AC converter is connected to the grid through an RL reactor, an RLC series filter to smooth current harmonics and a step-up $\Delta$-$Y_g$ transformer to insulate the PV farm from zero sequence current faults that might occur on the grid side. The system dimensioning and DC/AC controller tuning were presented in [28]. The AC/DC converter controls the PV voltage $v_{pv}$ via a perturb and observe MPPT algorithm [34], as well as the reactive power injected to the grid. The single-stage PV farm parameters are summarized in Table 1. The power-voltage characteristics of the PV panel are shown in Figure 2.

In Figure 1, $v_{pv}$ indicates the PV panel voltage whose reference value is generated by the maximum power point tracker (MPPT) algorithm; $C$ is the capacitor of the DC circuit; $i_{abc}$ is the converter AC current; $r_r$ and $L_r$ are the resistance and the inductance of the tie reactor; $R_f$, $L_f$ and $C_f$ are the resistance, the inductance and the capacitance of the filter. Both active and reactive power are considered positive when flowing from the PV system to the grid.

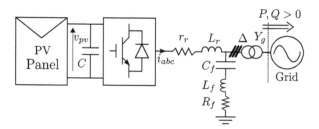

**Figure 1.** Overview of the single-stage PV system.

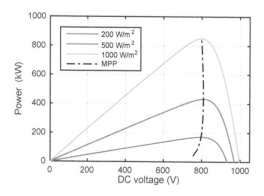

**Figure 2.** Power-voltage characteristics of the single-stage PV farm PV panel at 25 °C.

**Table 1.** Single-stage PV farm system parameters.

| Parameter | Value | Unit | Parameter | Value | Unit | Parameter | Value | Unit |
|---|---|---|---|---|---|---|---|---|
| Rated power | 850 | kW | $v_\Delta$ | 380 | V | $R_f$ | 0.5 | $\Omega$ |
| $v_{pv}$ @1000 W/m$^2$ | 798 | V | f | 50 | Hz | $L_f$ | 397.8 | µH |
| C | 87.8 | mF | $r_r$ | 1 | m$\Omega$ | $C_f$ | 0.64 | µF |
| | | | $L_r$ | 54.1 | µH | | | |

| Component | Reference | Series modules | Parallel modules | Total |
|---|---|---|---|---|
| PV module | Kyocera Solar KD205GX-LP | 30 | 138 | 4.140 |
| DC/AC converter | ABB 5SNA1600N170100 IGBT | 1 | 2 | 12 |

## 2.2. Two-Stage PV Farm Model

A two-stage PV farm is a topology where a combination of DC/DC converter(s) and DC/AC converter(s) is used to interface the PV panel to the electric grid [32,35]. Here, again, the DC/AC converter is usually a VSC. The DC/DC converter topology varies depending on the requirements, but its main role is always to decouple the panels' output voltage from the grid voltage, at the cost of higher power losses.

In this work, we adopt the two-stage PV farm shown in Figure 3, which is based on [31,32]. The DC/AC converter is a two-level three-phase VSC, and the DC/DC converter is a boost converter. Again, the DC/AC converter is connected to the grid through an RL reactor, an RLC series filter and a step-up Δ-$Y_g$ transformer. The system dimensioning and the DC/AC converter controller tunings were presented in [28]. The AC/DC converter controls the DC voltage $v_{dc}$ and the reactive power injected to the grid. The DC/DC converter controls the PV voltage $v_{pv}$ via a perturb and observe MPPT algorithm [34]. The two-stage PV farm parameters are summarized in Table 2. The power-voltage characteristics of the PV panel are shown in Figure 4.

In Figure 3, $v_{pv}$ is the PV panel voltage whose reference value is generated by the maximum power point tracker (MPPT) algorithm; $v_{dc}$ is the output voltage of the DC/DC; $C_{pv}$ and C are the DC circuit capacitors; $L_{DC}$ is the DC/DC converter inductor and $r_{DC}$ its resistance; $i_{abc}$ is the converter AC current; $r_r$ and $L_r$ are the resistance and the inductance of the tie reactor; $R_f$, $L_f$ and $C_f$ are the resistance, the inductance and the capacitance of the filter. Both active and reactive power are considered positive when flowing from the PV system to the grid.

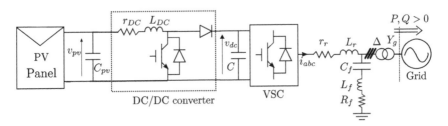

**Figure 3.** Overview of the two-stage PV system. VSC, voltage source converter.

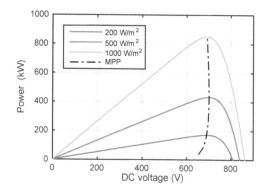

**Figure 4.** Power-voltage characteristics of the two-stage PV farm PV panel at 25 °C.

**Table 2.** Two-stage PV farm system parameters.

| Parameter | Value | Unit | Parameter | Value | Unit | Parameter | Value | Unit |
|---|---|---|---|---|---|---|---|---|
| Rated power | 850 | kW | $v_\Delta$ | 380 | V | $f$ | 50 | Hz |
| $v_{pv}$ @ 1000 W/m$^2$ | 691 | V | $r_{DC}$ | 1 | m$\Omega$ | $R_f$ | 0.5 | $\Omega$ |
| $C, C_{pv}$ | 87.8 | mF | $L_{DC}$ | 5.0 | mH | $L_f$ | 397.8 | µH |
| $r_r$ | 1 | m$\Omega$ | $L_r$ | 54.1 | µH | $C_f$ | 0.64 | µF |

| Component | Reference | Series modules | Parallel modules | Total |
|---|---|---|---|---|
| PV module | Kyocera Solar KD205GX-LP | 26 | 159 | 4.134 |
| DC/DC converter | ABB 5SNA1600N170100IGBT | 1 | 1 | 1 |
| DC/AC converter | ABB 5SNA1600N170100IGBT | 1 | 2 | 12 |

## 3. Methodology

For a given PV farm topology, the goal of this work is to evaluate the reactive power support capability and the associated losses during daytime for a wide range of operating points. Later on, the technical costs can be derived. We assume that priority is given to active power generation; therefore, only the remaining converter capability can be used for reactive power support.

### 3.1. Flowcharts

The flowcharts presented in Figures 5 and 6 summarize the proposed methodology for evaluating the losses and the reactive power capability for single-stage and two-stage PV farms, respectively. In the following paragraphs, these flowcharts are detailed.

For the single-stage PV farm, the goal is to cover the whole range of operating points $(G, Q_{ref})$ where $G$ is the solar irradiance and $Q_{ref}$ is the required reactive power support. To initiate the flowchart of Figure 5, the irradiance is set to zero, while $Q_{ref}$ is set to $Q_{min}$. Then, a simulation for the operating point $(G, Q_{ref})$ is performed. The next step consists of checking the controls: if the controller references are properly tracked, then the operating point is within the reactive power support capability area, and the power losses are evaluated. Otherwise, if the system controller references are not properly tracked, one of the converter capability limits is violated, and the operating point is outside of the capability area. After the losses evaluation, $Q_{ref}$ is increased by $\Delta Q$ until $Q_{max}$ is reached. To restart the evaluation for a new set of operating points, the irradiance $G$ is increased by $\Delta G$, and the procedure is repeated until the maximum local irradiance $G_{max}$ is reached. It is assumed in the simulations that the PV panel temperature is constant at 25 °C.

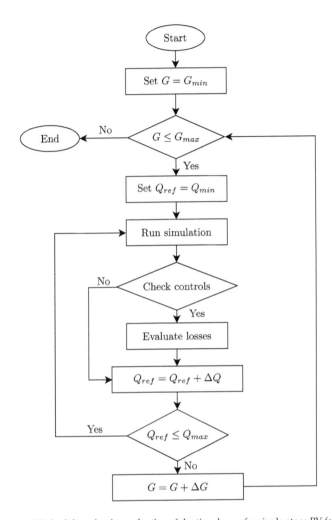

**Figure 5.** Methodology for the evaluation of daytime losses for single-stage PV farms.

For the two-stage PV farm, the procedure is similar to the previous one, but because of the extra degree of freedom offered by the use of the DC/DC converter, the flowchart presents an extra loop shown in red in Figure 6. Indeed, the DC voltage applied to the DC/AC converter $v_{dc}$ is not tied to the maximum power point (MPP) voltage, meaning that it can be set arbitrarily by the PV farm operator to reduce the DC/AC converter losses [36], for example. Therefore, the goal here is to cover the whole range of operating points $(G, v_{dc}, Q_{ref})$. Initially, the DC/DC converter output voltage $v_{dc}$ is set to $v_{dc\ min}$, then the procedure is similar to the one of the single-stage PV farm. To move on to the next operating point, $v_{dc}$ is increased by $\Delta v_{dc}$, and the procedure is repeated until $v_{dc\ max}$ is reached. The minimum DC circuit voltage is defined as [37]:

$$v_{dc\ min} = x \frac{2\sqrt{2}}{\sqrt{3}} v_\Delta \tag{1}$$

335

where $x$ is taken as 1.15 [38] and $v_\Delta$ is the RMS line-to-line voltage at the connection of the VSC output. The maximum DC voltage $v_{dc\ max}$ is taken close to the PV panel open circuit voltage for the maximum local irradiance at $-10\ ^\circ$C [39].

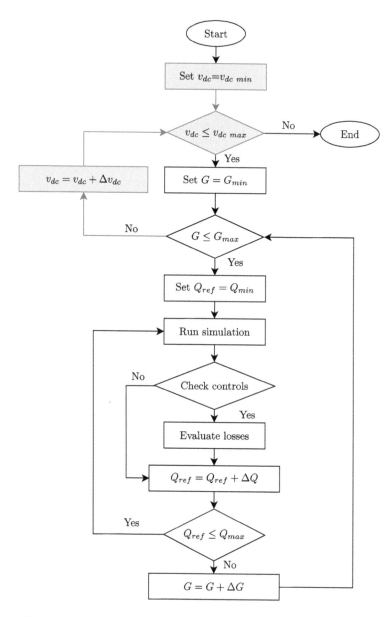

**Figure 6.** Methodology for the evaluation of daytime losses for two-stage PV farms.

### 3.2. PV System Losses

For a given operating point, the PV system losses $P_{total}$ are calculated as:

$$P_{total} = (P_{DC/DC}) + P_{DC/AC} + P_{reactor} + P_{filter} + P_{transformer} \tag{2}$$

$P_{reactor}$ and $P_{filter}$ are the Joule losses of the air-cored reactor and of the air-cored filter:

$$P_{reactor} = 3r_r I_r^2 \tag{3}$$

$$P_{filter} = 3R_f I_f^2, \tag{4}$$

where $I_r$ is the reactor current and $I_f$ is the filter current. $r_r$ and $R_f$ are the reactor resistance and filter resistance, respectively.

$P_{transformer}$ is the transformer losses, which can be divided into iron losses $P_{iron}$ and copper losses $P_{copper}$:

$$P_{transformer} = P_{copper} + P_{iron} \tag{5}$$

with:

$$P_{copper} = 3r_h I_h^2 + 3r_l I_l^2 \tag{6}$$

$$P_{iron} = 3\frac{V_g^2}{R_p} \tag{7}$$

where $I_h$ and $I_l$ are the high and low voltage winding currents and $V_g$ is the line-to-neutral voltage applied to the transformer high voltage windings. $r_h$, $r_l$ and $R_p$ are the high and low voltage windings' resistances and the iron losses' resistance [40], respectively.

For the DC/DC converter, $P_{DC/DC}$ can be written as:

$$P_{DC/DC} = P_r + P_{IGBT} + P_{diode} \tag{8}$$

where $P_r$ is the DC/DC converter air-cored inductor Joule loss, $P_{IGBT}$ is the IGBT loss and $P_{diode}$ is the diode loss. $P_r$ is calculated as:

$$P_r = r_{DC} I_{DC}^2 \tag{9}$$

where $r_{DC}$ is the air-cored inductor resistance and $I_{DC}$ is the DC bus current.

The losses of the semiconductor devices $P_{IGBT}$, $P_{diode}$ and the losses of the DC/AC converter $P_{DC/AC}$ do not have a closed formulation. The numerical evaluation of the converter losses for a wide range of operating points using a detailed switch model proved to be unfeasible due to the requirements of simulation steps in the order of nanoseconds. Therefore, alternative models were investigated [36,41–45]. The evaluation of these losses was carried out using the look-up table approach proposed by [45] with the same approximation as [28–30]: the junction temperature was assumed constant at 125 °C, so that only manufacturer data are considered instead of experimental data. The look-up table approach consists of comparing the switch currents' values pre and post the switching transitory. This comparison determines whether the switch was turned on or turned off, and the losses of the respective transitory are obtained through the look-up table. Furthermore, if the switch current is greater than zero, than the respective conduction losses are calculated.

### 3.3. Technical Costs

According to [30], the technical costs of operating a PV farm as a STATCOM during daytime "are associated with the electric power that must be bought from the grid to supply the difference between the system losses with and without reactive power support". Therefore, for a given irradiance, by subtracting the PV system losses without reactive power support $P_{total}(Q = 0)$ from the losses with reactive power support $P_{total}(Q \neq 0)$, one can derive the technical costs:

$$\text{Technical Costs} = P_{total}(Q \neq 0) - P_{total}(Q = 0) \tag{10}$$

By setting $Q_{min} = 0$ p.u. and $Q_{max} = 0$ p.u., one can evaluate $P_{no\ support}$ and by setting $Q_{min} = -1.0$ p.u. and $Q_{max} = +1.0$ p.u., one can obtain the daytime reactive power support capability map and the associated $P_{with\ support}$ of the PV systems.

## 4. Operation without Reactive Power Support

This section presents the results for losses without reactive power support (needed to derive the technical costs) and shows how the addition of the second stage impacts the losses.

### 4.1. Single-Stage PV Farm Losses without Reactive Power Support

For the single-stage PV farm, the DC voltage is tied to the irradiance following the MPP as shown in Figures 2 and 7. As a result, the losses are only a function of the solar irradiance. For nighttime operation, when the irradiance is 0 W/m$^2$, the DC voltage must be set to $v_{dc\ min}$ given by Equation (1). For both test systems, this value is 720 V. Figure 8 details the losses per component, while Figure 10 summarizes the losses as a percentage of the rated power . It can be seen that converter losses are predominant and exhibit an approximately linear behavior with respect to the solar irradiance.

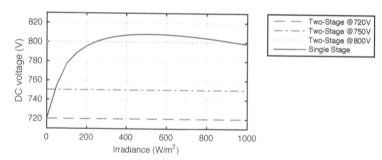

**Figure 7.** PV farm operating DC voltage $v_{dc}$ as a function of the solar irradiance.

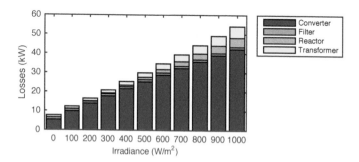

**Figure 8.** Single-stage PV farm losses per component, without reactive power support. The losses without reactive power support are approximately linear as a function of the irradiance.

### 4.2. Two-Stage PV Farm Losses without Reactive Power Support

For the two-stage PV farm, the DC voltage can be chosen by the PV farm operator. Three different output DC/DC converter voltages $v_{dc}$ are considered here: 720 V is the minimum voltage for the correct operation of the DC/AC converter; 750 V is the minimum voltage for which the DC/AC converter is able to generate 1.0 p.u. of reactive power support; and 800 V is the voltage at the maximum power point for the standard test conditions (25 °C and 1000 W/m²) for the single-stage PV farm. Figure 9 details the losses per component.

Figure 10 summarizes the losses as a percentage of the rated power. We observe the same trends as for the single-stage case. The use of the second stage leads to extra losses, but they can be minimized by operating at the lowest DC/DC converter output voltage possible (given by (1)). In the next section, the usefulness of using the second stage will be clarified.

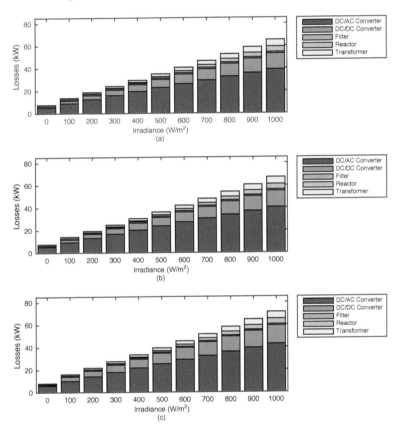

**Figure 9.** Two-stage PV farm losses per component for various $v_{dc}$, without reactive power support. (**a**) At 720 V; (**b**) at 750 V; (**c**) at 800 V. The losses without reactive power support are approximately linear as a function of the irradiance.

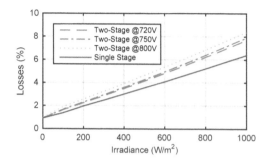

**Figure 10.** PV farm losses, without reactive power support, as a percentage of the rated power.

## 5. Operation with Reactive Power Support

This section presents the results regarding reactive power support capability, the evaluated losses for daytime operation with reactive power support and the operation map with the associated technical costs.

### 5.1. Reactive Power Support Capability Area

We remind that the reactive power support is provided by the DC/AC converter. The PQ-diagram for the 850 kVA DC/AC converter used in both test systems is shown in Figure 11. The circle of radius 1.0 p.u. represents the AC current converter limit, while the arcs represent the AC voltage limit. The AC voltage limit is tied to the DC voltage: a higher DC voltage enables a higher reactive power support capability because the converter is able to synthesize a higher AC voltage [46]. The operating point of the DC/AC converter must be inside those limits.

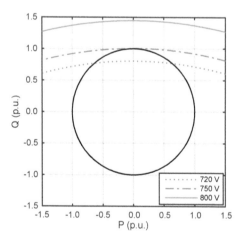

**Figure 11.** PQ-diagram for the 850 kVA DC/AC converter of both PV farms.

As we focus our analysis on PV farms, in addition to the DC/AC converter limits, one should consider the limits linked to the irradiance. The possible operating points of the DC/AC converter of a PV farm is then better visualized by plotting an GQ-diagram as shown in Figure 12.

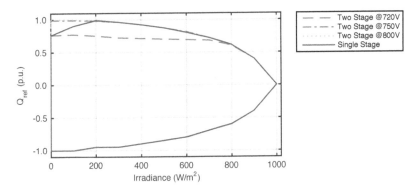

**Figure 12.** Reactive power support capability area of both PV farms.

For the single-stage PV farm, the reactive power support capability is reduced for irradiance values below 200 W/m$^2$. This is because the MPPT algorithm generates a low voltage reference $v_{dc\ ref}$ for low irradiance values (according to Figure 7). For irradiance values above 200 W/m$^2$, the reactive power support is only limited by the AC current converter limit. To extend the reactive power support for low irradiance, one could consider a manual control of the DC voltage, but the system would not operate anymore at the MPP.

For the two-stage PV farm operating at $v_{dc}$ equal to 720 V, the reactive power support is limited compared to the single-stage PV system. However, for $v_{dc}$ equal to 750 V or more, within the converter AC current limit, there is no reactive power support limitation even for low irradiance values. Therefore, the second stage, by decoupling the input DC voltage of the DC/AC converter from the MPP voltage, allows one to extend the reactive power support capability at low irradiance.

### 5.2. PV Farm Losses with Reactive Power Support

The losses per component can be obtained similarly as before for various reactive power references. At 500 W/m$^2$, for example, Figures 13 and 14 show the losses for the PV systems as a function of the reactive power references. We observe that the losses increase approximately symmetrically with respect to $|Q_{ref}|$ and that the converter losses are always predominant.

### 5.3. Technical Cost for Reactive Power Support

Figure 15 shows the technical costs for 500 W/m$^2$ as an example. Note that because of the AC current limit, one cannot provide 1 p.u. of reactive power at this operating condition. The technical costs of operating the single-stage PV farm and the two-stage PV farm with $v_{dc}$ set to 800 V are approximately equal. Moreover, the cost of operating the two-stage PV farm with $v_{dc}$ set to 720 V or 750 V is slightly higher than the cost of operating it at 800 V. Similar results have been obtained for all irradiance values (not shown here).

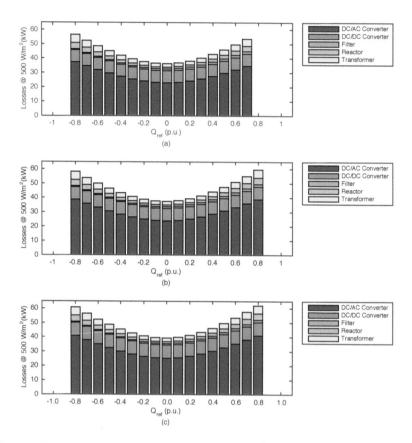

**Figure 13.** Two-stage PV farm losses per component at 500 W/m$^2$ for various $v_{dc}$, with reactive power support. (**a**) At 720 V; (**b**) at 750 V; (**c**) at 800 V. The losses with reactive power support are approximately a quadratic function of the reactive power reference value.

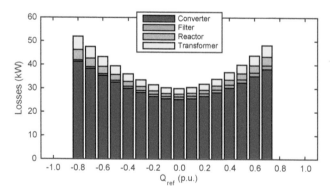

**Figure 14.** Single-stage PV farm losses per component at 500 W/m$^2$, with reactive power support. The losses with reactive power support are approximately a quadratic function of the reactive power reference value.

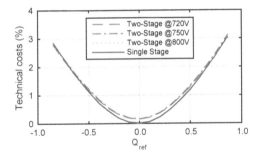

**Figure 15.** Technical costs of providing reactive power support at 500 W/m² for both PV farms, as a percentage of the rated power. Despite losses from two-stage PV farms being higher, the technical costs are similar for both topologies.

By evaluating the technical costs for each point inside the reactive power capability area, one can obtain the maps of Figure 16. One can note that even if the losses of the two-stage PV farm are higher than the ones of the single-stage system, the technical costs of providing reactive power support are similar: they reach 5.1% for the single-stage system and 5.6% for the two-stage system for the worst operating point. These maps underline another point: the cost of providing reactive power is higher at low irradiance than at high irradiance.

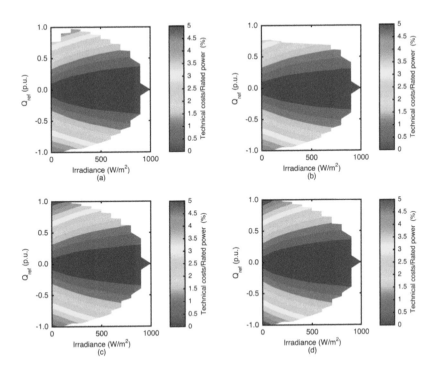

**Figure 16.** Reactive support capability areas with associated technical costs: (**a**) Single-stage; (**b**) Two-stage at 720 V; (**c**) Two-stage at 750 V; (**d**) Two-stage at 800 V. At 1000 W/m², there is no reactive power capability since full converter capacity is being used to inject active power.

*5.4. Reactive Power Support Economics*

The maps presented in the previous subsection can then be used to estimate the profits of a PV farm providing reactive power support to the grid. To illustrate, we have considered a PV farm located in the Brazilian northeastern region. The hourly average solar irradiance data for a random day was obtained using the Vortex Solar Series satellite data package, and the reactive power dispatch was obtained from [47]. These data are shown in Figure 17. The base value considered for the solar irradiance is 1000 W/m$^2$ and for reactive power is 850 kVAr.

The Brazilian system operator pays for the reactive power support as an ancillary service. The standard remuneration for this service is 6.88 BRL/MVArh [48] (1 USD = 3.92 BRL). Moreover, at the last energy auction, the average price of the energy sold by PV farms was 145.68 BRL/MWh, which we have considered as the cost for energy losses to estimate the expense generated by the reactive power support. Considering the scenario presented in Figure 17, the evaluation of the reactive power support economic feasibility is presented in Table 3.

The reactive energy demand is obtained by integrating the reactive power dispatch in Figure 17. The technical costs are obtained from the maps in Figure 16 and are shown in Figure 18. By integrating the data in Figure 18, the technical costs can be obtained in MWh. From the results presented exclusively for the reactive power support operation, the two-stage PV farm operating at 720 V is the most profitable for this specific dispatch and daily irradiance cycle.

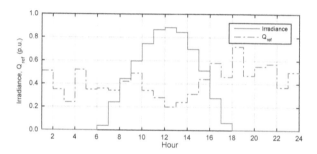

**Figure 17.** Hourly average solar irradiance data and reactive power dispatch for estimation of the reactive power support revenue. The base value considered for the solar irradiance is 1000 W/m$^2$ and for reactive power is 850 kVAr.

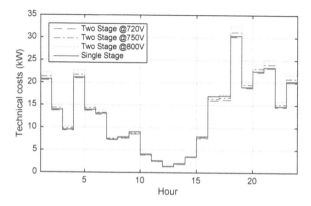

**Figure 18.** Technical costs for the test systems following the reactive power dispatch and daily irradiation cycle presented in Figure 17.

**Table 3.** Reactive power support economic feasibility evaluation. Revenue, expenses and profit are expressed in BRL.

| Test System | Reactive Power (MVArh) | Technical Costs (MWh) | Revenue | Expenses | Profit |
|---|---|---|---|---|---|
| Single-stage | 8.59 | 0.046 | 59.06 | 46.53 | 12.16 |
| Two-stage @720V | 8.59 | 0.045 | 59.06 | 45.57 | 13.12 |
| Two-stage @750V | 8.59 | 0.047 | 59.06 | 47.18 | 11.51 |
| Two-stage @800V | 8.59 | 0.049 | 59.06 | 48.93 | 9.75 |

As the reactive power markets develop, more complex analysis should be made by the PV farm operators. With dynamic reactive power pricing, more options are available for the PV farm operator to expand the product portfolio. The results presented in this work are the first steps towards analysis that do not feature active power priority, but a compromise between the active and reactive power prices looking to maximize the PV farm profits.

## 6. Conclusions

As energy mixes worldwide move away from the conventional carbon-emitting sources, PV farms arise as an environmentally-friendly option. In this work, we have presented a step-by-step methodology to evaluate the reactive power support capability and the associated technical costs for single- and two-stage PV farms operating in the daytime. The methodology consists of sweeping a wide range of operating points to help PV farm owners plan their bidding strategies for the reactive power markets that are under discussion. It was shown that the use of the two-stage PV farm allows an improved reactive power support for irradiance values lower than 200 W/m². For higher irradiance values, the converter AC current is the limiting factor if the priority is given to active power generation. It was underlined that even if the losses are higher for the two-stage topology in comparison to single-stage PV farms, the technical costs of providing reactive power are quite similar. They reached 5.1% and 5.6% for the single-stage and two-stage PV farms, respectively. Based on the operation maps with reactive power support, it was demonstrated how the profits can be evaluated for a specific dispatch according to the current Brazilian ancillary services policy. These results are interesting to PV farm owners, and during the planning phase, they could be used to determine the best topology: if the revenue provided by reactive power support compensates the technical costs and the extra losses linked to the DC/DC converter, then the two-stage PV farm should be selected.

**Author Contributions:** Luís F. N. Lourenço proposed the test system, ran the simulations and processed the results; Luís F. N. Lourenço, Renato M. Monaro, Maurício B.C. Salles, José R. Cardoso and Loïc Quéval discussed and validated the methodology; Luís F. N. Lourenço, Renato M. Monaro, Maurício B.C. Salles, José R. Cardoso and Loïc Quéval analyzed and discussed the results; Luís F. N. Lourenço, Renato M. Monaro, Mauricio B. C. Salles and Loïc Quéval wrote the paper.

**Funding:** This work was supported by São Paulo Research Foundation (FAPESP) grant # 2014/05261-0, by the Brazilian National Research Council (CNPq) grant # 2017-1/06628,and by the Coordination for the Improvement of Higher Education Personnel (CAPES).

**Acknowledgments:** The authors would like to thank the Brazilian National Research Council (CNPq) and the Coordination for the Improvement of Higher Education Personnel (CAPES) for their financial support.

**Conflicts of Interest:** The authors declare no conflict of interest.

## References

1. SolarPower Europe. *Global Market Outlook for Solar Power 2015–2019*; Technical Report; Euoropean Photovoltaic Industry Association: Bruxelles, Belgium, 2015.
2. Quitmann, E.; Erdmann, E. Power system needs—How grid codes should look ahead. *IET Renew. Power Gener.* **2014**, *9*, 3–9. [CrossRef]

3.  Industrial Standards Committee. *IEEE Standard for Interconnecting Distributed Resources with Electric Power Systems*; IEEE Standard 1547; Institute of Electrical and Electronics Engineers: New York, NY, USA, 2003.

4.  European Standard. *Voltage Characteristics of Public Distribution Systems*; EN Standard 50160; EN: Bruxelles, Belgium, 2004.

5.  Technischen Richtlinie Erzeugungsanlagen am Mittelspannungsnetz. Richtlinie für Anschluss und Parallelbetrieb von Erzeugungsanlagen am Mittelspannungsnetz. Available online: http://www.mega-monheim.de/assets/bdew_rl_ea-am-ms-netz_juni_2008_end.pdf (accessed on 12 June 2018).

6.  Gil, J.B.; San Román, T.G.; Rios, J.A.; Martin, P.S. Reactive power pricing: A conceptual framework for remuneration and charging procedures. *IEEE Trans. Power Syst.* **2000**, *15*, 483–489.

7.  Thomas, R.J.; Mount, T.D.; Schuler, R.; Schulze, W.; Zimmerman, R.; Alvarado, F.; Lesieutre, B.C.; Overholt, P.N.; Eto, J.H. Efficient and reliable reactive-power supply and consumption: Insights from an integrated program of engineering and economic research. *Electr. J.* **2008**, *21*, 70–81. [CrossRef]

8.  Chattopadhyay, D.; Chakrabarti, B.B.; Read, E.G. A spot pricing mechanism for voltage stability. *Int. J. Electr. Power Energy Syst.* **2003**, *25*, 725–734. [CrossRef]

9.  Amjady, N.; Rabiee, A.; Shayanfar, H. Pay-as-bid based reactive power market. *Energy Convers. Manag.* **2010**, *51*, 376–381. [CrossRef]

10. Hinz, F.; Moest, D. Techno-economic Evaluation of 110 kV Grid Reactive Power Support for the Transmission Grid. *IEEE Trans. Power Syst.* **2018**. [CrossRef]

11. Gabash, A.; Li, P. Active-reactive optimal power flow in distribution networks with embedded generation and battery storage. *IEEE Trans. Power Syst.* **2012**, *27*, 2026–2035. [CrossRef]

12. Varma, R.K.; Khadkikar, V.; Seethapathy, R. Nighttime application of PV solar farm as STATCOM to regulate grid voltage. *IEEE Trans. Energy Convers.* **2009**, *24*, 983–985. [CrossRef]

13. Varma, R.K.; Rahman, S.A.; Mahendra, A.; Seethapathy, R.; Vanderheide, T. Novel nighttime application of PV solar farms as STATCOM (PV-STATCOM). In Proceedings of the IEEE Power and Energy Society General Meeting, San Diego, CA, USA, 22–26 July 2012; pp. 1–8.

14. Varma, R.K.; Das, B.; Axente, I.; Vanderheide, T. Optimal 24-hr utilization of a PV solar system as STATCOM (PV-STATCOM) in a distribution network. In Proceedings of the IEEE Power and Energy Society General Meeting, San Diego, CA, USA, 24–29 July 2011; pp. 1–8.

15. Varma, R.K.; Rahman, S.A.; Vanderheide, T. New control of PV solar farm as STATCOM (PV-STATCOM) for increasing grid power transmission limits during night and day. *IEEE Trans. Power Deliv.* **2015**, *30*, 755–763. [CrossRef]

16. Mulolani, F.; Armstrong, M.; Zahawi, B. Modeling and simulation of a grid-connected photovoltaic converter with reactive power compensation. In Proceedings of the 9th International Symposium on Communication Systems, Networks & Digital Signal Processing (CSNDSP), Manchester, UK, 23–25 July 2014; pp. 888–893.

17. Romero-Cadaval, E.; Francois, B.; Malinowski, M.; Zhong, Q.C. Grid-connected photovoltaic plants: An alternative energy source, replacing conventional sources. *IEEE Ind. Electron. Mag.* **2015**, *9*, 18–32. [CrossRef]

18. Albuquerque, F.L.; Moraes, A.J.; Guimarães, G.C.; Sanhueza, S.M.; Vaz, A.R. Photovoltaic solar system connected to the electric power grid operating as active power generator and reactive power compensator. *Sol. Energy* **2010**, *84*, 1310–1317. [CrossRef]

19. Samadi, A.; Ghandhari, M.; Söder, L. Reactive power dynamic assessment of a PV system in a distribution grid. *Energy Procedia* **2012**, *20*, 98–107. [CrossRef]

20. Blaabjerg, F.; Teodorescu, R.; Liserre, M.; Timbus, A.V. Overview of control and grid synchronization for distributed power generation systems. *IEEE Trans. Ind. Electron.* **2006**, *53*, 1398–1409. [CrossRef]

21. Gabash, A.; Li, P. Active-reactive optimal power flow for low-voltage networks with photovoltaic distributed generation. In Proceedings of the 2012 IEEE International Energy Conference and Exhibition (ENERGYCON), Florence, Italy, 9–12 September 2012; pp. 381–386.

22. Ullah, N.R.; Bhattacharya, K.; Thiringer, T. Wind farms as reactive power ancillary service providers—Technical and economic issues. *IEEE Trans. Energy Convers.* **2009**, *24*, 661–672. [CrossRef]

23. Jung, S.; Jang, G. A Loss Minimization Method on a reactive power supply process for Wind Farm. *IEEE Trans. Power Syst.* **2017**, *32*, 3060–3068. [CrossRef]

24. Zhang, B.; Hu, W.; Hou, P.; Tan, J.; Soltani, M.; Chen, Z. Review of Reactive Power Dispatch Strategies for Loss Minimization in a DFIG-based Wind Farm. *Energies* **2017**, *10*, 856.

[CrossRef]

25. Gabash, A.; Li, P. Evaluation of reactive power capability by optimal control of wind-vanadium redox battery stations in electricity market. *Renew. Energy Power Qual. J.* **2011**, 1–6. [CrossRef]

26. Gabash, A.; Li, P. On variable reverse power flow—Part I: Active-Reactive optimal power flow with reactive power of wind stations. *Energies* **2016**, *9*, 121. [CrossRef]

27. Gabash, A.; Li, P. On variable reverse power flow—Part II: An electricity market model considering wind station size and location. *Energies* **2016**, *9*, 235. [CrossRef]

28. Lourenço, L.F.N. Technical Cost of Operating a PV Installation as a STATCOM during Nighttime. Master's Thesis, Universidade de São Paulo, São Paulo, Brazil, 2017.

29. Lourenço, L.F.N.; Salles, M.B.C.; Monaro, R.M.; Quéval, L. Technical Cost of Operating a Photovoltaic Installation as a STATCOM at Nighttime. *IEEE Trans. Sustain Energy* **2018**. [CrossRef]

30. Lourenço, L.F.N.; Salles, M.B.C.; Monaro, R.M.; Quéval, L. Technical cost of PV-STATCOM applications. In Proceedings of the 2017 IEEE 6th International Conference on Renewable Energy Research and Applications (ICRERA), San Diego, CA, USA, 5–8 November 2017; pp. 534–538.

31. Blaabjerg, F.; Chen, Z.; Kjaer, S.B. Power electronics as efficient interface in dispersed power generation systems. *IEEE Trans. Power Electron.* **2004**, *19*, 1184–1194. [CrossRef]

32. Kouro, S.; Leon, J.I.; Vinnikov, D.; Franquelo, L.G. Grid-connected photovoltaic systems: An overview of recent research and emerging PV converter technology. *IEEE Ind. Electron. Mag.* **2015**, *9*, 47–61. [CrossRef]

33. Yazdani, A.; Di Fazio, A.R.; Ghoddami, H.; Russo, M.; Kazerani, M.; Jatskevich, J.; Strunz, K.; Leva, S.; Martinez, J.A. Modeling guidelines and a benchmark for power system simulation studies of three-phase single-stage photovoltaic systems. *IEEE Trans. Power Deliv.* **2011**, *26*, 1247–1264. [CrossRef]

34. De Brito, M.A.; Sampaio, L.P.; Luigi, G.; e Melo, G.A.; Canesin, C.A. Comparative analysis of MPPT techniques for PV applications. In Proceedings of the 2011 International Conference on Clean Electrical Power (ICCEP), Ischia, Italy, 14–16 June 2011; pp. 99–104.

35. Huang, L.; Qiu, D.; Xie, F.; Chen, Y.; Zhang, B. Modeling and Stability Analysis of a Single-Phase Two-Stage Grid-Connected Photovoltaic System. *Energies* **2017**, *10*, 2176. [CrossRef]

36. Blaabjerg, F.; Jaeger, U.; Munk-Nielsen, S. Power losses in PWM-VSI inverter using NPT or PT IGBT devices. *IEEE Trans. Power Electron.* **1995**, *10*, 358–367. [CrossRef]

37. Hansen, A.D.; Michalke, G. Modelling and control of variable-speed multi-pole permanent magnet synchronous generator wind turbine. *Wind Energy* **2008**, *11*, 537–554. [CrossRef]

38. Liserre, M.; Blaabjerg, F.; Dell'Aquila, A. Step-by-step design procedure for a grid-connected three-phase PWM voltage source converter. *Int. J. Electron.* **2004**, *91*, 445–460. [CrossRef]

39. Fronius International GmbH Solar Energy Division. *Sizing the Maximum DC Voltage of PV Systems*; Fronius International GmbH Solar Energy Division: Wels, Austria, 2015.

40. Martinez, J.A.; Mork, B.A. Transformer modeling for low-and mid-frequency transients—A review. *IEEE Trans. Power Deliv.* **2005**, *20*, 1625–1632. [CrossRef]

41. Blaabjerg, F.; Pedersen, J.K.; Sigurjonsson, S.; Elkjaer, A. An extended model of power losses in hard-switched IGBT-inverters. In Proceedings of the Conference Record of the 1996 IEEE Industry Applications Conference Thirty-First IAS Annual Meeting (IAS'96), San Diego, CA, USA, 6–10 October 1996; Volume 3, pp. 1454–1463.

42. Rajapakse, A.; Gole, A.; Wilson, P. Electromagnetic transients simulation models for accurate representation of switching losses and thermal performance in power electronic systems. *IEEE Trans. Power Deliv.* **2005**, *20*, 319–327.

43. Wong, C. EMTP modeling of IGBT dynamic performance for power dissipation estimation. *IEEE Trans. Ind. Appl.* **1997**, *33*, 64–71. [CrossRef]

44. Drofenik, U.; Kolar, J.W. A general scheme for calculating switching-and conduction-losses of power semiconductors in numerical circuit simulations of power electronic systems. In Proceedings of the 2005 International Power Electronics Conference (IPEC 9205), Niigata, Japan, 4–8 April 2005; pp. 4–8.

45. Munk-Nielsen, S.; Tutelea, L.N.; Jaeger, U. Simulation with ideal switch models combined with measured loss data provides a good estimate of power loss. In Proceedings of the Conference Record of the IEEE Industry Applications Conference, Rome, Italy, 8–12 October 2000; Volume 5, pp. 2915–2922.

46. Cole, S. Steady-State and Dynamic Modelling of VSC HVDC Systems for Power System Simulation. Ph.D. Thesis, Katholieke Universiteit Leuven, Leuven, Belgium, 2010.

47. Haghighat, H.; Kennedy, S. A model for reactive power pricing and dispatch of distributed generation. In Proceedings of the IEEE Power and Energy Society General Meeting, Providence, RI, USA, 25–29 July 2010; pp. 1–10.
48. Tarifas de Energia de Otimização e de Serviços Ancilares para 2018 (Optimization and Ancillary Services Tariffs for 2018). Available online: http://www.aneel.gov.br/sala-de-imprensa-exibicao-2/-/asset_publisher/zXQREz8EVlZ6/content/fixadas-as-tarifas-de-energia-de-otimizacao-e-de-servicos-ancilares-para-2018/656877?inheritRedirect=false (accessed on 7 June 2018).

MDPI

St. Alban-Anlage 66

4052 Basel

Switzerland

Tel. +41 61 683 77 34

Fax +41 61 302 89 18

www.mdpi.com

*Energies* Editorial Office

E-mail: energies@mdpi.com

www.mdpi.com/journal/energies

Lightning Source UK Ltd.
Milton Keynes UK
UKHW020914180223
417109UK00006B/66